Genetic Algorithms and Remote Sensing Technology for Tracking Flight Debris

Maged Marghany
Syiah Kuala University, Indonesia

A volume in the Advances in Mechatronics and Mechanical Engineering (AMME) Book Series

Published in the United States of America by
IGI Global
Engineering Science Reference (an imprint of IGI Global)
701 E. Chocolate Avenue
Hershey PA, USA 17033
Tel: 717-533-8845
Fax: 717-533-8661
E-mail: cust@igi-global.com
Web site: http://www.igi-global.com

Library of Congress Cataloging-in-Publication Data

Names: Marghany, Maged, author.
Title: Genetic algorithms and remote sensing technology for tracking flight debris / by Maged Marghany.
Description: Hershey, PA : Engineering Science Reference, 2020. | Includes bibliographical references and index. | Summary: "This book explores the use of genetic algorithms in the context of the analysis of remotely sensed images. This is illustrated with a case study of the M370 crash"-- Provided by publisher.
Identifiers: LCCN 2019035517 (print) | LCCN 2019035518 (ebook) | ISBN 9781799819202 (h/c) | ISBN 9781799819219 (s/c) | ISBN 9781799819226 (eISBN)
Subjects: LCSH: Aircraft accidents--Investigation--Data processing. | Marine debris--Remote sensing--Data processing. | Genetic algorithms. | Malaysia Airlines Flight 370 Incident, 2014.
Classification: LCC TL553.5 .M325 2020 (print) | LCC TL553.5 (ebook) | DDC 363.12/46501519625--dc23
LC record available at https://lccn.loc.gov/2019035517
LC ebook record available at https://lccn.loc.gov/2019035518

This book is published in the IGI Global book series Advances in Mechatronics and Mechanical Engineering (AMME) (ISSN: 2328-8205; eISSN: 2328-823X)

British Cataloguing in Publication Data
A Cataloguing in Publication record for this book is available from the British Library.

For electronic access to this publication, please contact: eresources@igi-global.com.

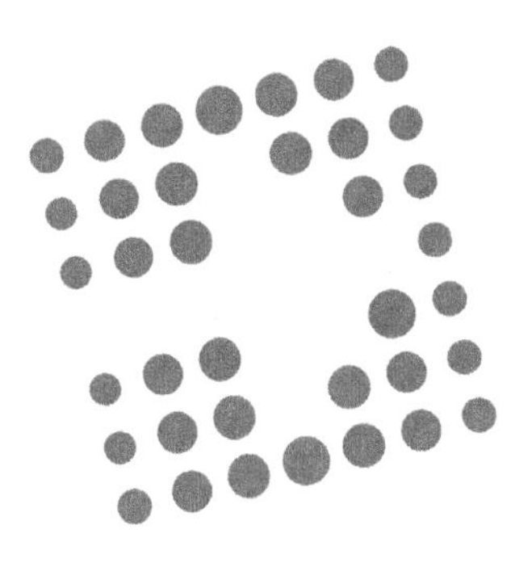

Advances in Mechatronics and Mechanical Engineering (AMME) Book Series

J. Paulo Davim
University of Aveiro, Portugal

ISSN:2328-8205
EISSN:2328-823X

Mission

With its aid in the creation of smartphones, cars, medical imaging devices, and manufacturing tools, the mechatronics engineering field is in high demand. Mechatronics aims to combine the principles of mechanical, computer, and electrical engineering together to bridge the gap of communication between the different disciplines.

The **Advances in Mechatronics and Mechanical Engineering (AMME) Book Series** provides innovative research and practical developments in the field of mechatronics and mechanical engineering. This series covers a wide variety of application areas in electrical engineering, mechanical engineering, computer and software engineering; essential for academics, practitioners, researchers, and industry leaders.

Coverage

- Nanomaterials and nanomanufacturing
- Biologically Inspired Robotics
- Control Methodologies
- Tribology and surface engineering
- Computer-Based Manufacturing
- Micro and nanomechanics
- Computational Mechanics
- Vibration and acoustics
- Bioengineering Materials
- Sustainable and green manufacturing

IGI Global is currently accepting manuscripts for publication within this series. To submit a proposal for a volume in this series, please contact our Acquisition Editors at acquisitions@igi-global.com or visit: https://www.igi-global.com/publish/.

The Advances in Mechatronics and Mechanical Engineering (AMME) Book Series (ISSN 2328-8205) is published by IGI Global, 701 E. Chocolate Avenue, Hershey, PA 17033-1240, USA, www.igi-global.com. This series is composed of titles available for purchase individually; each title is edited to be contextually exclusive from any other title within the series. For pricing and ordering information please visit https://www.igi-global.com/book-series/advances-mechatronics-mechanical-engineering/73808. Postmaster: Send all address changes to above address.

Titles in this Series

For a list of additional titles in this series, please visit:
https://www.igi-global.com/book-series/advances-mechatronics-mechanical-engineering/73808

Modeling and Optimization of Solar Thermal Systems Emerging Research and Opportunities
Jagadish (National Institute of Technology Raipur, India) and Agnimitra Biswas (National Institute of Technology Silcha, India)
Engineering Science Reference • © 2020 • 170pp • H/C (ISBN: 9781799835233) • US $185.00

Airline Green Operations Strategies Emerging Research and Opprtunities
Yazan Khalid Abed-Allah Migdadi (Qatar University, Qatar)
Engineering Science Reference • © 2020 • 215pp • H/C (ISBN: 9781799842552) • US $185.00

Recent Technologies for Enhancing Performance and Reducing Emissions in Diesel Engines
J. Sadhik Basha (International Maritime College Oman, Sohar, Oman) and R.B. Anand (National Institute of Technology, Tiruchirappalli, India)
Engineering Science Reference • © 2020 • 298pp • H/C (ISBN: 9781799825395) • US $215.00

Practical Approach to Substrate Integrated Waveguide (SIW) Diplexer Emerging Research and Opportunities
Augustine Onyenwe Nwajana (University of East London, UK) and Kenneth Siok Kiam Yeo (Universiti Teknologi Brunei, Brunei)
Engineering Science Reference • © 2020 • 171pp • H/C (ISBN: 9781799820840) • US $195.00

Diverse Applications of Organic-Inorganic Nanocomposites Emerging Research and Opportunities
Gabriele Clarizia (Institute on Membrane Technology, National Research Council, Italy) and Paola Bernardo (Institute on Membrane Technology, National Research Council, Italy)
Engineering Science Reference • © 2020 • 237pp • H/C (ISBN: 9781799815303) • US $195.00

Handbook of Research on Artificial Intelligence Applications in the Aviation and Aerospace Industries
Tetiana Shmelova (National Aviation University, Ukraine) Yuliya Sikirda (Flight Academy of National Aviation University, Ukraine) and Arnold Sterenharz (EXOLAUNCH GmbH, Germany)
Engineering Science Reference • © 2020 • 517pp • H/C (ISBN: 9781799814153) • US $295.00

Design and Optimization of Sensors and Antennas for Wearable Devices Emerging Research and Opportunities
Vinod Kumar Singh (S. R. Group of Institutions Jhansi, India) Ratnesh Tiwari (Bhilai Institute of Technology, India) Vikas Dubey (Bhilai Institute of Technology, India) Zakir Ali (IET Bundelkhand University, India) and Ashutosh Kumar Singh (Indian Institute of Information Technology, India)
Engineering Science Reference • © 2020 • 196pp • H/C (ISBN: 9781522596837) • US $215.00

701 East Chocolate Avenue, Hershey, PA 17033, USA
Tel: 717-533-8845 x100 • Fax: 717-533-8661
E-Mail: cust@igi-global.com • www.igi-global.com

Dedicated to My Mother Faridah, Captain Zaharie Ahmad Shah, COVID-19 Victims, and Nikola Tesla and Richard Feynman, who taught me that the real professor survives even after his death, whereas the fake professor does not exist after his death.

Table of Contents

Preface

Up to today, no one can tell where Flight MH370 went. Even with advanced remote sensing sensors and communication technologies, MH370 cannot be traced since March 8, 2014 up until this moment. The main claim of the horror of tragic MH370 is based on Captain Zaharie Ahmad Shah 's committed suicide. Is it a precise conclusion that Captain Zaharie was behind the tragedy of MH370? It is such a critical question and requires a deep benevolence of the aerodynamic mechanisms. Regardless of advanced remote sensing technology, remote sensing scientists could not detect an accurate geographical location for MH370's vanishing. In this regard, the MH370 vanishing did not leave evidence beyond it to explain the mechanism of that tragedy ending compared to other flight crashes. In this view, the vanishing of flight MH370 is what could conceivably swap 30 years of biased productivity evolution in aeronautics transportation and localizing.

Conversely, there has been countless satellite data that has shown numerous objects belonging to the flight MH370. One of these satellites is a Thai satellite, which had detected 300 floating objects in the Indian Ocean, approximately 200 kilometers from the worldwide search region for the missing Malaysia Airlines MH 370 at 10 am Perth local time on the 24th of March 2014. In this context, the THEOS satellite payload aspects each excessive resolution in panchromatic mode and large field of view in multispectral mode and has been man-made to Thailand's precise requests with an international imaging capability. In addition, it incorporates 2m resolution for black and white images and 15m resolution of the panchromatic image.

The foremost question can be raised up "what marvelous sensors can be used to reveal and discover flight MH370 debris"? The high-resolution sensors both on board of satellite or airborne can observe and perceive the flight MH370 debris. Even HF ground radar can discover any overseas objects drifting in the coastal zone. The widespread techniques of object computerized detection through the use of high-resolution microwave satellite data with 1 m two as in the spot mode of each RADARSAT-2 SAR, TerraSAR-X satellite data is additionally required. The RADARSAT-2 SAR satellite has a synthetic aperture Radar (SAR) with a couple of polarization modes, such as a completely polarimetric mode in which HH, HV, VV and VH polarized information are acquired. Its perfect resolution is 1m in Spotlight mode (3m in Ultra-Fine mode) with 100 m positional accuracy requirement. Moreover, RADARDSAT-2 SAR Scan Narrow SCNB beam is its and a high revisit duration of 7 days. Additional has nominal close to and far resolutions of 7 m. If the size of the flight debris, for instance, is 24 m, the potential it could definitely be detected in RADARDSAT-2 SAR Scan Narrow. Nonetheless, none of HF ground radar and satellite radar data delivered any information regarding the missing flight MH370.

This book will attempt to answer a critical question " Did flight MH370 plunge into the Southern Indian Ocean or not? In this view, the book will explore what could have happened to Malaysia Airlines

Flight MH370 debris and flaperon due to the impacts of ocean dynamic fluctuations. The scope of the book is restricted to remote sensing data available for the MH370. The first three chapters of the book discuss the theories beyond MH370 missing, novel theories behind flap failure impact on the MH370 crash, and the mathematical theory of flight MH370 ditching, respectively. The explicit concern of the diverse optical satellite data which tracked the possible MH370 debris across the Southern Indian Ocean has been addressed in this book.

This book has also introduced the multiobjective genetic algorithms as a possible solution for tracking MH370 debris and deciding whether the flight ditched into the Southern Indian Ocean or not. Chapter four has addressed the basic of the genetic algorithm and multi-objective genetic programming. Conversely, this book implemented the genetic algorithm as advanced image processing tools to investigate the existing of MH370 debris. Further, chapter five also uses the multiobjective genetic algorithm to simulate the actual MH370 route, in addition, to accurately route if Captain Zaharie Ahmad Shah 's was willing to commit suicide.

Unfortunately, major Malaysian oceanographers and remote sensing researchers are not familiar with the Indian ocean dynamic or specifically the Southern Indian Ocean. In fact, Malaysian researchers knowledge is restricted partially along the coastal waters of Peninsular Malaysia. It is mainly focused on marine biology, marine chemistry and marine geology without a perfectly tolerant understanding of the physical and ocean dynamic. Further, the remote sensing researchers also are not able to address the mystery of MH370. In spite of plenty of Malaysian researchers in remote sensing and oceanography fields are titled by professors, they are not able since the missing of MH370 to deliver any sort of solution or conclusion. Under these facts, those researchers are mainly relying on the Australia scientists to model the trajectory movements of debris and flaperon across the Southern Indian Ocean. In this regard, chapter six addresses the principle of optical remote sensing while chapter seven reviews the optical satellite images used to track MH370 debris such as Gaofen-1satellite. In continuing with chapter seven, chapter eight proposes a possible automatic detection algorithm that could be useful to detect MH370 debris such as Otsu thresholding algorithm and deep belief nets algorithm. More advanced algorithms based on multiobjective genetic algorithm and Pareto optimization algorithm are addressed in chapter nine. However, chapter ten reviews the fundamentals of radar altimeter as a keystone to understanding the impact of ocean dynamic features on the stability of MH370 across the Indian ocean.

In this view, more advanced work on the Indian Ocean wave dynamics is presented from chapter eleven to chapter thirteen. This book also presented a new technique which is based on the altimeter interferometry to monitor the impact of the significant wave height fluctuations on the MH370 debris and flaperon as revealed in chapter eleven. Unlike other studies, the book does not only concern the surface trajectory model but involves other dynamic ocean components such as Rossby wave, eddies, velocity potential, and vorticity as presented in chapter tweleveth. These were taken into account to answer the critical question

"How did flaperon wash up on Réunion Island?"

The book also introduces a new technique to model the trajectory drift of the MH370 debris based on the amalgamation between Volterra-Lax-Wendroff algorithm and multiobjective algorithm based on the Pareto optimal solution as explained at chapter thirteen. Finally, chapter fourteen involves the genetic algorithm based on the Pareto optimization to answer the significant question of "Why MH370 could not plunge into the Southern Indian Ocean?" Lastly, this book attempts to answer the requested question by the public of where is MH370? The answer to this question is addressed from the point of view of the conspiracy theories based on the Pareto optimization in the last chapter. In addition, the book also addresses the benefit of sea operation for searching MH370 approximately 1500 km off Perth, Australia. The answers to these critical questions are addressed scientifically in the last chapter.

Maged Marghany
Syiah Kuala University, Indonesia

Chapter 1
Speculations Behind the Disappearance of Flight MH370

ABSTRACT

Several theories and conspiracies have flooded the media regarding the vanishing of Malaysia Airline 370 (MH370) in the Indian Ocean on March 8, 2018. This chapter aims at reviewing these theories to investigate the vanishing of MH370. Futhermore, this chapter also discloses doubts about tracking MH370 as approximately the signal information concerning BTO and BFO is non-existent. This chapter concludes with some theories such as aliens and the Bermuda Triangle that are not scientifically grounded. This chapter suggests a serious concern about the development of ocean dynamic studies as the keystone to investigating the vanishing of MH370 in such an extremely dynamic ocean as the Indian Ocean.

INTRODUCTION

It is a challenging mission to comprehend the speculations behind the disappearance of Malaysia Airlines Flight 370 (MH370) since some of these assumptions are not scientifically grounded, for instance, those concerning aliens and the Bermuda Triangle. In this scenario, it is not a logical assumption that aliens have hijacked MH370 and besides, the Bermuda Triangle is not geographically located near the South China Sea nor the Indian Ocean. It is just a media brainwash of global audiences.

In this view, both theories are not inadequate in unraveling the mystery of MH370 vanishing. Modern physics and advanced data analysis are necessary to comprehend this catastrophic event. There was continuous research, both hypothetically and experimentally by the Australian Transport Safety Bureau (ATSB), to comprehend precisely the reason for the airplane's disappearance. In that sense, the mystery of MH370 remains unsolved. This chapter is devoted to delivering novel findings and speculations behind airplane vanishing mechanisms.

DOI: 10.4018/978-1-7998-1920-2.ch001

What Happened on March 8, 2014?

Despite the superior space, marine, and communication technologies, the mystery of the missing flight cannot be clarified. Despite twelve countries that joined in for the search and rescue efforts of the missing flight, it is extremely complicated to investigate its mystery vanishing from the Malaysian secondary ground radar, northwest of Penang Island, Malaysia (Figure 1). The aircraft was shown flying at a cruise altitude of 350 m and was traveling at an actual airspeed of 872.26 km/h. Later analysis estimated that flight MH370 had 41,500 kg (91,491.84 lb) of fuel when it disappeared from secondary radar (Bureau, 2014; Bureau,2015; Tyler, 2016).

Figure 1. Location of the disappearance of MH370 on the radar screen

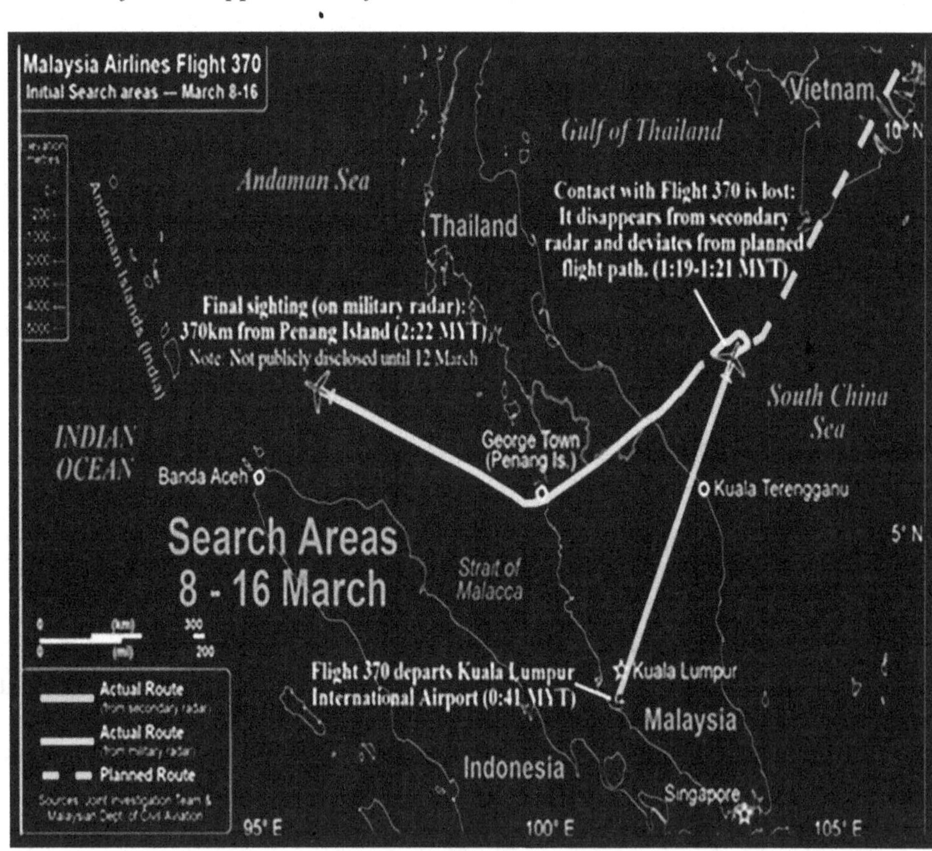

The initial critical question is what was happening before March 8th 2014? On March 7th 2014 approximately at 16:42, flight MH370 departed from Kuala Lumpur International Airport (KLIA) constrained for Beijing; China under normal circumstances. In this understanding, the Mode S transponder system on-board the airplane was reacting as probable to debrief from the ATC Secondary Surveillance Radar till it disappeared on the ATC radar screen at 17:21:13 UTC. In this view, no message or signal was acknowledged from the airplane to account for a system catastrophe (**Bureau, 2015;** The Star online, 2016). Consequently, at 17:07:29 UTC on-board Aircraft Communications Addressing and Reporting

System (ACARS) has collected six reports which were produced at five-minute intervals from 16:41:43 UTC to 17:06:43 UTC. In these regards, this information delivered the aircraft position and motion, for instance, latitude, longitude, altitude, air temperature, airspeed, wind direction, wind speed, and true heading. During the cruise, the ACARS location reports are programmed to be conveyed at thirty-minute intervals. Conversely, at 17:37 UTC, the succeeding scheduled report was not delivered. At 17:19:30 UTC, the crew of MH370 has the last recorded radio transmission whilst the airplane was directed to communicate to Vietnamese ATC on departure from the Malaysian Flight Information Region (Figure 1). Subsequently, Vietnamese ATC with the efforts of several countries, including Malaysia has been failed at 17:39:06 UTC to establish any accurate position of MH370 (**Bureau,2015 and** Fox, 2016).

On the contrary, at 17:21:13 UTC, the Malaysian military radars were successively talented to reveal major radar backscatters combined with MH370 diverging from MH370 instantaneously later the demise of Secondary Surveillance Radar by requiring a left turn to end up traveling in a South Westerly direction. In this view, radar backscatters indicate the airplane roving back across Malaysia before turning near Penang Island and traveling in a North Westerly direction up the Straits of Malacca at 18:22:12 UTC (Green, 2014; **Bureau,2015;**Team, 2015; Fox, 2016). In this understanding, the timing measurement, which is known as the Burst Timing Offset (BTO), and the frequency measurement, the Burst Frequency Offset (BFO), was used to estimate the trajectory movement of MH370 route. Table 1 reveals that Aircraft Earth Station (the aircraft satellite communications unit) (AES) obtained information about BTO from time of 18:25 UTC to 18:28 UTC. However, BFO information did not available at 18:25 UTC but is presented at 18:28 UTC. This delivers uncertainties among the obtained information of BFO and BTO(Davey et al., 2016). Furthermore, Ground Earth Station (the ground component of the satellite communications system) (GES) did not obtain any information about BTO at time 23:15 UTC(**Bureau,2015).**

Table 1. Briefing on SATCOM data available for MH370 (Davey et al., 2016)

Incident	Period (UTC)	BTO	BFO
Aircraft departed KLIA	16:42	Available	Available
Final ACARS transmission	17:07	Available	Available
AES instigated start	18:25	Available	Not-available
AES access request	18:28	Available	Available
Unreciprocated ground to air handset call	18:39	Not-available	Available
GES instigated handshake	19:41	Available	Available
GES instigated handshake	20:41	Available	Available
GES instigated handshake	21:41	Available	Available
GES initiated handshake	22:41	Available	Available
Unreciprocated ground to air handset call	23:15	Not-available	Available
GES instigated handshake	00:10	Available	Available
AES instigated start	00:19	Available	Not-available

Nonetheless, the abortive communication data log comprises frequency, but doesn't time metadata. Consequently, MH370 could be logged on the SATCOM network. In this circumstance, a query was recieved from the ground station, which is known as handshake messages. These handshake retorts comprise both timing and frequency metadata. As reported by Davey et al., (2016), handshake messages occurred from 19:41:00 UTC (Table 1) to the last SATCOM transmission received from the aircraft at 00:19:29 UTC without answering from a satellite telephone call. In this ambiguity scenario, the conclusion was drawn as there are no signal responses from MH370 owing to fuel exhaustion and the succeeding triggering of the secondary power unit (Green 2014; Iannello,2014; **Bureau,2015**). It can reveal that AES did not receive BFO at 00:19 UTC (Davey et al., 2016).

Briefly, the complete timeline scenario of that March 8, 2014 event till March 2016 is compiled and addressed as follows (Bureau 2015):

1.19am

From the cockpit was MH370's final voice contact with air traffic control: "Good night. Malaysian three seven zero."

1.21am

Just two minutes later, the plane's transponder stops communicating with radar and air traffic control systems.

The key information such as speed and altitude no longer available to ground control.

1.22am

MH370 disappears from Thai military radar, which had detected the aircraft making a sharp left turn that sees the plane now flying in a south-westerly direction.

1.30am

Civilian radar based in Malaysia loses contact with MH370 over the Gulf of Thailand, between Vietnam and Indonesia.

1.37am

A scheduled ACARS transmission is not received.

2.15am

After multiple attempts to contact the flight fails, the aircraft is detected by Malaysian military radar near the island of Pulau Perak – hundreds of kilometers off course.

2.40am

Malaysia Airlines learn from air traffic control MH370 is missing from radar.

The airline begins searching for the plane by contacting various air traffic control stations and other aircraft.

3.45am

Malaysia Airlines issues a "code red" alert, indicating a crisis that requires an emergency response.

5.30am

An initial search-and-rescue operation is launched in the South China Sea and the Gulf of Thailand, near where the aircraft lost contact with air traffic control.

6.30am

MH370 does not arrive at Beijing International Airport as scheduled.

7.24am

Nearly an hour after the flight was meant to land, Malaysia Airlines issues a public statement announcing MH370 is missing.

"Malaysia Airlines confirms that flight MH370 has lost contact with Subang Air Traffic Control at 2.40 am, (8 March 2014)," the statement reads. "Flight MH370, operated on the B777-200 aircraft, departed Kuala Lumpur at 12.41am on 8 March 2014.

"MH370 was expected to land in Beijing at 6.30 am the same day.

"The flight was carrying a total number of 227 passengers (including 2 infants), 12 crew members.

"Malaysia Airlines is currently working with the authorities who have activated their Search and Rescue team to locate the aircraft."

Satellite data later reveals the aircraft is still in the air at this point.

8.11am

British satellite telecommunications company Inmarsat receives a "ping" at 8.11am that appears to be MH370.

It is detected during routine, automated communications and goes unnoticed for several days. It's the final time any electronic signal is detected from the aircraft, about seven hours and 30 minutes after take-off.

9.15am

The aircraft does not respond to the next automated communication attempt with Inmarsat's satellites. Aircraft and vessels from multiple countries in the region are now searching for the plane based on where MH370 lost contact. In the days after MH370's disappearance, police raid Captain Zaharie Ahmad Shah's home and seize a flight simulator.

March 15

After a week searching in South China Sea, Malaysian Prime Minister Najib Razak publicly reveals MH370 flew for several hours after losing contact with air traffic control. The initial search is ended and new zone along two possible flight arcs is established.

March 18

A surface search of the southern Indian Ocean begins about 2500km south-west of Perth. Australia agrees to lead the search.

March 24 2014

Malaysian Prime Minister Najbi Razak announces MH370 is presumed to have crashed somewhere in the search area.

"This is a remote location, far from any possible landing sites," he said.

"It is therefore with deep sadness and regret that I must inform you that, according to this new data, Flight MH370 ended in the southern Indian Ocean."

A relative of one of the Chinese passengers aboard MH370 collapses in grief after being told of the latest news.

April – May 2014

Multiple searches are now underway: an acoustic search to detect pings from the aircraft flight recorders ends on April 14, the ocean surface search on April 28 and a seafloor sonar search end on May 28.

October 2014 - June 2015

A new underwater search begins after a lengthy bathymetric survey of 208,000km2 of the ocean floor.

Autonomous underwater vehicles are used in the search mission.

The underwater search zone followed a possible arc of the doomed plane.

The search runs until June 2015 but fails to find any debris from MH370.

March 8, 2015

A year after the plane's disappearance, relatives of some Chinese passengers protest outside the Malaysian embassy in Beijing.

July 29, 2015

Sixteen months after MH370 vanished, a flaperon – a section of wing – is found on a beach on the island of Reunion, around 4000km west of the initial search area.

It is the first confirmed piece of wreckage of MH370.

Ocean drift modelling showed the Reunion find was consistent with possible MH370 debris.

December 2015 – March 2016

More airline parts are found.

Some are confirmed to belong to MH370, while others are deemed only "highly likely" due to a lack of identifying features (Bureau, 2015).

What are the New Findings?

In the four years since the disappearance of Flight 370, numerous new theories approximating the fate of the missing aircraft, have been taking the internet by storm. On March 8, 2014, the Boeing 777 aircraft, which had 239 people on board, vanished on the path to Beijing from Kuala Lumpur above the Indian Ocean. To date, can anyone articulate how and where MH370 vanished? The Star online (2016) reported that Peter McMahon, an Australian mechanical engineer made the shocking claims that the missing flight has been tracked down on Google Earth riddled with bullet holes, and suggested a location near Round Island, north of Mauritius (Figure 2).

Figure 2. The claim that MH370 crashed north of Mauritius

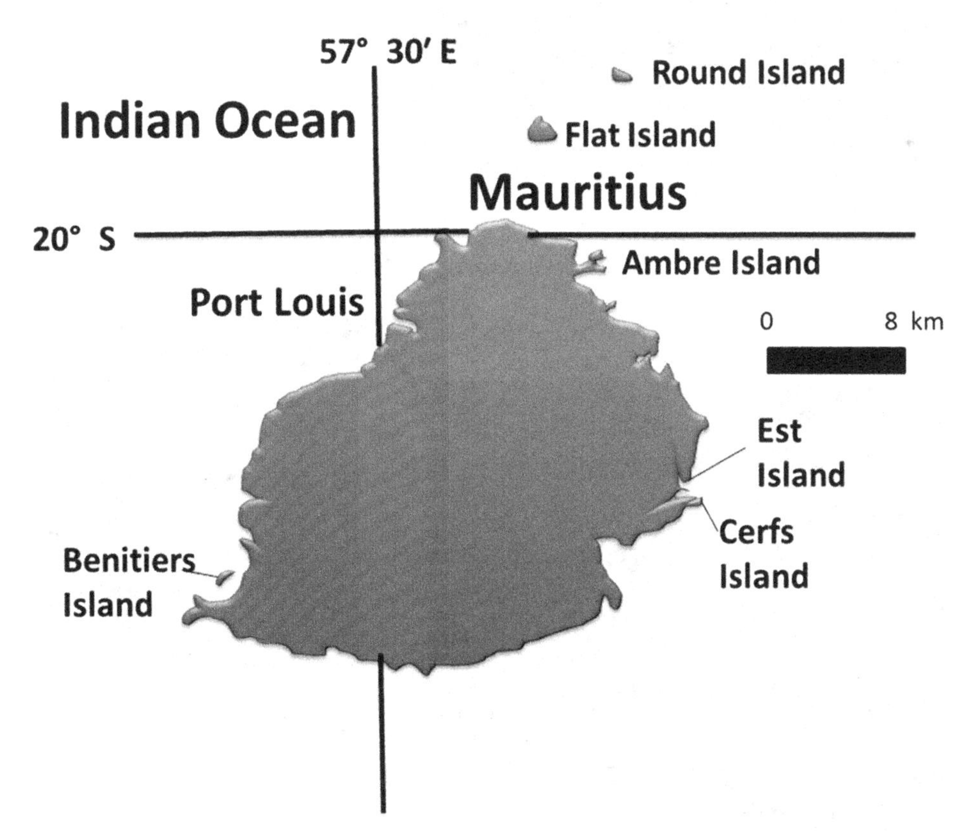

Conversely, the disappearance of Flight 370 remains mysterious. In this view, MH370 investigators argued that the images supposedly belonging to the missing flight (Figure 3a) were taken years earlier, before the flight disappeared. They explained that those images (Figure 3b) were captured on November 6, 2009, more than four years before the flight disappeared (Green, 2014; **Bureau,2015;**Team,2015).

This explanation can be true as long as the experts had never found any MH370 that flew corpses as passengers. In this concern, there are many questions raised that require scientific responses to resolve the mystery of MH370 vanishing. Did the suspected objects on satellite images belong to the missing flight? The first 'location' is near the round island that can be found on an image that was dated as June

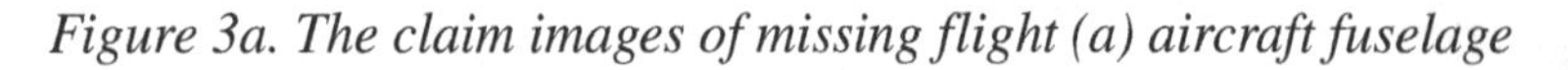

Figure 3a. The claim images of missing flight (a) aircraft fuselage

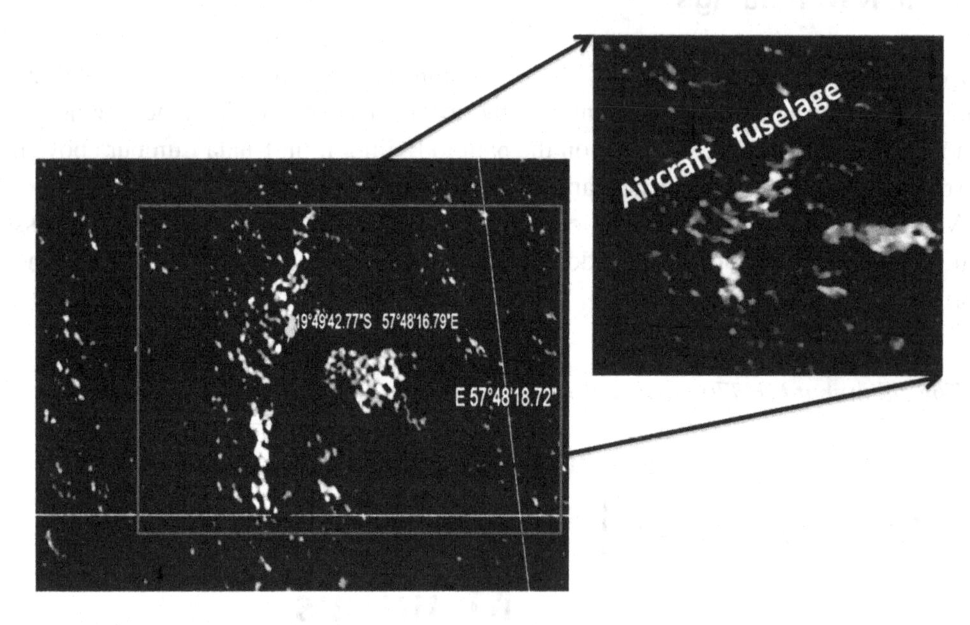

Figure 3b. Part of the front cabin, in the south of Rodrigues Island

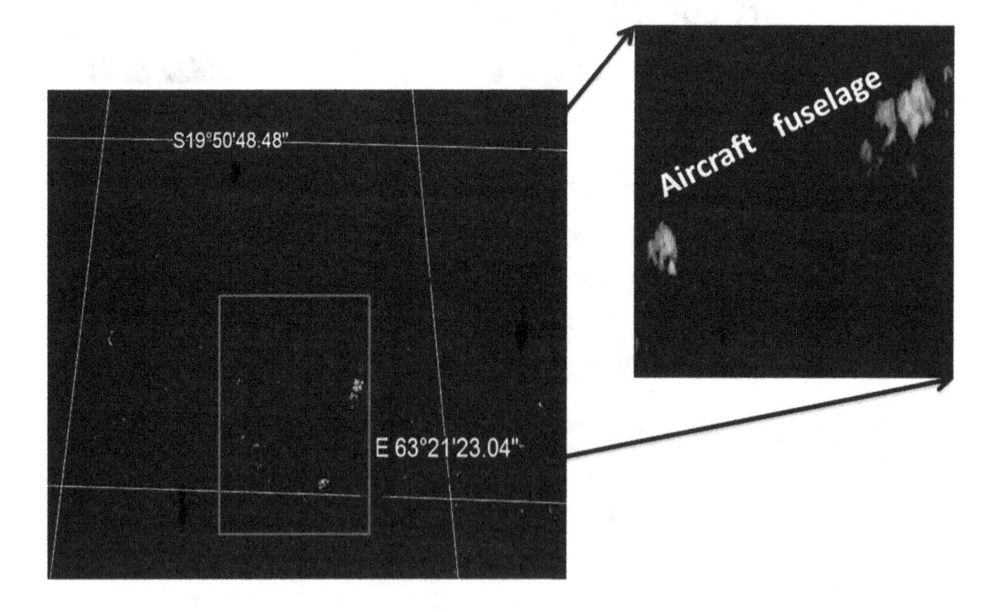

11, 2009, specifically, five years before the flight disappeared. Besides, this site is located between 19°49'42.77"S and 57°48'16.79"E, which is approximately 1 mile northeast of Round Island. Consequently, these random patterns are shown in both images dated June 11, 2009, and October 27, 2014, respectively, that present the ocean wave propagations. Moreover, the deepwater far north coastal zone of Mauritius never allows the airplane fuselage to remain longer on the water surface (Fox, 2016). In other words, the flight MH370 must sink deeper that that. The collapse of recent findings to solve the mystery of MH370 vanishing will raise the following critical questions.

THE SEARCH AND RESCUE: WHAT HAPPENED NEXT?

Consequently, the Malaysian navy radar confirmed that MH370 traveled over the Malacca Straits, while the blue circle (Figure 4) shows the disappearance of MH370 from the radar screen, near Penang Island (Figure 4).

Figure 4. Last appearance of MH370 on the radar screen

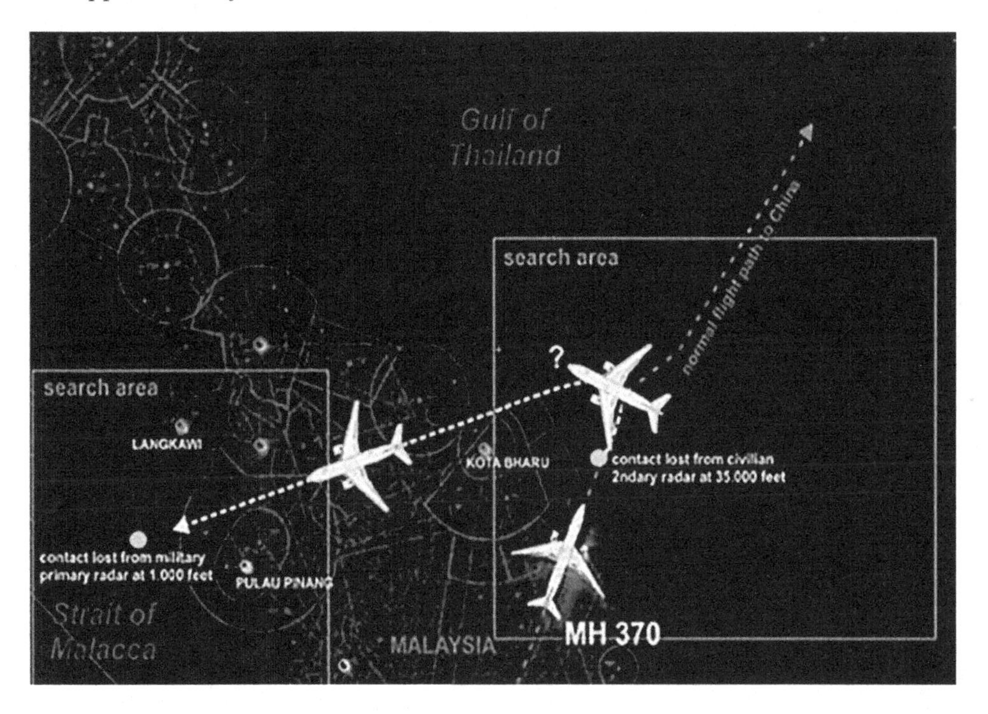

The Malaysian navy radar confirmed that the airplane changed rapidly its path to Beijing. It was suggested that the airplane continued flying for more than 6 hours until it lost contact with air control. In this view, the fuel must runoff for those 6 hours of concern. In fact, at 01:06 MYT, the total fuel remaining was about 43,800 kg (96,600 lb) (Team, 2015 and Davey et al., 2016). In this concern, MH370 could not fly to the southern Indian Ocean as the required adequate fuel was more than 43,800 kg. In other words, the airplane was geared up with sufficient fuel for 6 hours, beginning in 1642. It is expected that there would be extra fuel for an emergency. In this regard, MH370 was able to fly until 0019 (approximately 7.5 hours). Therefore, Yu (2015) stated that the aircraft must have been flying economically – at a speed that minimized fuel consumption at a low altitude.

WHEN WAS DEBRIS FROM THE PLANE DISCOVERED?

The briefing confirmation of MH370 debris has been reported as follows.

In July 2015, the flaperon was found in Reunion and in May 2016, a wing fragment was found in Mauritius. Moreover, in June 2016, a wing flap was found on Pemba Island, off the coast of Tanzania. Besides that, from the wreckage there was found smaller pieces of wing or tail. However, the fuselage of the aircraft has not been found.

Later in December 2015, a flap track fairing was found on a Mozambique beach. Also in February 2016, a horizontal stabiliser is found on a Mozambique beach. Finally, in March 2016, a section of the cabin bulkhead was found in Mauritius and an engine cowling bearing a Rolls-Royce mark that matches the type used by Malaysian Airlines.

In more detail, a week after the disappearance of flight MH370, satellite images of promising debris suggested that airplanes crashed in the Indian Ocean, south-west of Australia. Nevertheless, the exploration of the black box of MH370 in the confirmed search area is not considered valid as long as the airplane was not discovered. This is proof that the search area carried in the wrong location. Consequently, on July 29, 2015, more than a year after the disappearance airplane, debris was discovered by volunteers cleaning a beach in St Andre, Reunion. As a result, investigators verified the debris came from the missing flight. It did not, conversely, assist to trace the airplane as it had flowed into the Indian Ocean. Later, on October 7, 2017, it was declared that two wing flaps located in Mauritius did belong to MH370. Finally, in August 2017, Australia broadcasted satellite imagery reported that 12 artificial MH370 debris, which found floating near the search area (Green, 2014 and **Bureau,2015**). The satellite images were detained by a French military satellite two weeks after the disappearance of Flight MH370, nevertheless never release to the public. In this regard, why the French authority did not share this information with the Australian?

WHAT HAS MH370'S DEBRIS REVEALED?

The Australian transport safety Bureau reported that the aircraft was not ditched accurately in the Southern Indian Ocean after the flaperon has split away from MH370. If the pilot turned into planning, a managed ditching of the aircraft, the wing flaps might have been configured for landing. However, MH370 seemed to be out of control during its flying. In this debate, if someone had been controlling the airplane, it could have covered a great distance than believed. In this concern, the MH370 investigators could survey the wrong search area. A deep-sea sonar signal cannot the fuselage or black box on the missing flight. Conversely, there is more than 20 MH370 debris confirmed. These wreckages have washed ashore on coastlines all over the Southern Indian Ocean (Davey et al., 2016).

WHAT ARE THE PRESUMPTIONS BEHIND THE MISSING MH370?

We live in an age and era, where technology tracks every tiny change in the Earth. Why is then it is extremely challenging to retrieve a missing MH370? The Malaysia Airlines flight went missing on March 8, 2014, and every new piece of information seems to shroud the disappearance of flight in more mystery. Theories are involved hijacking, sabotage and the possibility of the pilot suicide or pilot error (The Star online, 2016 and Saroni et al., 2019). With no accurate facts, these kinds of details are becoming almost conceivable. With no mayday call, no data and no wreckage, there are too many theories to follow. This is the perplexing question that the world is still struggling in the four years since the disappearance of Flight 370. In other words, there are numerous causes for MH370 vanished, nevertheless, the details are imperfect. The investigators, therefore, are just struggling to investigate how and where MH370 vanished?

Among the rejected theories is mechanical failure or fire. Indeed, a fire would have had to be vastly adequate to not offer the crew any time to communicate their emergency (Wright 2010;Davey et al., 2016; Saroni et al., 2019). Furthermore, the mechanical failure did not allow the airplane to continue flying until its fuel run out. Consequently, it also rejects the theory of flight decomposers. In this circumstance, MH370 must be visible on radar and keeping automatically flying toward Beijing.

Shoot-Down Theory

The South China Sea is a navy conflict zone between U.S. and Chinese army Forces. Tensions have elevated over the numerous years after it becomes learned that China had located surface-to-air-missiles on one of the Parcel Islands. Since the incident, U.S. Pacific Command chief Adm. Harry Harris has seen that China is militarizing the region. In this regard, certainly, one of China's fantastically advanced islands within the northern part of the South China Sea, Wooden Island, has been geared up with surface-to-air missiles and fighter aircraft. These actions have come simply as many defense analysts, which have anticipated for years and are likely a demonstration of factors to return for China's other island outposts in the course of the South China Sea (Cawthorne, 2014; Fox, 2016; Saroni et al., 2019).

Particularly, the Chinese activity is around Scarborough Shoal in the northern part of the Spratly archipelago, approximately 125 miles (200 km) west of the Philippine base of Subic Bay (Tyler,2016). In this view, Cawthorne (2014), suspected that when the jet was shot down at some point of a US-Thai Joint Strike Fighter jet training exercise, searchers deliberately have been sent off target as a part of a sophisticated cowl-up. Further, Cawthorne (2014), confirmed that New Zealand oil rig worker Mike McKay also claimed to have seen a burning airplane going down in the Gulf of Thailand. However, there is no any MH370 debris has been reported in the Gulf of Thailand. The Cawthorne 's book received a massive grievance, in particular from the Australian authority, where it became characterized for this reason: "Cawthorne undoes anybody's properly paintings with the aid of retrieving each obsolete and discredited non-reality from the trash, slapping the whole lot among the covers." Spouse and children of those aboard Flight 370 criticized the e-book as "untimely and insensitive. Consistent with Cawthorne (2014), four Malaysian policemen were reported about a low-flying aircraft, believed to be the missing flight MH370, was lodged by eyewitnesses from Tumpat and Bachok, which was headed north towards the South China Sea. Based on the reports, the plane was sighted between 1.30 a.m. and 1.45 a.m. A bus driver, who gave his voluntary statement on Sunday, 8 March 2014 said he saw a low-flying plane at Penarik, Setiu, Terengganu, at about 3.45 am the same day flight MH370 went missing (Iannello, 2014 and Saroni et al., 2019).

Finally, it was also reported that the co-pilot attempted to make a midnight call from his mobile phone before the plane vanished, but the call ended abruptly. In this regard, the shutdown was occurring at the speed of light due to internal or external exploration, which ended the call rapidly. However, the exploration would have left a radar signature and MH370 debris would float in water.

Cyber Attacks

It has been speculated that novel technology is known as cyberjaked used to hijack the airplane remotely. In other words, hackers may have changed the airplane's speed, direction, and altitude employing radio signals in the airplane's flight control system. Whether existing security on commercial flights is sufficient to prevent such an attack is also a matter of debate, although Boeing has dismissed the possibility. Nevertheless, all the Boeing airplane control systems are precisely protected and a hacker can't gain either external nor internal access processes (Iannello, 2014; Yu, 2015; Wang, 2016; Morrell, and Klein, 2018; Network, 2018; Saroni et al., 2019). Finally, if MH370 was hijacked, it would not have vanished from radar. The certain assumption is MH370 vanished somehow cannot yet recognize.

Cracking of the Fuselage Skin Lower SATCOM Antenna

It is suggested that the satellite communication (SATCOM) antenna adapter dominated by cracking of the fuselage skin. In other words, MH370 had lost communication with SATCOM because of an explosion triggering a hovel in the fuselage. Conversely, the SATCOM system can deliver the exact location of the missing flight over the southern Indian Ocean (Davey et al., 2016).

MH370 Behind Terrain

It advocated that the airplane was flown at a lower elevation of 5000 ft as confirmed by a radar signal. In this view, the radar detectors were not able to spot the aircraft. In other words, the radar range signals were blocked by terrain topography and the flight was invisible (Figure 5).

Figure 5. Concept of train impact on radar signal

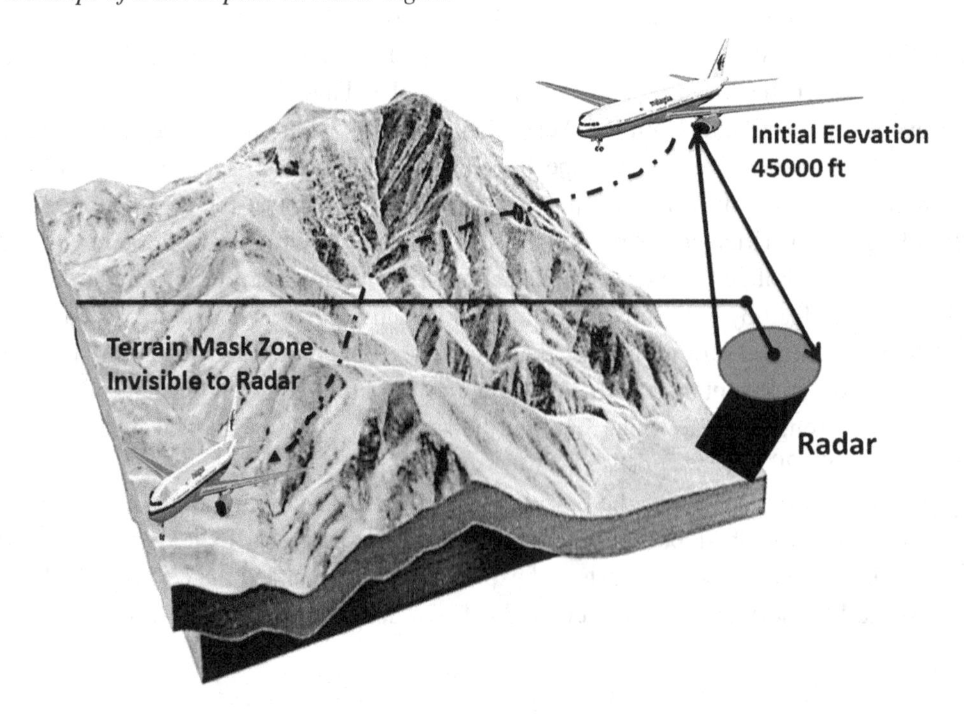

Besides, if the flight made a U-turn across the complicated terrain, it could be crashed into the mountains and heavy forest covers (Cawthorne, 2014; Green, 2014; **Bureau,2015**).

Pilot Suicide

One of the confusing conspiracy theories behind the missing MH370 is pilot suicide. This theory is based on the route of MH370 was recorded by Malaysian military radar. The theory suggests that Captain Zaharie dipped the wing of the missing flight MH370 over his hometown of Penang in northwest Malaysia to have a last look before he committed suicide.

In this evidence, experts concluded that Captain Zaharie committed suicide and killed everybody else on board, and he did it deliberately.

On the contrary, the experts do not understand deeply the Malaysian culture, which prevents the Malaysian from such a dramatic event. If Captain Zaharie wants to commit suicide, he must decide on a short flight path to end his life. This will discuss extremely in chapter 5.

The experts additionally concluded, based on the physical evidence recovered with the aid of reliable investigators, that the aircraft did not strike the water in a sharp dive as previous reviews counseled. However, the airplane was controlled until plunged into the southern Indian Ocean, near-coastal zone of Perth, Australia.

Diego Garcia Conspiracy

Several articles and a book have expressed many exceptional unproven theories. No phase of the aircraft has been officially found. The modern-day unverified theory, written with the aid of former Proteus Airlines CEO Marc Dugain and published through Paris Match, claims the Boeing 777-200 ER can also have been hijacked by using a "remote control system" and perchance shut down utilizing U.S. forces near Diego Garcia in the Indian Ocean. The U.S. has denied the plane came down near the British island.

WHAT HAS THE FINAL REPORT SAID?

Despite the search area was extended to intact 25,000-kilometer square in the southern Indian Ocean, the Australian Transport Safety Bureau (ATSB) is unable to determine the location and the fuselage of the vanished MH370 or to answer how did it end up in the Indian Ocean?. Therefore, the report cannot answer why did the plane divert from its planned route. Further, the underwater bathymetry was surveyed most of the suspected areas, which are simulated based on the MH370'route. However, the search was ended up with no findings. Conversely, it is intolerable to survey all the Indian Ocean due to an extremely expensive cost and bad weather conditions. On January 31, 2018, the searching ship disappeared from the radar for 80 hours (Team, 2015). It is still unclear what did cause that. Consequently, the Australian-led underwater search for MH370 is officially suspended.

SEARCH AREA

Flight Routes

What was the accurate flight route? It is a critical question which is required logical order way of thinking, which can assist to detect accurately the search area. In this understanding, prior to the search area, the flight paths must be clarified. This can assist to investigate how did the MH370 vanish. Figure 6 reveals the number of the possible routes of MH370 to the southern Indian Ocean, and three boxes indicate where MH370 likely ended. On the word of Green, (2014), these flight paths differ based on different projections of the aircraft's speed (Saroni et al., 2019).

Figure 6. Flight paths from Malaysia to Indonesia

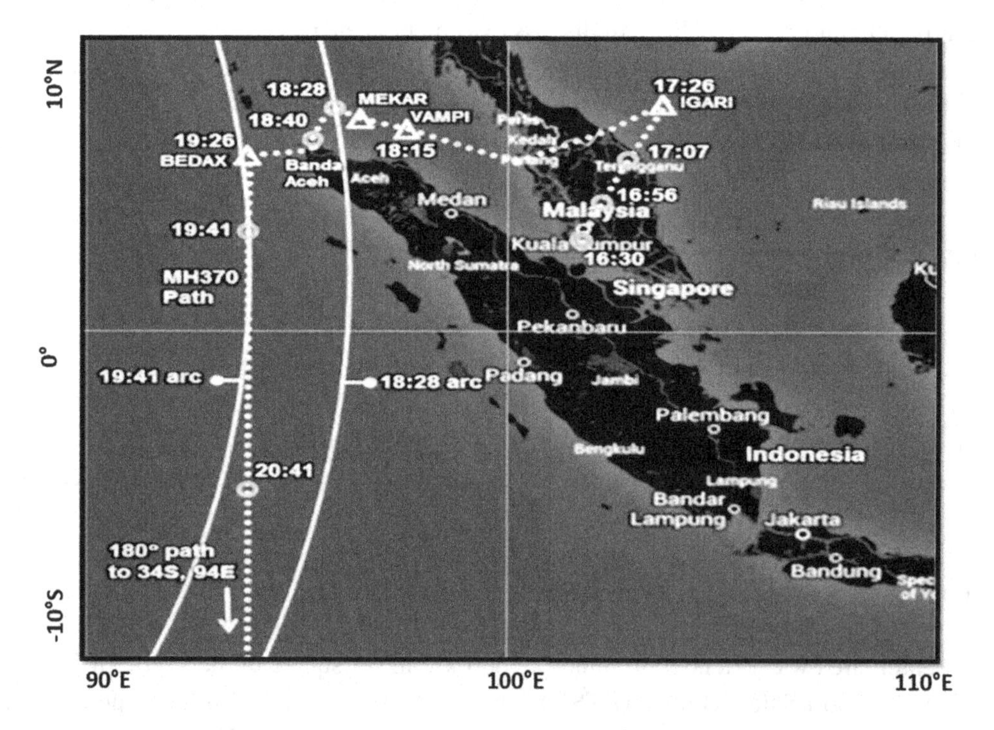

It is strange that the proposed airspeed is extremely low. Typical cruise airspeed of 777 is about 490 knots, not 350 or 320 knots. If we assume that the airspeed was actually 450 knots, the possible path of the aircraft is very different and leads to a very different fact. In the Burst Offset Frequency chart from 19:41 UTC until the final ping at 00:11 UTC the track lines for MH370 show increasing frequency, which means the flying was towards the satellite (Figure 7). Consequently, the Burst Offset Frequency was assumed to rectify the flight track. In this view, track curves must be to the west (6th Arc) not east (7th Arc) (Figure 8). If the track curves were fundamentally wrong, how could this ping chart have used to plot the final search area of the missing flight MH370?

Figure 7. Ending path of MM370 over the Indian Ocean

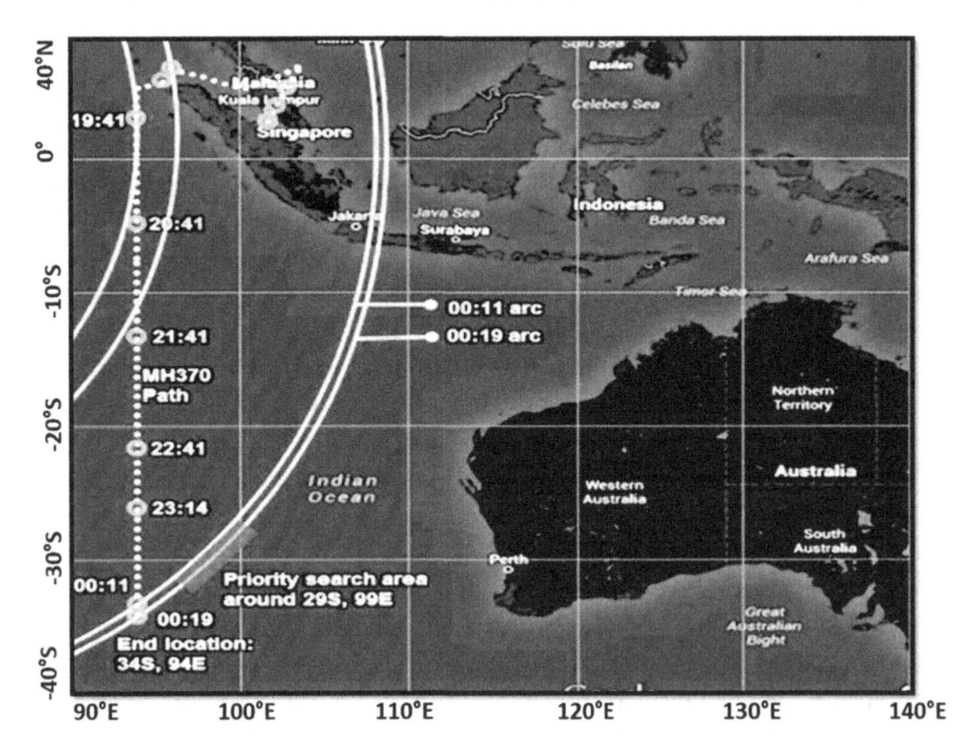

Figure 8. Arcs number 6 and number 7

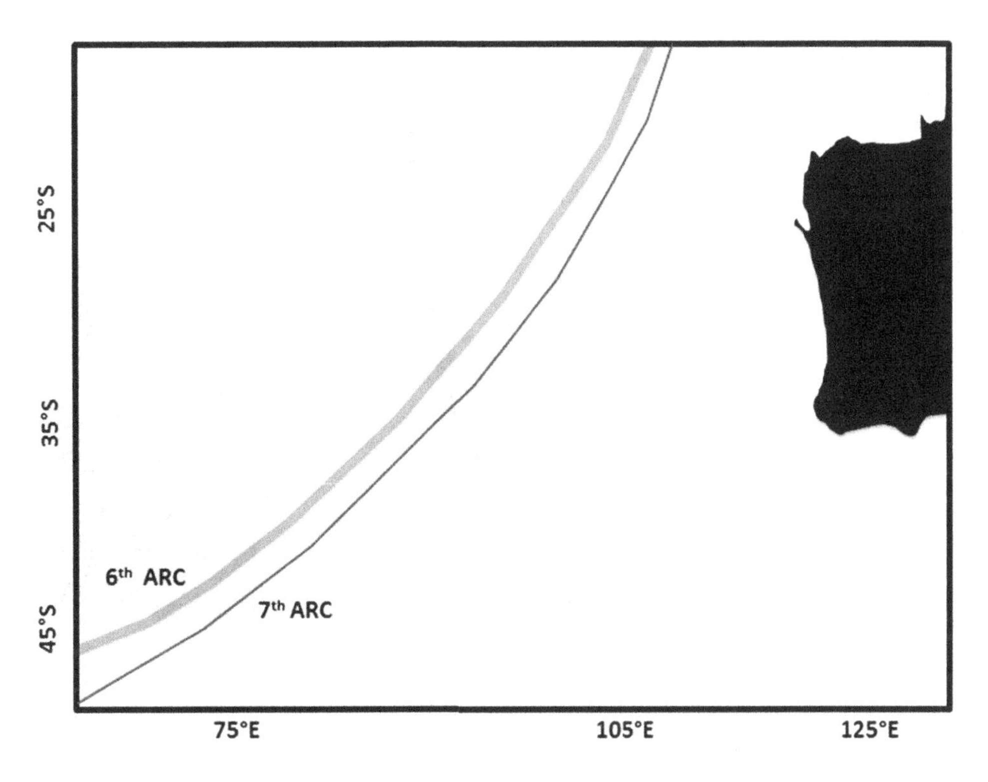

Frequency Offset

The "frequency offset" is a Doppler-effect-derived value. It is related to the speed of the aircraft towards/ away from the satellite, not its distance; the rising line is consistent with an aircraft slowly turning away from the satellite. On the contrary, it is also consistent with the aircraft performing several more improbable maneuvers(Monks, 2014). Also, the time that it acquires the signal to be referred and sensed, through the satellite, to the ground station can be depleted to determine the distance of the aircraft from the satellite (**Bureau,2015**).

The maximum range cruise (MRC) routine was estimated for a turn soon subsequently the last radar contact near the 1st arc of 1828 and at the time of the 1st telephone call about 1840. These scenarios were used to appraise the MRC frontier as they denoted both an early and a late turn (Figure 9). In this understanding, the direction and velocity were reserved so that all tracks transacted the burst frequency offset (BTO) curvatures at the epochs of the handshakes. Consequently, the paths that delineated, and were within, the MRC frontier which was a far the 7th arc. In this view, arcs were the paths for a turn near the 1st arc of 1828 were 35,000 ft, 30,000 ft, and 25,000 ft, respectively (Figure 10) and for the arc of 1840 were about 40,000 ft, 35,000 ft, 30,000 ft, and 25,000 ft, respectively (Green, 2014 and Iannello, 2014). Therefore, the endpoints of the MRC boundary are combined with the routes that dismissed very near the 7th arc. To the south of the arc of 1828 was about 40,000ft and to the north of the arc of 1840 was approximately 20,000ft. These calculations were considered under constant airspeed, but ground speed varied because of the wind impacts.

Figure 9. Maximum range cruise (MRC) routine

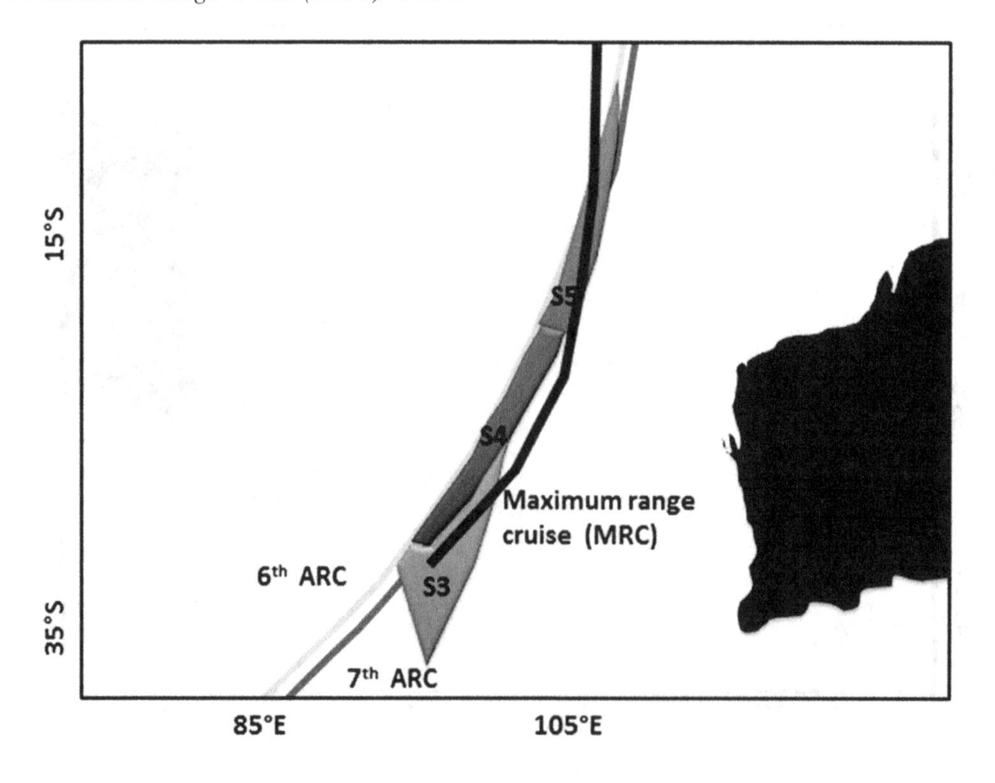

Figure 10. Burst frequency offset (BTO)

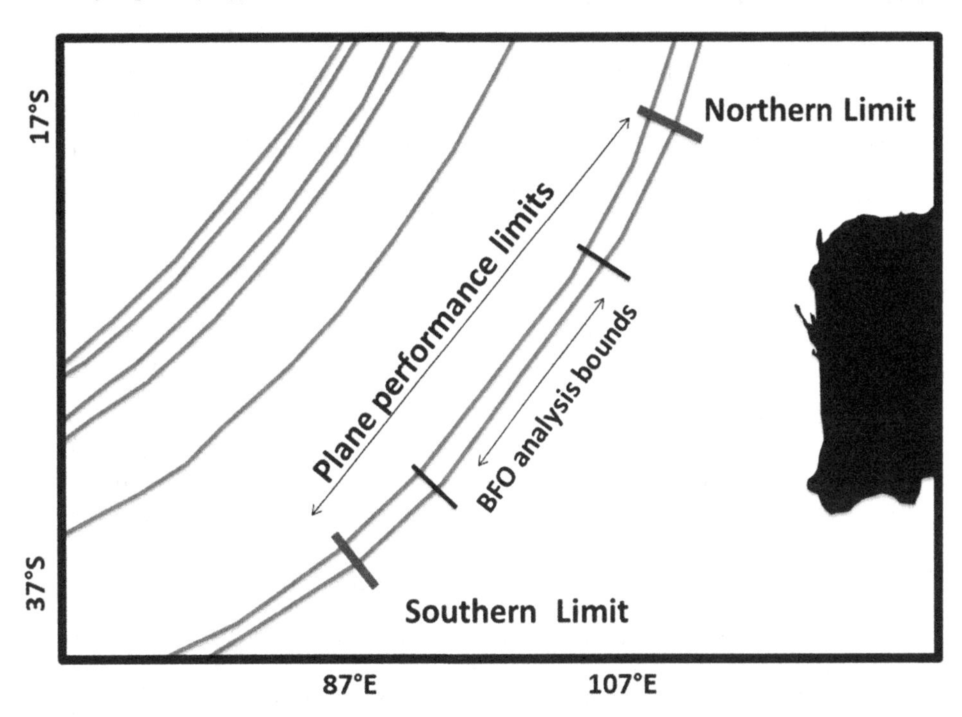

Some southern routes from this analysis were beyond the expected routine boundaries of the gray area which is involving acceptance for the MRC occurrence frontier approximation of the airplane. The routine boundaries were fairly confident. However, the results south of the southern routine boundary were not counted reasonable.

Truly, the MH370 could be overturned for the northern bearing. Nevertheless, the southern route of the MH370 is considered a reliable guess. In this regard, the southern route is based on a measure of the Doppler signal effects, which were sent to the satellite (Green, 2014 and Iannello, 2014).

The main question can be raised about how the flight path is represented in two arcs? However, these two arcs are not considered as the MH370 routes, but they are represented as the last satellite ping that occurred along the arcs. How can anybody create a computer model with just this information? Or does it mean that the models have more information available than the general public?

How the Search Area was Identified?

Using various assumptions, the flight path was simulated using the combinations of aircraft altitudes, speeds, and headings generate candidate paths and calculated the BFO values at the arc crossings for these paths. These values, compared with the recorded BFO values, provided a measure of statistical consistency. The latest refinements to the SATCOM system model affected the calculations of the BFO, therefore the various analyses were required to be repeated. The analysis methods were categorized into two classes:

1- Constrained autopilot dynamics - The aircraft was assumed to have been flown using one of the autopilot modes. Each of these modes is modeled to generate candidate paths. The satellite data (BFO, BTO) values were then calculated for these flight paths and compared against the recorded values.
2- Data error optimization - The candidate path was broken up into steps using successive recorded BTO/BFO values. At each step, speed and/or heading values were varied to minimize the error between the calculated BFO of the path and the recorded value. Paths were not constrained by the aircraft autopilot behavior.

End-Of-Flight Scenarios

Figure 11 presents the end MH370 along the 7th arc over the west coast of Perth, Australia. To estimate and have confidence in a reasonably wide search area, it is important to understand the aircraft system status at the time of the SATCOM transmission from the aircraft at 0019.29 UTC (Table 1) (log-on request), and the variations in aircraft behavior and its trajectory that were possible from that time.

Figure 11. The scenario of MH370 vanishing along the search area

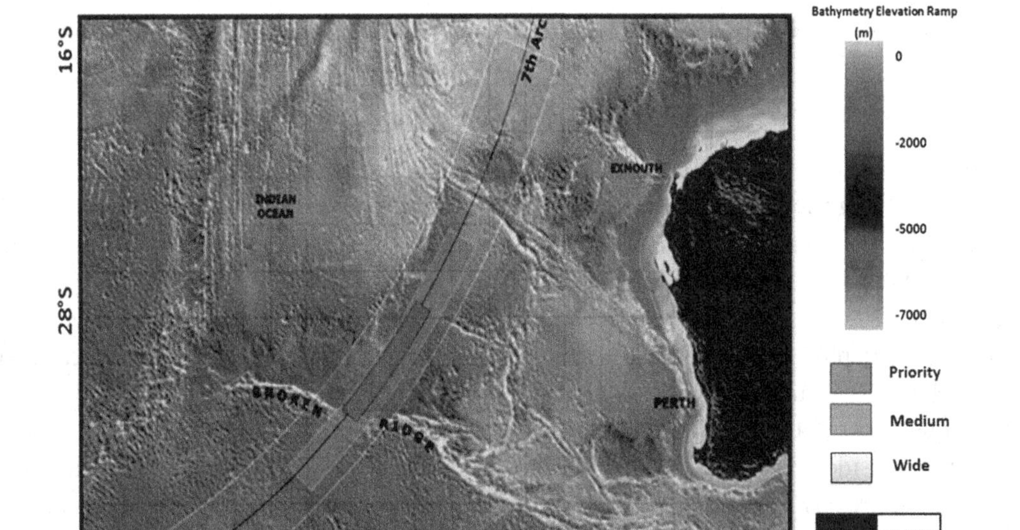

The log-on request recorded at the final arc occurred very near the estimated time of fuel exhaustion. The recorded BFO values indicated that the aircraft had descended at that time. In Particular, aircraft systems, which involve the electrical and auto flight system were precisely functioning. However, the simulator activities involved fuel exhaustion of the right engine followed by a flame out of the left engine with no control inputs. This scenario resulted in the aircraft entering a descending spiraling with a low bank angle and left turn. Then, the aircraft plunged into the water in a relatively short distance after the last engine flameout. Nonetheless, when consideration of the arc tolerances, a record of messages and simulator activities are combined, it indicates that the aircraft may be located within relative proximity to the arc. Whilst the systems analysis and simulation activities are operating, based on the analysis to date, the search area width described in the ATBS's report remains reasonable with the underwater search to commence at the 7th arc and progress outwards both easterly and westerly (Davey et al., 2016).

However, a lot of conflicting and missing information remains. The arcs shown in Figures 7 to 10 are based on the estimation made at the time between the ping going out and the one received by satellite. In this understanding, the 0.001-second error of time delay would produce a location error of about 1000 miles. Besides, ping arcs at 1:11, 2:11, 3:11 and 4:11, respectively are not delivered. Finally, there is a conflict about the content of 1:07 ACARS transmission. It is worth mentioning that the Malaysian Airline did not contract for ADS-C information. On the other hand, it is also mentioned that the 1:07 transmission included it (Iannello, 2014; Wright 2014; **Bureau,2015**).

DYNAMIC OCEANOGRAPHY AND MH370

Dynamic oceanography models the dynamic relationship between the ocean's physical properties, the atmosphere, and the coast and seafloor. The ocean's physical properties include temperature, salinity, and density variations (BoM, 2012 and **Bureau,2015**). The ocean's dynamic components are waves, currents, and tides. Physical oceanographers additionally investigate how the ocean interacts with the Earth's atmosphere to generate climate and weather systems. Under this circumstance, physical oceanography theories and models must extremely be instigated to resolve the mystery of flight MH370.

The observational procedures which are taught to undergraduates of physical oceanography do not work in this case. Indeed, standard and modified models are required to verify the information from the *Inmarsat* satellite. Many researchers just use physical oceanography models and do not understand how the models operate. Physical oceanography scientists who have been correctly trained have deep background knowledge of physics, mathematics, numerical analysis, finite element models, and modeling and can verify information delivered by any source.

To date, some marine institutes with a low quality of physical oceanography research are unable to explain, scientifically, how flight MH370 vanished in the southern Indian Ocean. It is believed that researchers are not even able to solve the 1-D mathematical coastal water flow equation! The main physical and dynamical oceanography work is based on the blind use of software, models and ocean observations without understanding the physics and mathematics beyond them. Ocean data observations do not represent the end work of scientists - they need to go beyond in situ data collection by developing models and solving boundary condition problems.

For instance, the turbulent water flow throughout the further south of Africa - known as the Agulhas Current - is a portion of a superior "ocean conveyor belt" which distributes the water across the globe. The Agulhas Current is based on currents, wind, and water density changes. Scientists have defined a new finding, named the Agulhas leakage. This represents the increments of water flow from the southern Indian Ocean into the Atlantic Ocean. It may well be that flight MH370 debris moved to the Atlantic Ocean through the Agulhas leakage.

Modern physical oceanography studies predict that the ocean conveyor belt will be slowed down, something which could cause dramatic changes to weather and climate patterns, and consequently the sea level could rise due to the melting of ice caps. This melting could reduce the salinity of ocean water and change its biogeochemical matrix. Consequently, these circumstances could affect the trajectory movement of MH370 debris across the ocean.

CONCLUSION

This chapter has reviewed the theories and conspiracies behind the crashing of MH370 into the Indian Ocean on March 8, 2014. It can be demonstrated that the complicated challenge of understanding these theories is that some of these theories are not scientifically grounded. The chapter also reveals uncertainties of tracking MH370 as some of the signal information regarding BTO and BFO are not available. This can lead to misunderstanding the scenario of MH370 crashing. The significance of dynamic oceanography in understanding and tracking the MH370 vanishing in the Indian Ocean can be ignored.

Along with the above perspective, the next significant view must be the focus on novel theories that can explain the dynamic impact of flap separation from the fuselage on the MH370 vanishing. The accurate explanation can lead to understanding perfectly the scenario of MH370 crashing in the Indian Ocean. The next chapter will deal with the impact of flap dynamic failure on the MH370 crash scenario.

REFERENCES

BoM, A. (2012). APS1 Upgrade of the ACCESS-G Numerical Weather Prediction System. *NMOC Operations Bulletin*.

Bureau, A. T. S. (2014). *MH370: Flight Path Analysis Update*. Australian Transport Safety Bureau.

Bureau, A. T. S. (2015). *MH370 search area definition update.* ATSB transport safety report, External Aviation Investigation AE-2014-054.

Cawthorne, N. (2014). *Flight MH370-The Mystery*. Kings Road Publishing.

Davey, S., Gordon, N., Holland, I., Rutten, M., & Williams, J. (2016). *Bayesian Methods in the Search for MH370*. Springer Singapore. doi:10.1007/978-981-10-0379-0

Fox, N. (2016). *US Navy deploys several ships to South China Sea as tensions rise*. https://www.foxnews.com/world/2016/03/04/us-navy-deploys-several-ships-to-south-china-sea.html

Green, J. J. (2014). *Boeing rules out cyber sabotage connection to missing plane.* WTOP-FM.

Iannello, V. (2014). *MH370 Scenario with a Landing At Banda Aceh*. http://jeffwise.net/mh370-scenario-with-a-landing-at-banda-aceh-by-victor-iannello-august-23-2014//

Monks, S. (2014). Flight MH370: International law and how we use it. *Austl. Int'l LJ*, *21*, 101.

Morrell, P. S., & Klein, T. (2018). *Moving boxes by air: the economics of international air cargo*. Routledge. doi:10.4324/9781315180632

Network, A. S. (2018). *Criminal Occurrence description*. Academic Press.

Saroni, A. N., Abd Samat, M. A., & Ibrahim, J. (2019). The case study of emergency response plan (ERP) implementation during the Malaysia Airlines flight MH370 disappearance. *Malaysian Journal of Computing*, *4*(2), 270–277. doi:10.24191/mjoc.v4i2.4809

Team, M. (2015). Factual Information Safety Investigation for Mh370–Malaysia Airlines MH370 Boeing B777-200ER (9M-MRO) 08 March 2014. *Malaysian ICAO Annex, 13*.

The Star Online. (2016). *Missing MH370: Cops get reports on low-flying aircraft and loud noise*. https://www.thestar.com.my/news/nation/2014/03/12/cops-get-reports-on-lowflying-aircraft-and-loud-noise/

Tyler, R. (2016). *Surface To Air Missiles Arrive On China's Island Outpost In The South China Sea*. https://foxtrotalpha.jalopnik.com/surface-to-air-missiles-arrive-on-chinas-island-outpost-1759519656

Wang, Z. (2016, May). Searching for the Accident Plane. In *2016 2nd Workshop on Advanced Research and Technology in Industry Applications (WARTIA-16)*. Atlantis Press. 10.2991/wartia-16.2016.142

Wright, T. (2010). The 727 That Vanished One more intrigue that seemed to have 9/11 written all over it. *Air & Space Smithsonian*, 48.

Wright, T. (2014, July 30). Unlawful interference not ruled out in MH370 disappearance. *Financial Times*.

Yu, Y. (2015). The Aftermath of the Missing Flight MH370: What Can Engineers Do? *Proceedings of the IEEE*, *103*(11), 1948–1951. doi:10.1109/JPROC.2015.2479336

Chapter 2
Novel Theories Behind Flap Failure Impact on Flight MH370 Crash

ABSTRACT

This chapter is devoted to delivering the fundamentals of aerodynamics. In this view, a perfect understanding of the principles of aerodynamics can assist to investigate how the MH370 crashed in the Indian Ocean. Bernoulli's principle is the keystone to understanding the aerodynamic mechanism of MH370. This chapter addresses a new theory of the MH370 crash based on the turbulence impacts on the airframe. In the circumstance of the MH370 air crash, the dynamic processes are addressed to reveal how the MH370 was uncontrolled.

INTRODUCTION

Understanding aerodynamic principles can assist in the investigation of aircraft crashes. Rationally, the aircraft crashes because of one of two reasons: (i) there is either pilot error or (ii) technical error involved. The malfunctioning of the stabilizers and the flaps on the plane's wings are major technical errors. In this sense, the deep details and comprehensive perception of plane flying and crashing mechanisms are required. This chapter is devoted to explicating the different dynamical scenarios that can cause Flight MH370's vanishing into the southern Indian Ocean. Further, this chapter will answer the question as to how did Flight MH370 end up in the Indian Ocean?

PRINCIPLES OF FLIGHT FLYING

It is well established that Bernoulli's Principle is a keystone for dynamic flying. More specifically, an airplane's wing is designed in a manner that triggers the air drifting over the wing to travel a farther distance than that existing beneath the wing at the matching time. In this sense, this drives a faster

DOI: 10.4018/978-1-7998-1920-2.ch002

airflow at the top of the wing than its bottom (Figure 1). Consequently, the air on the bottom wields higher pressure on the wing than the air on the top. Then the airplane is driving up because of the higher pressure below the wing. The counteracting force from above the airplane is much less (Wegener **1997**).

Figure 1. Bernoulli's principle for aerodynamic

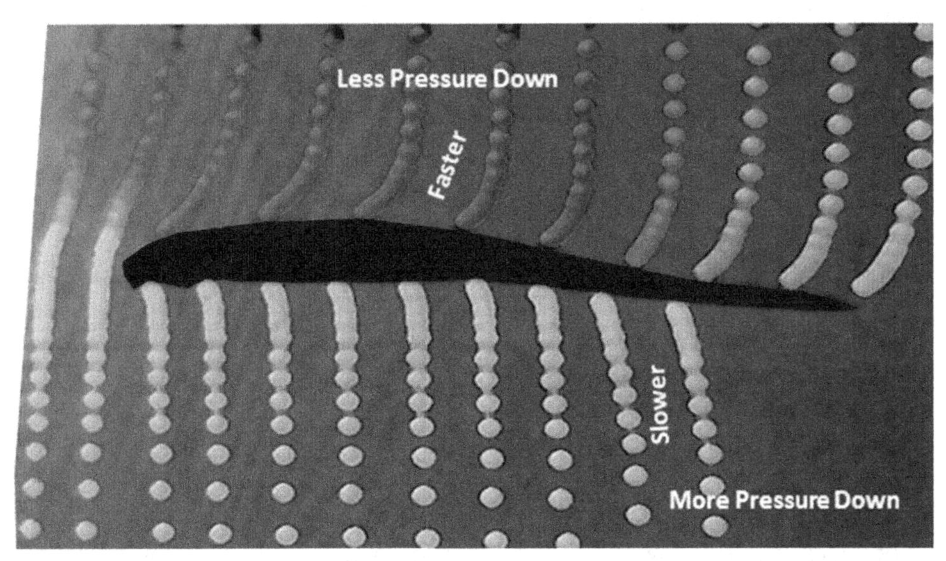

Wegener (1997)and Anderson and Eberhardt (1999) stated that the flaps and stabilizers adjust the heaviness alterations between the areas overhead and beneath the wing, triggering the plane to revolve or move up and down. Defective flaps or stabilizers will instigate the pilot to misplace control of the aircraft, predictably consequential in a crash.

As said by Cook (2012), all the forces and moments operating on an aeroplane are a result of pressure influences. In this regard, the pressure is acting normal to the aeroplane surfaces. Therefore, the shear force is acting along the aeroplane surfaces. Thus, both forces are formed by the airflow pattern configuration over the aircraft body. In this understanding, the moments (torques) are a product of the forces and, as well as a measure of rigid body motion analysis. Along with Anderson and Eberhardt (1999), Aircrafts can commonly be considered as rigid bodies once exploring the dynamics of their mobility. The lift force "holding" a plane up is generated by airflow over the wings. This airflow is only possible if the plane moves relative to the air, hence lift is only possible if the plane moves relative to the air. And the relative airspeed must be large enough for sufficient lift to be generated (Wegener, 1997and Anderson and Eberhardt 1999).

Another common technical error in aeroplanes is an engine failure. When a plane's engine fails, it loses its forward thrust and is slowed down by intense air resistance. In order for Bernoulli's Principle to work on an object as massive as an aeroplane, there must be a large speed involved. As the plane is slowed, the air on top of the wing does not travel as fast and pressure increases. This starts to equalize the pressure above and below the wing, and the plane starts to drop. Engine failures can be brought about by many different reasons, such as faulty wiring or poor quality fuel (Goel and Gupta, 1984).

Airfoils

In aerodynamics, plane wings are known as airfoils. They have a cambered profile which permits them to create lift, even for angles of attack (α) equal to zero (Wegener, 1997) (Figure 2.).

Figure 2. Plane lift with α equals zero

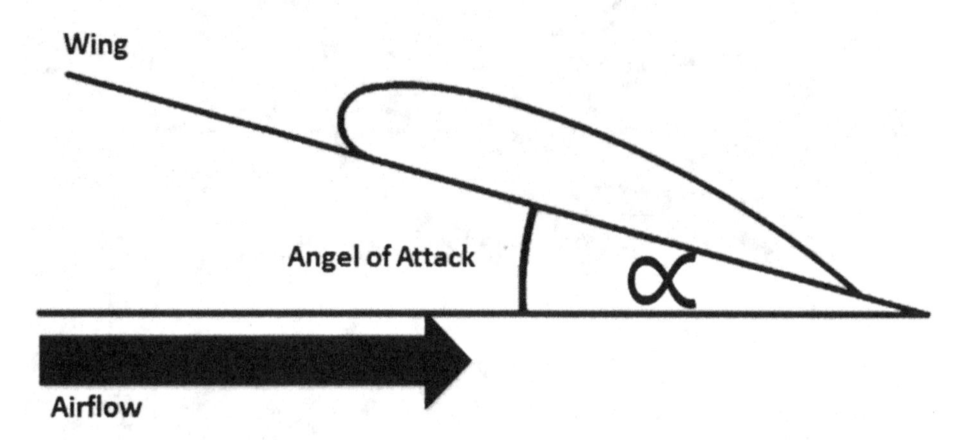

The angle of attack specifies how tilted the wing is to conduct to the looming air. So as to create a lift, or aloft dynamism operating on the wing, Newton's third law articulates that there must be an equal force acting in the opposite direction. If we can employ a potency on the air, so as it is directed down (Mamou and Khalid 2007), the air will bring to bear an aloft force back on the wing. Figure 3 depicts how the Coanda influence directs the airflow for diverse angles of attack.

Figure 3 spectacles that growing the angle of attack amplifies how much the air is bounced downwards. If the angle of attack is excessively large, the airflow will no longer tail the camber of the wing. In this circumstance, a small vacuum will be created just behind the wing. As the air hastens in to bridge this gap, termed cavitation. More specifically, it instigates dense pulsations on the wing. In this regard, cavitation impressively reduces the effectiveness of the wing. Hence, aeroplane wings are commonly slanting (Figure 4). This wing usefully leads the airflow descending, which in turn pushes up on the wing, creating lift.

This technique of determining lift, therefore, is termed momentum change. Further approaches computing the same lift exploit the alteration in pressure forces overhead and beneath the wing. Either technique is precise on its own, nonetheless never complements the two approaches mutually.

As well as creating lift on an aeroplane, Bernoulli's principle and the Coanda effect perform an important function in the manoeuvre of a propeller. More specifically, the blades of the propeller seem like an airfoil or wing. Principally, a propeller blade is a wing turned on its side. Just as wings touring forward are lifted aloft, a rotating propeller blade is forced or strapped frontward. Therefore, a propeller blade also has something that wings don't: they are warped. If a propeller turns very slowly, the twist of the blade triggers it to exchange the air consistently and move it backward. Moreover, the propeller blades are located at an angle. This is known as the propeller pitch (Figure 5). The larger the pitch of the propeller, the more air it can move. To maintain an even flow rate as much as possible, the hub pitch

Figure 3. Coanda Effects

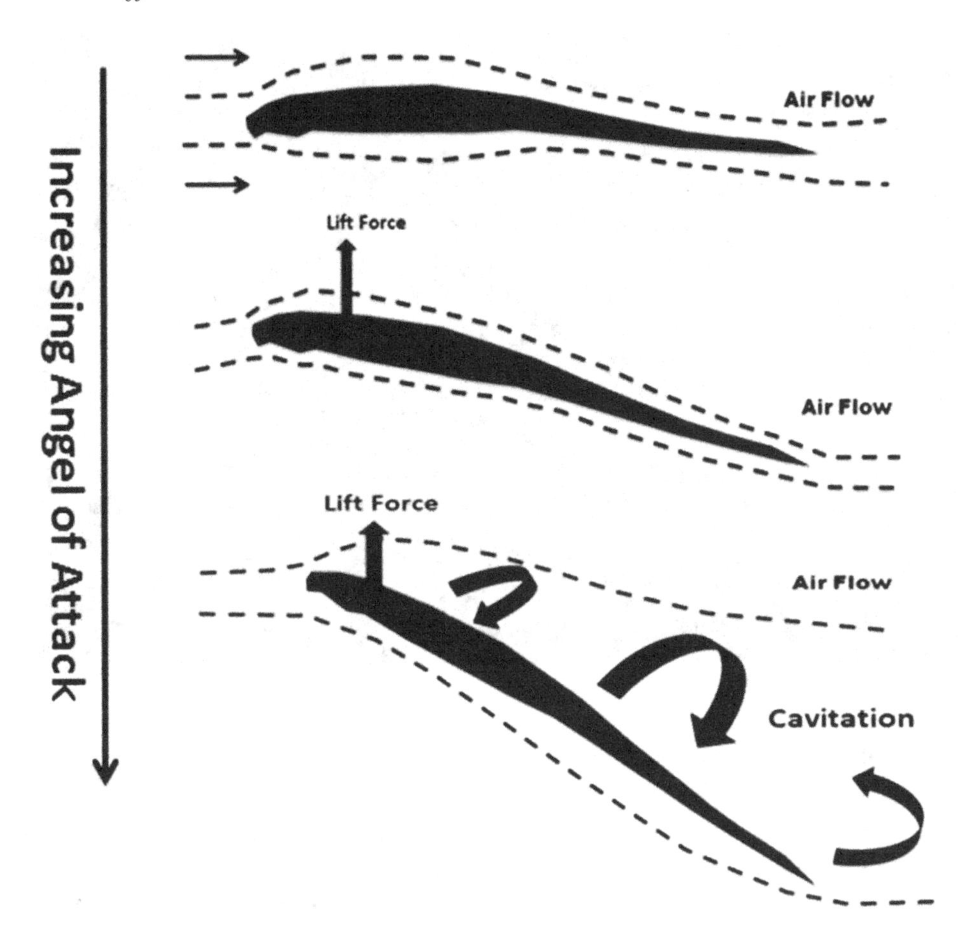

Figure 4. Aeroplane slanting

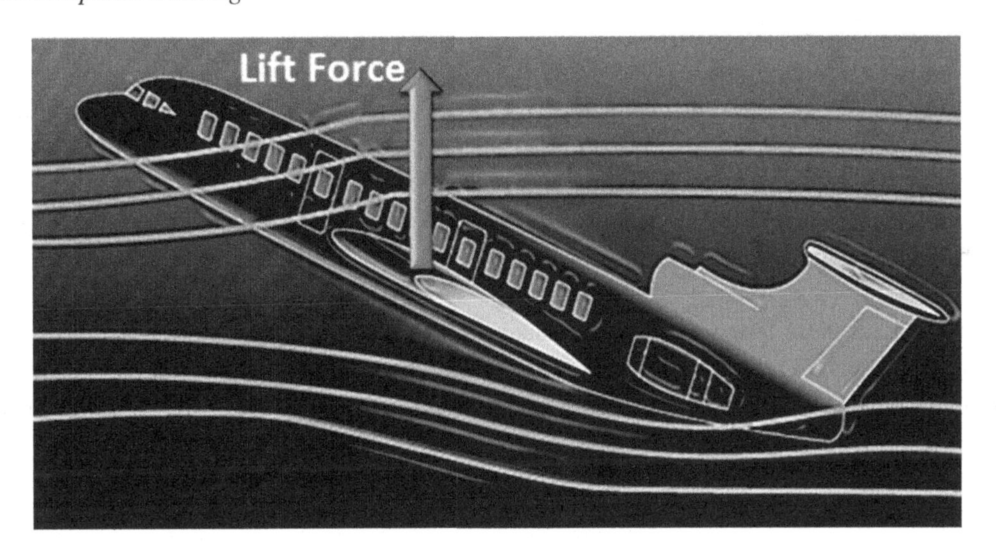

Figure 5. Propeller pitch

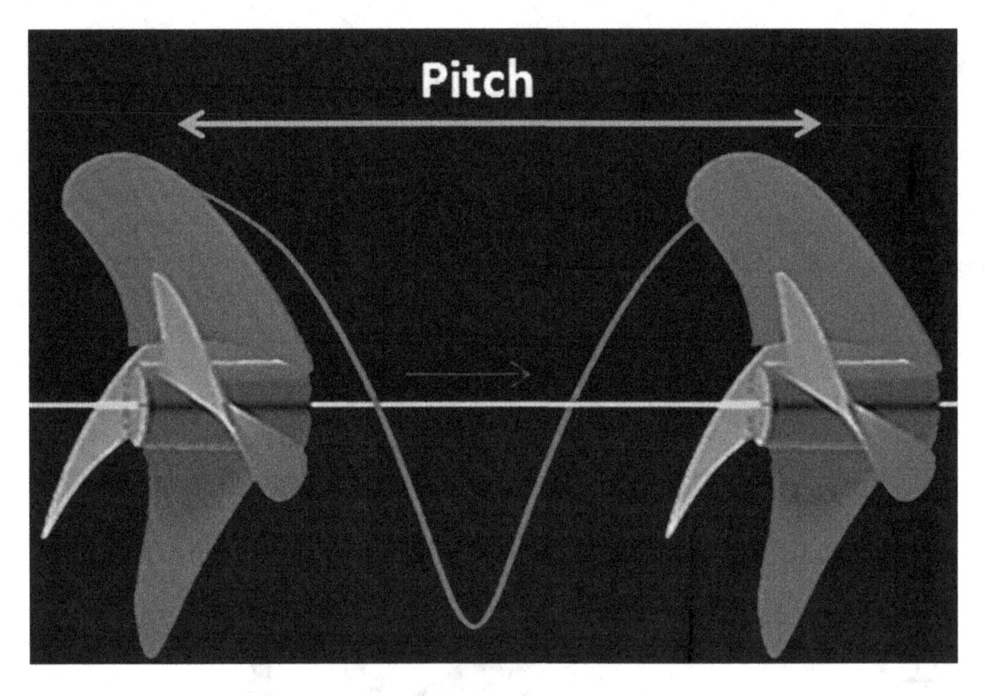

(pitch at the root of the propeller) has to be very sharp whereas the propeller tips have to be practically flat! This will assist protect an even flow of air over the duct.

The existence of air viscosity is what permits an airfoil to produce a lift. The perceptive behind this is complicated and involves rather complex mathematics. However, the air basically flows around an airfoil because of the airfoil 's lower half is a function of a greater viscosity force than the top one. Accordingly, a lift force is produced. This is perhaps the slightest agreed, and supreme questioned, the phase of how aircraft fly (Figure 6) (Anderson and Eberhardt 1999).

Forces Acting On An Airplane

It is well known that the aircraft is influenced by five forces which are drag, thrust, lift, weight, and gravity. More specifically, the lift force (*L*) is expressed as upright to the velocity (*V*) of the plane relative to the air. The drag force (*D*) is termed as analogous to the velocity (*V*). The thrust force (*T*), consequently is in a similar direction as velocity (*V*) (Figure 7). The weight (*W*) of the plane, however, directs straight down in the route of gravity (Goel and Gupta, 1984; Wegener 1997; Ringrose, 1997;Anderson and Eberhardt 1999).

In this regard, all these forces must be in balance since the aeroplane's motion is constant and its acceleration then equals zero. In other words, the thrust force created by the aircraft engines must equal the drag force triggered by air resistance. Further, the lift force formed by the aircraft wings must equal the aircraft weight (Ringrose, 1997).

Similar forces are acting on an aeroplane body through both takeoff or landing manoeuvres (Figure 8). The aeroplane's weight can be expressed mathematically as (Wegener, 1997):

Figure 6. Air viscosity variations around the aeroplane.

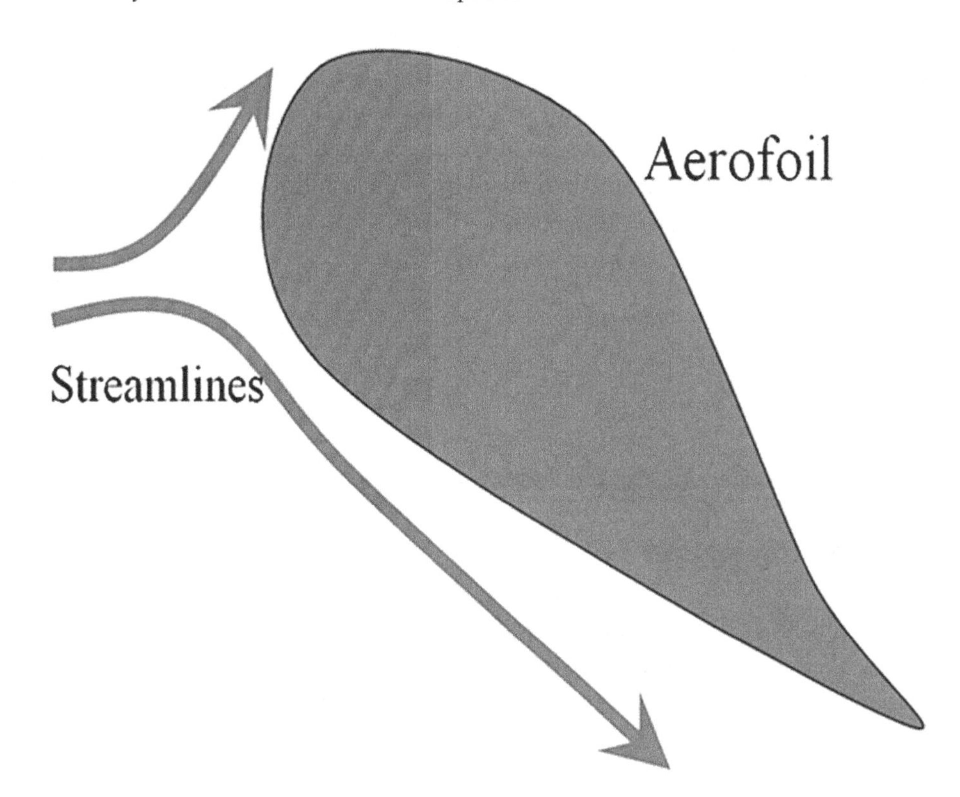

Figure 7. Forces acting on an aeroplane

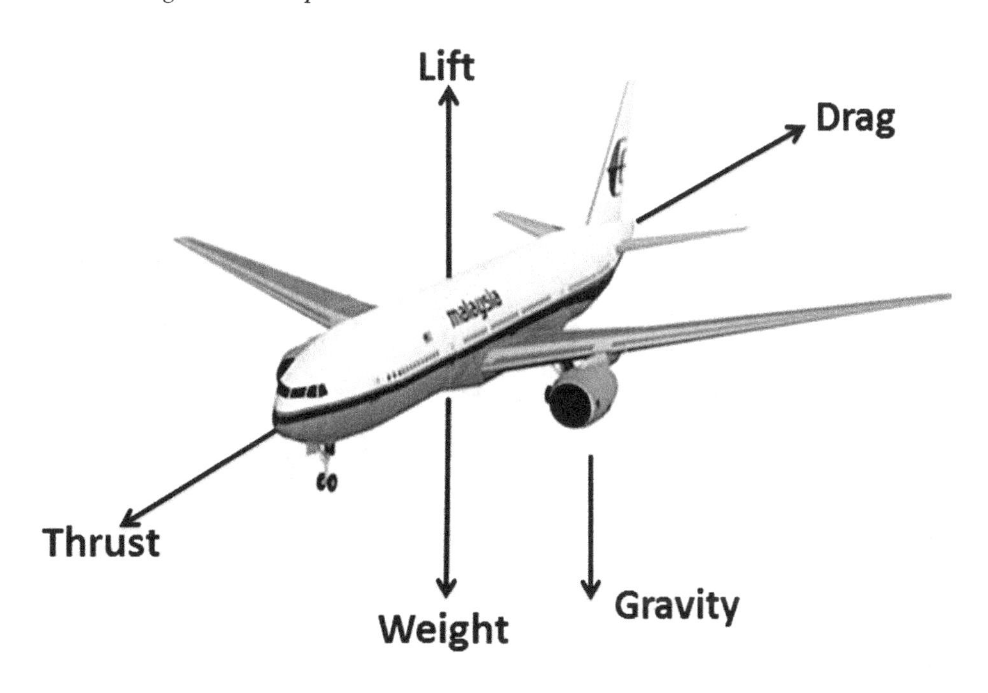

$$W = mg, \tag{2.1}$$

where m is the mass of the plane and g is the acceleration due to gravity, where g = 9.8 m/s^2. If an aeroplane is moving at constant velocity with respect to the ground, then all the forces acting on the plane must be balanced. In other words, the sum of the forces equals zero in both vertical and horizontal directions. These can be expressed in a mathematical formula as follows in both vertical and horizontal directions, respectively (Bearman and Zdravkovich, 1978):

$$L\cos\theta + T\sin\theta - D\sin\theta - W = 0 \tag{2.2}$$

$$T\cos\theta - D\cos\theta - L\sin\theta = 0 \tag{2.3}$$

where θ is a direction in concerning vertical and horizontal directions.

Maneuvering and Navigation

Aircraft operates their navigation route and orientation corresponding to the direction of air flow by attuning the physical components on the outside of the aeroplane. Consequently, these physical components are known as control surfaces and involve ailerons, elevators, rudders, spoilers, flaps, and slats. In

Figure 8. Forces acting on airplane maneuvers

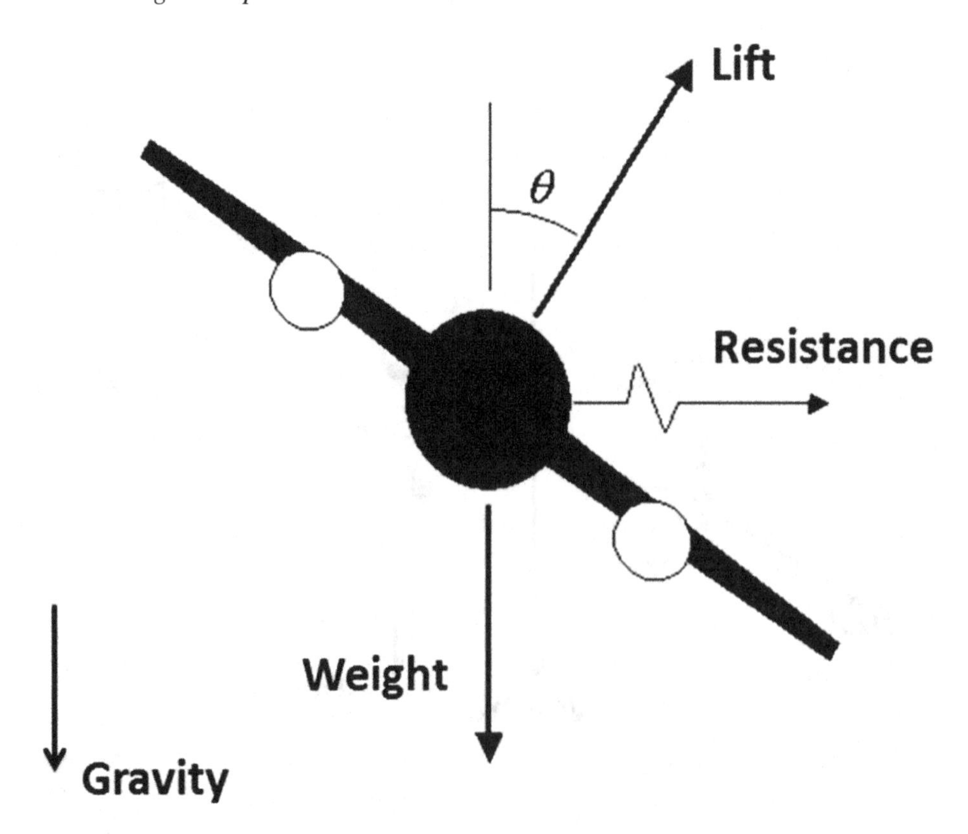

other words, these physical components which adjust the airflow pattern across the plane, triggering the aircraft to regulate its orientation and aircraft route (Repperger and Morris 1989). In retrospect, the main mechanical operators used to adjust the flight path are either pitching, rolling, or yawing, or a sequence of these (Figure 9) (Eberhardt, 2000).

Figure 9. Mechanical operators for adjusting the flight path

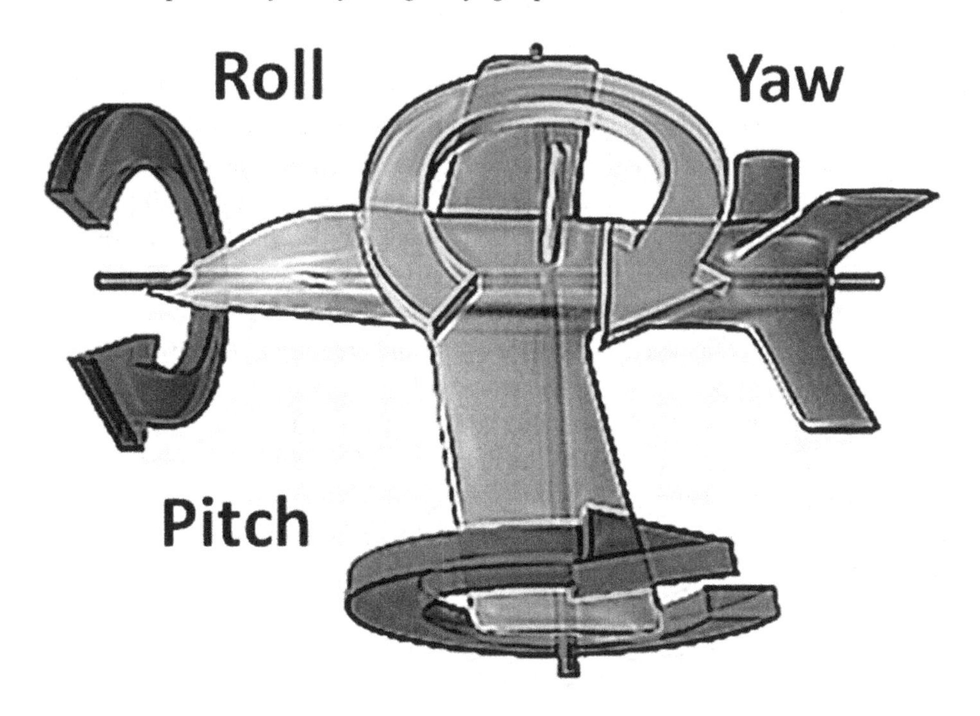

For instance, for a plane to perform a horizontal turn, its body orientation must be tilted (roll) such that the resulting aerodynamic forces enable the plane to go around a turn (Figure 8). Crosswise forces allowing a turn are solitarily probable by tilting the aeroplane. In this view, the lift force (*L*) has a lateral component required to equalize the centripetal acceleration created during the turn which is given by the second Newton 's Law as (Anderson and Eberhardt,1999):

$$L\sin(\theta) = \frac{mV^2}{R} \tag{2.4}$$

where θ is the bank angle, m is the mass of the plane, V is the velocity of the plane concerning ground, and R is the radius of the turn (Goel and Gupta, 1984; Wegener 1997; Ringrose, 1997;Anderson and Eberhardt 1999). Then in the vertical direction, the force balance is given by

$$L\cos(\theta) = mg \tag{2.5}$$

The radius of the turn has obtained the combination of equations 2.4 and 2.5

$$R = \frac{V^2}{g \tan\theta} \tag{2.6}$$

If the wind is blowing in the different direction than the pilot wants, the pilot must fly at a certain angle relative to the air such that the plane's flight with respect to the ground is in the desired direction (Ringrose, 1997).

Consistent with the above perspective, the airplanes will often hover in high elevation jet streams because of the existence of fast-flowing air currents. In this circumstance, a plane can be " conveyed " by a jet stream. Consequently, the flight time dramatically reduces when the jet stream is flowing in the projected route of travel. Of further advantage, the high altitude flight also diminishes the drag force since air has a much lower density at high altitude. In other words, the drag force is proportional to density, and lower drag force results in grander fuel effectiveness since less engine thrust is required to overwhelm the drag force(Eberhardt,2000).

The wing of an aeroplane is shaped to control the speed and pressure of the air flowing around it. The air moving over the curved upper surface of the wing will travel faster and thus produce less pressure than the slower air moving across the flatter underside of the wing. This difference in pressure creates lift which is a force of flight that is caused by the imbalance of high and low pressures (Bearman and Zdravkovich 1978 and Kusunose et al., 2006).

Supersonic Flight

Subsonic velocity is less than the speed of sound which is about 340 m/s. In this regard, all commercial aeroplanes hover at subsonic velocities. Under this circumstance, the aeroplane‘s aerodynamic system is designed to fit this speed. The airfoil shapes, for instance, are designed for subsonic aircraft. In this manner, the speed of sound must be relatively smaller than the aircraft speed relative to the surrounding air stream(Mizobata et al., 2014).

This entails a noteworthy modification in the aerodynamic device from that of aeroplane hovering at subsonic momentums. The elementary cause for this is that air conveys "data" in the configuration of compression waves at a confident rate of about 340 m/s, which fluctuates with both air density and temperature. These gravity waves are owing to the tangible movement of matters in an exchange with the air force (Anderson and Eberhardt, S. 1999 and Mizobata et al., 2014).

An aircraft, for instance, traveling across the air generates turbulence in the air precisely in connection with the body. These turbulences are in the form of gravity waves, then moving away from the body, in all tracks, over the air, at the velocity of sound. For an aeroplane traveling at less than the speed of sound, the turbulences roving upstream of the flight do so at a speed faster than the aircraft. Consequently, the air upstream of the flight can "respond" rapidly enough to adapt the route of the aircraft across the air. Henceforth, the airflow across an aircraft causing at subsonic velocities can slickly drift around the aeroplane (Mizobata et al., 2014).

However, for an aircraft moving at supersonic velocity, the air upstream of the aircraft cannot react rapidly to modify for the existence of the aircraft when it travels across the air. Consequently, air piles up contrary to the aircraft in a forceful and abrupt routine. This ensures the generation of shock waves, which is perceived as a sonic explosion (Eberhardt, 2000).

The presence of these shock waves, therefore, require an altered wing pattern than that castoff for aircraft hovering at subsonic momentums. Archetypally, supersonic wing patterns have a delta form, which produces satisfactory lift, whereas also deteriorating the shock waves, and as a result, this relegates the drag force (Wegener, 1997 and Kusunose et al., 2006) (Figure 10).

Figure 10. Shock waves along the wing and front tips

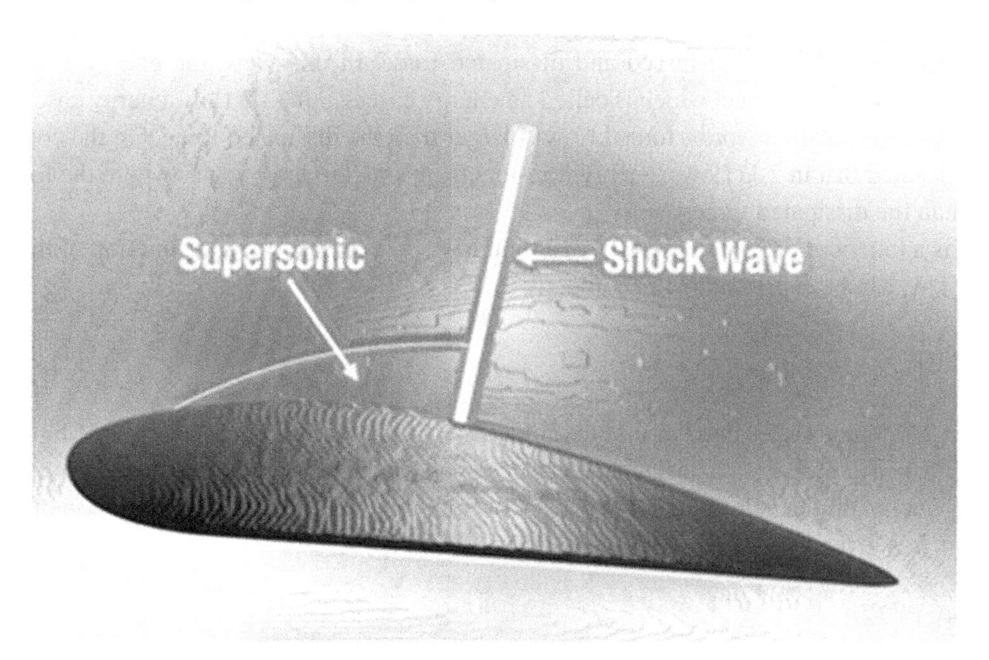

New Theory of Flight MH370 Crash

The principles of a flight aerodynamic can assist to deliver a new theory to understand Flight MH370 event crash. In this view, we attempt to establish a new crash theory based on turbulence impacts on the airframe. This theory has never been addressed in any of the early discussions regarding Flight MH370 vanishing.

Turbulence Impact

In physical terms, the fluid elements brake faster than accelerate, in a large set of flow,which is well known for driving in dense traffic. In this context, the geometric signature of erratic "flight–crash" experiences, correlated with fast particle deceleration. In this sense, it provides a technique to enumerate irreversibility in a turbulent flow. Specifically, the third moment of the power instabilities alongside a flight path, non dimensionalized by the energy flux. This displays an extraordinary strength regulation as a function of the Reynolds number in two and three spatial dimensions, respectively. This establishes a relation between the irreversibility of the system and the range of active scales. The irreversibility concept can describe precisely the dynamic of the Flight MH370 event crash. It has been acclaimed for its capacity for the quantifying and evolution of unsteady-state or nonequilibrium systems.

In the circumstance of fluid turbulence, a representative example of Flight MH370 is extremely nonequilibrium. The motion of Flight MH370 provides a perfect manifestation of time irreversibility. Explicitly, Flight MH370 tended to lose kinetic energy faster than it gained it. In this regard, Flight MH370's presence of rare "flight–crash" events, where loose moving airframe suddenly decelerates into a region where is no resistance to slow Flight MH370. Curiously, the statistical significance of this event ascertains a computable relation between the quantity of irreversibility and turbulence quantity.

The turbulence is dominated by two properties: (i) forced to flow; and (ii) dissipated force. The steady-state is balanced between forced and dissipated force. In this view, the energy is transferred through scales at an average rate, which is called an energy cascade. In 3-D flow, energy cascades from large to small scales.in other words, forced flow is larger than the dissipated force. On thc contrary, Xu et al., (2014) stated that in 2-D flows, energy transfers from small to large scales where the forced flow is smaller than the dissipated force.

Let $E(t)$ is a kinetic energy per unit mass of Flight MH370 velocity $V(t)$ as a function of the fluctuation time t. This can be expressed mathematically as:

$$E(t) = 0.5V^2(t) \tag{2.7}$$

Equation 2.7 demonstrates that there were moments of velocity differences along the Flight MH370 path $V(t) - V(0) = \partial V$, whose statistical characteristics are a function of time transformation $E(t) \rightarrow -t$. In the point of view of the statistic energy *differen+ce* $W(\tau)$, Flight MH370 crash is a function of the stress moment which is given by

$$W(\tau) = E(t+\tau) - E(t) \tag{2.8}$$

Equation 2.8 reveals that $-\langle W^3(\tau) \rangle$ increases as τ^3 approaches short time, then reduces at transitional times, and remnants, positive over the entire variety of turbulence dynamical time dimensions. Let M_r is the third moment, which is identified as

$$M_r = -\langle F^3 \rangle \varepsilon^{-3} \tag{2.9}$$

Being that F is the power and is ε energy cascades. In both 2-D and 3-D, M growths with the Reynolds number R_λ. Hence, it occurs with the separation of scales between forcing and dissipation energy. In 3-D, the $M_r \propto \left(\frac{F}{d}\right)^{2/3} \propto R_\lambda$ where $\left(\frac{F}{d}\right)^{2/3}$ is known as the Taylor-scale Reynolds number for 3D turbulence. In this understanding, the irreversibility increases also with this Reynolds number approximately as $M_r \propto R_\lambda$. In this regard, the scaling of the power recommends that such Flight MH370 crash event stipulates the foremost impact to the moments of the energy deviations and power. Therefore, this can be determined by the correlation between the Flight MH370 velocity V and its trajectory movements δV. Under this circumstance, the inertial time intervals for Flight MH370 crash can be casted by

$$W(\tau) \approx \frac{V\sqrt[3]{V(\varepsilon\tau)}}{\tau} \tag{2.10}$$

Equation 2.10 suggests that the Flight MH370 vanishing time was very short as turned into the non-equilibrium state due to turbulence instability variations through its structures. In other words, the finite-time average power of $W(\tau)$ increases when τ decreases and saturates at a short time τ. In this view, the saturation power $\langle p^2 \rangle \propto \varepsilon^2 R_\lambda^{4/3}$ to perform Flight MH370 crash event.

This implication leads Flight MH370 to turn into the turbulent flow (Figure 11) where the force distribution of equations 2.1 to 2.3 changes. Then the total balance of the forces is broken, and Flight MH370 became non-stationary and jolting (Granichin et al., 2017). Consequently, the moment of the force triggers the irregular rotation of Flight MH370 around its center of mass changing the angular momentum of the plane. To a large extent, the irregular rotation of Flight MH370 is connected to the influence of quaking.

Figure 11. Unsteady flow across the Flight MH370

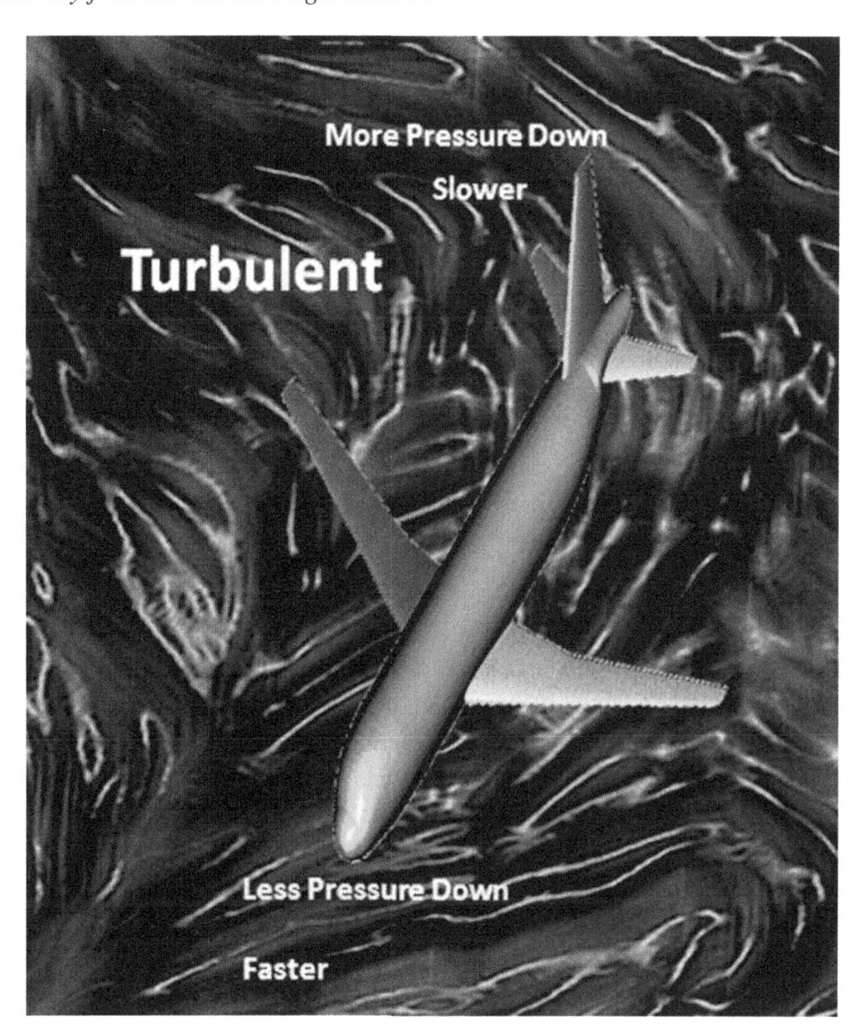

Borders of the Flight Envelope

Let assume that the flight MH370 was encountered an unsteady state pattern. In this view, it must hover at high speed. Under this circumstance, the buffet boundary represents a serious boundary for aircraft, as structural fatigue, maintenance, and passenger security were at a risk. Therefore, the flight envelope concept based on the zonal detached eddy simulations (ZDES) is used to explain the unsteady state of flight MH370 flying (Figure 12). There are three possible phases are involved in the framework of ZDES. Phase 1 involves flows where the separation is generated by a comparatively sudden deviation in the geometry of flaps. In this regard, phase 2 concerns once the locality of separation is caused by a stress gradient on a mildly arched plane. Consequently, phase 3 is retained the flows when the separation is robustly shaped by the diminuendos of the external boundary layer. In these concerns, Deck et al., (2017) reported that ZDES can predict the flight MH370 's scenarios which are befalling at the borders of the flight envelope. These are included buffet onset and dive.

Figure 12. Eddy around the Flight MH370 wings

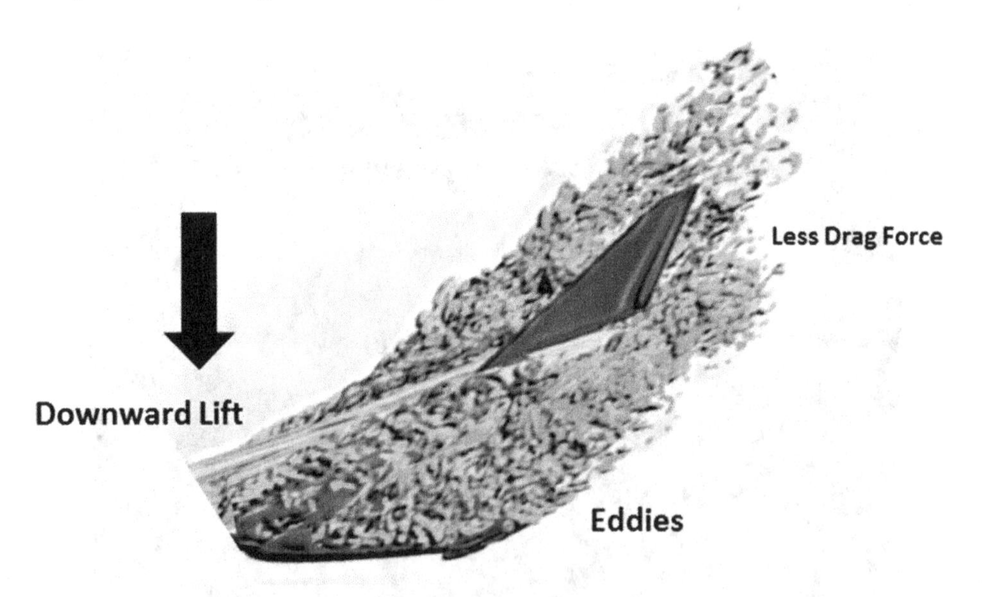

The term buffet can be used to designate the aero-dynamics instabilities and can create buffeting. In this understanding, it could have a trivial influence on the aerodynamic behavior of flight MH370. The buffeting phenomenon acts at high lift coefficients when the MH370 's Mach number or the angle of attack decreases. This phenomenon restrains the flight MH370 envelope. In contrast, Flight MH370 was dominated by a small angle of attack due to the unsteady state which was characterized by eddies around the aeroplane frame (Figure 13). In other words, the angels of attack are a function of the Reynold number. Indeed, the lower Reynold number is generated by less opposing gradient alongside the low-pressure side of the airfoil. In this circumstance, the delay separation occurs and operates at smaller angles of attack.

It is well known that the air accelerates as it travels over the top of a wing - it's a basic part of Bernoulli's lift. Subsequently, if the flight is hovering near the speed of sound, for instance, Mach 8, the air

Figure 13. Flight MH370 angle of attack decreases due to eddy viscosities

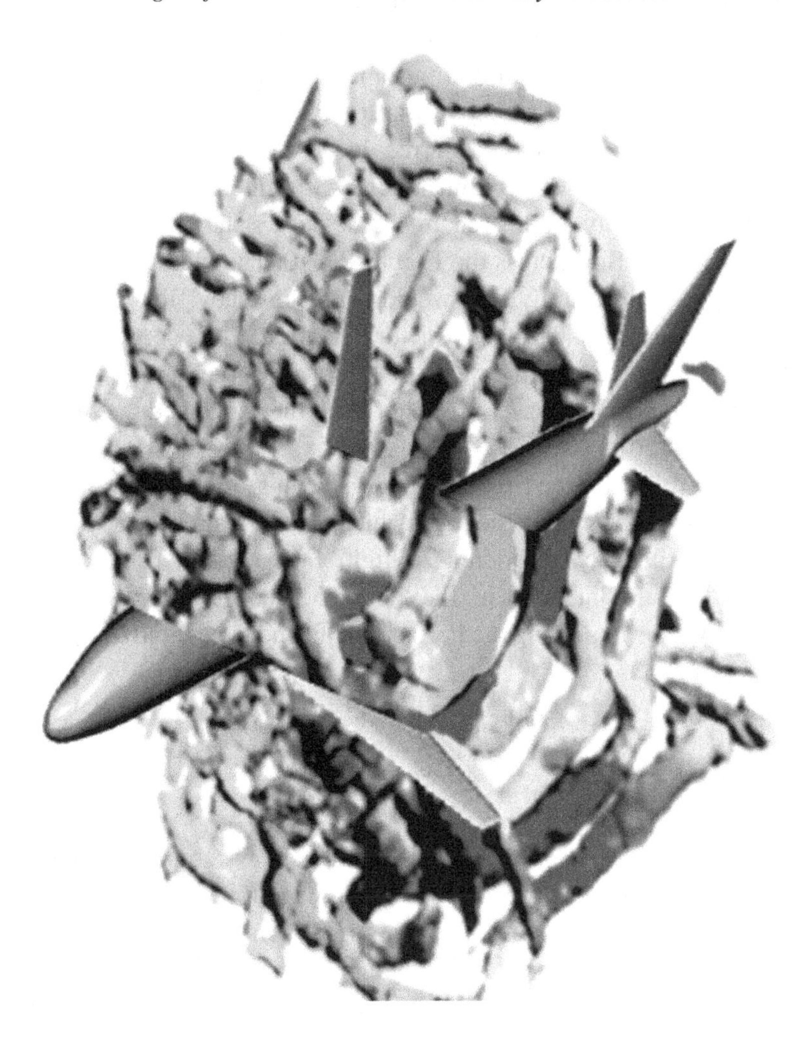

is flowing over the wing could speed up to Mach 1. This is known as the supersonic flow. In this regard, the critical Mach number is the speed where air flowing over the wing first reaches Mach 1(Caruana et al., 2005).

What is the dilemma with that? The airflow doesn't continue supersonic incessantly - it speeds up, surpasses Mach, and then decelerates back down to subsonic velocity. In this context, the more supersonic air moves over the wing, the faster flying. Nevertheless, *once the air decelerates down* below Mach 1, it creates a shock wave. As the air flows along the wing, it radiates out pressure waves – which travel at the rapidity of sound. That means that the pressure waves cannot travel forward through the supersonic airflow. Instead, they build up into a massive pressure or shock wave above the flight MH370.

That shock wave creates masses of drag. The air flowing over the wingspans a massive pressure boundary, which sucks the energy out of the airflow - causing drag. As well as, the air can lose extreme energy that it separates from the wing, causing more drag. This drag is called wave drag. The inquest into the accident determined that the major cause of the accident was a tire blow out which resulted in the impact of a piece of the tire into the bottom of the left-wing structure. It was eventually determined that the combined effects of mechanical reaction in the wing skin and a shock wave, set-up in the fuel tank within the wing due to the impact, appear to have ruptured the tank outward.

A sudden decompression of the flow caused transonic flows which are intersected by shock waves. Consequently, the boundary layers are perturbed by transonic flows which formed a complex interaction. Under this circumstance, the flow separation occurred due to the reduction of the flow velocity. In this sense, the angle of attack and the flow Mach number increase the intensity of the shock waves which lead to the large scale of flight instability. In this regard, as the large scale of the shock wave is developed, the flow is turned into turbulent flow across the flight MH370. The pressure levels, and consequently the lift, fluctuate incredibly significantly. This would cause unsteady flow patterns across the flight MH370 which could lead to fatigue and trigger structural damage. More specifically, the unsteady state occurs in the aerodynamic interaction on the wing. In other words, there was unsteadiness between the shock, the incoming attached boundary layer upstream from the shock and the separation downstream from the shock (Caruana et al., 2005).

The dive is another possibility for the flight MH370 vanishing. In this view, the massive flow separations were produced by spoilers to cause the dive, which directs to an unsteady loading of the spoiler axes that requires to be quantitatively considered. At high Mach number, a shock-persuaded separation over the wing could exist and also interact with the spoiler. Moreover, the wake of the spoiler has to be accurately resolved, as it may be impinging the horizontal tail and trigger tail buffet leading to handling issues in some cases.

STALLING

Boundary layer separation in the airflow over airfoils and wings is commonly referred to as a stall. In the case of a stall, the airflow separates from the wing, the wing loses lift and control of the aircraft is lost for the duration of the stall. In this view, airflow normally moves uniformly through a jet engine. When it is disrupted or distorted as it enters the engine, there's a high risk for a compressor stall. The most common reasons for disrupted air entering the engine are foreign object damage, for instance, the pieces of Flight MH370 tire puncture, worn or dirty compressor components, in-flight icing, operations outside the design envelope, and improper engine handling.

High Speed

Flight MH370 encountered high speed which could reduce an angle of attack which created lost lift. In other words, the nose of flight MH370 was extremely in a steep angle which makes the plane lost lift and went into a stall. On the contrary, the expert pilots can manage the speed to lower the nose to recover from a stall. This statement indicates that the pilots were under critical circumstances which not allow them to take any further action.

In this circumstance, the strong vortices were generated downward of the nose and high pressure above the wind. This caused the flight to spin around it axes and turned the flight vertical or with the smallest angle of attack to land vertically.

Air separation and stalls inside the engine occurred at a blade's critical angle of attack, just like on the wing of the plane. Disrupted, turbulent airflow follows this airflow separation around the blades, entering the engine.

Wing Sweepings

Sweeping the wings also affects the stall pattern. The number of spanwise flow compounds as approach the wingtip, decreasing the wingtip's effective airspeed and thickening the boundary layer. This can cause the wingtip to stall before the wing root - meaning you lose aileron control at the onset of the stall. That could make for a wild ride. This turbulent airflow entering a turbine engine is bad news. When the airflow is disrupted, there is an excess of fuel compared to the amount of air required. This excess fuel is burning off and creates flames.

HOW DID FLAPERON EFFECT THE FLIGHT MH370?

The flaperon is solitary of the wing device surfaces of the aircraft. Additional resistor surfaces on the aircraft are the ailerons, used to move or bank the planes. Consequently, the flaps are used to diminish an aircraft speed for departure and landing. In this view, a flaperon commingles attributes of both flaps and ailerons.

The flaperon debris indicates that the wing exhibited impact damage. In this sense, the wing spar could be fractured in multiple places and buckled around its base by the narrow oblique angle of about 45° to 50°. This could create a fracture in the cable just inboard of the wing root which may be presented at "broom straw" signature. Finally, it suggests that the propeller must be remained attached to the engine.

Consistent with the above perspective, flaps are there primarily to increase lift at low speeds and assist by providing enough lift so the aircraft designer can lower the landing speed for safer handling and shorter landing/braking distance. Whilst they increase lift, they also increase drag so have a braking effect in themselves.

In theory, the aeroplanes can land without flaps, however, the final approach speed will be extremely high to maintain the correct rate of pitching. Further, there is more danger involved, for instance, tires will blowout and it is difficult corrections because of high speed and small distance.

Here is an aeroplane about to land and if we do not have flaps then it is going to have to fly to a certain speed which probably is reasonably fast so it will take a certain amount of distance before it comes down from its starting height with a glide slope. It is going to be relatively shallow.

It we put flaps on, all drag goes up and the plane will slow down and the lift will be increased up. In this understanding, the plane will end up with different approach slopes. If the flap is on, then the plane still decent at the same speed without taking a large distance to pitch. In other words, when the flap is on the descent distance will be extremely shorter than no flap on-field Pruett. The flaps enable the plane to make much steeper approaches without falling out of the sky.

If the plane is landing without flap on with the much steeper angle, two things dramatically happen. First, if it moves faster and then it is tried to slow down, it would stall, fall and break. If it simply puts a nose-down it would turn its height into speed. Therefore, it would be going so fast and it would glide a long way.

When the flaperon separated from the Flight MH370, it would create much drag, which would pull it backward. It would also spine along its axis because there is no balance between the forces act on the flight. This would increase the pressure upwards and by the way, the flight nose directed down and then rotated around its axis.

WHY FLIGHT MH370 WAS UNCONTROLLED

The flight MH370 seemed uncontrolled due to the serious destruction. One of those which never mention by the conspiracy theories is unbalanced of atmospheric pressure. In this view, the splitting of the flaperon from the wing caused unbalanced atmospheric pressure to act on the aircraft. Under this circumstance, the sudden death of the pilots resulted in a sudden decrease in the air in the pilot room. In this understanding, the flight would supreme hypothetically become unrestrained and subsequently crash down.

Falling Process

Hypothetically, the primary access site of the Flight MH370 to fall into the sea is predominantly exaggerated by the acting forces, for instance, the lift force, the drag force, etc., (Figure 14).

Figure 14. Falling Process of MH370

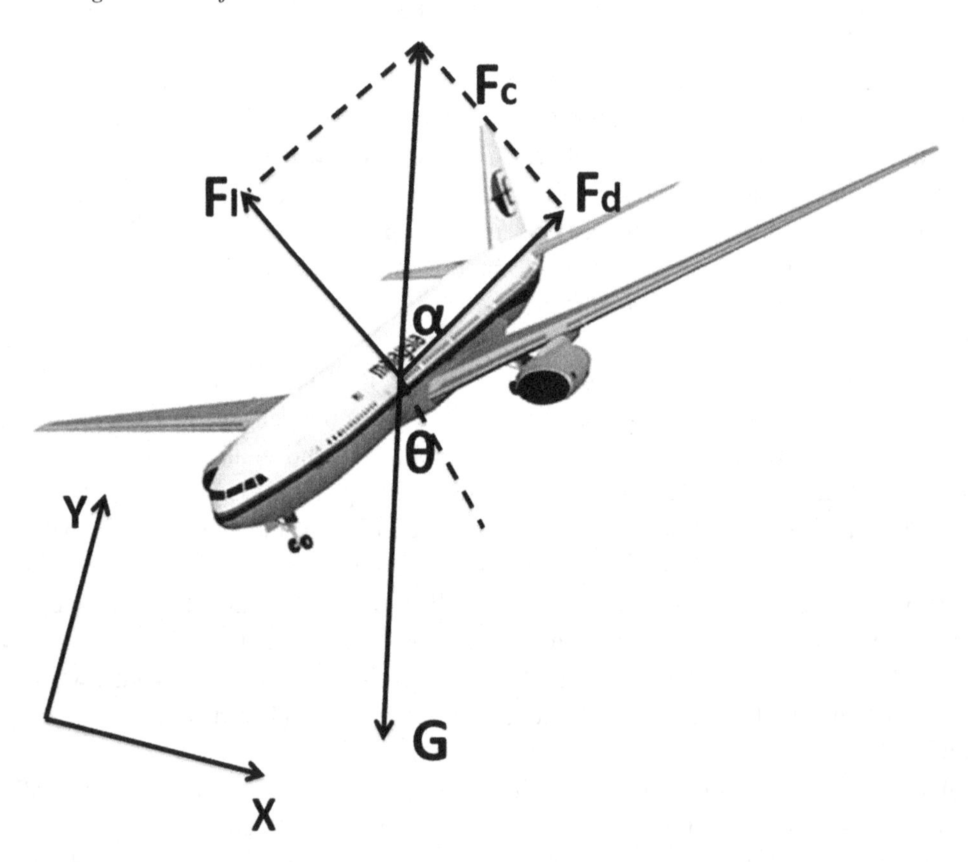

Let the lifting force F_l and the drag force F_d result in a composite force F_C. The two components of the kinetic energy are acting on the MH370 frame in x and y directions, respectively. This can be expressed mathematically as:

$$Mg\sin\theta - F_c\cos\alpha = M\frac{dV}{dt} \tag{2.11}$$

$$Mg\cos\theta - F_c\sin\alpha = MV\frac{d\theta}{dt} \tag{2.12}$$

being, M is the mass of the aeroplane (kg), F_C is the value of the composite force (N), and θ is the angle between the gravity and the normal line of the aeroplane's moving direction (degree) (He et al., 2016). Therefore, θ is specifying the gliding angle of the falling aeroplane. Further, α (degree) is the angle between F_c and F_d and can be obtained by calculating the ratio of F_l to F_d. Finally, V is the velocity of the aeroplane (m/s), and t is the time (s) (Green, 2014). The value of lifting and drag forces can be estimated respectively, by

$$F_l = 0.5\rho V^2 C_l S \tag{2.13}$$

$$F_l = 0.5\rho V^2 C_d S \tag{2.14}$$

where ρis air density (1.293 kg/m³), S is the action area of both airlift and drag forces during the falling process (m²) which are computed respectively, by

$$C_d = 0.12149 - 0.01714C^2{}_l + 0.00690C_l^3 \tag{2.15}$$

$$C_l = 1.1488 + 0.11832\alpha^2 + 0.00092\alpha^3 \tag{2.16}$$

Then the second derivatives of the falling process velocity components can be expressed mathematically by (He et al., 2016):

$$\ddot{x} = -\frac{0.5}{M}\rho SC\sqrt{\dot{x}^2 + \dot{y}^2}(\dot{y}\sin\alpha + \dot{x}\cos\alpha) \tag{2.17}$$

$$\ddot{y} = -g + \frac{0.5}{M}\rho SC\sqrt{\dot{x}^2 + \dot{y}^2}(\dot{x}\sin\alpha - \dot{y}\cos\alpha) \tag{2.18}$$

Both equations spectacle that the falling process is affected by air density, which is taken under different altitudes. It is worth mentioning that both lift and drag forces are increased as a function of growth. In the case of MH370 α is ranged between 1° to 2°. Under this circumstance, the maximum of both lift and drag forces are 0.108130 and 1.141748, respectively (Table 1) (Figure 15) (He et al., 2016).

Table 1. Impact of α lift and Drag Forces

α (deg)	Lift force	Drag Force
2	0.108130	1.141748

Figure 15. Lift and drag variations as a function of α (deg).

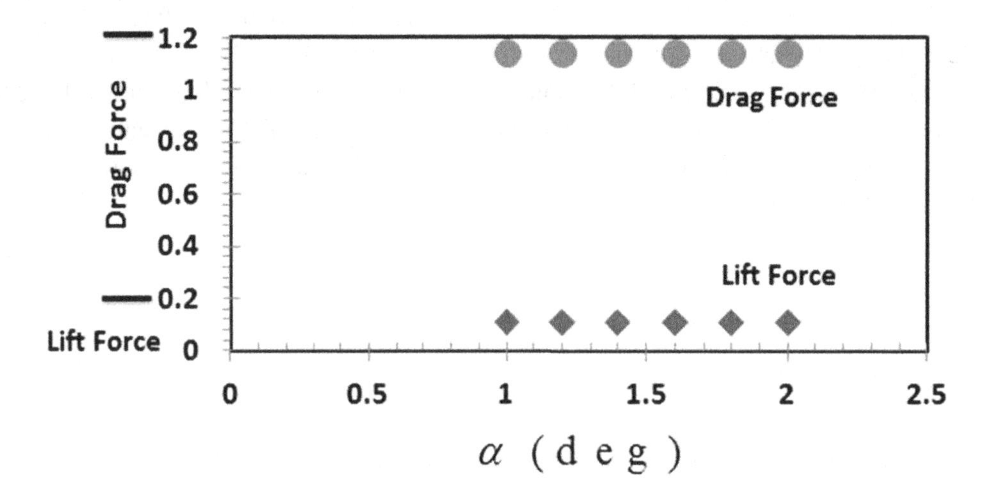

Figure 16 presents also the warning times which is the time elapsed from the instant of the crash detection on the impact of the aircraft with the ground/ocean. Mekki et al., (2017) stated that the statistic results indicate that the warning times are greater than 15 seconds in 75% of the crash cases, greater than 30 seconds in 59% of the cases, greater than 60 seconds in 34% of the cases and greater than 120 seconds in 23% of the cases. Thus, the protocol should ensure a very low delay, as much as possible, for storing the last recorded FDR data during the warning times.

Figure 16. Warning time for the air crash

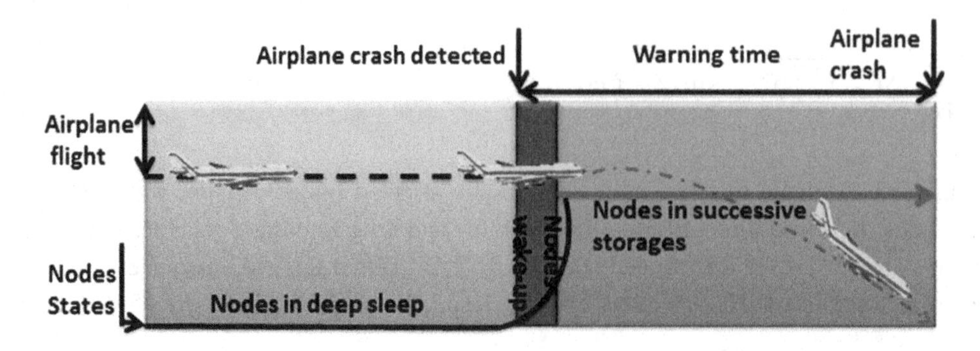

CONCLUSION

This chapter has been demonstrating some aspects of aerodynamic principles. One of the important aspects is Bernoulli's principle for aerodynamics. The understanding of aerodynamic principles can assist to determine the flap failure impact on the aerodynamic system. A new theory of the MH370 crash is delivered based on the turbulence impacts on the airframe. In the circumstance of the MH370 air crash, the dynamic processes have addressed to reveal how the MH370 was uncontrolled.

Based on the novel theories of aerodynamic principles, the next chapter will be able to address the accurate mathematical theory of MH370 ditching.

REFERENCES

Anderson, D., & Eberhardt, S. (1999). *How airplanes fly: A physical description of lift*. Sport Aviation.

Bearman, P. W., & Zdravkovich, M. M. (1978). Flow around a circular cylinder near a plane boundary. *Journal of Fluid Mechanics*, *89*(1), 33–47. doi:10.1017/S002211207800244X

Caruana, D., Mignosi, A., Corrège, M., Le Pourhiet, A., & Rodde, A. M. (2005). Buffet and buffeting control in transonic flow. *Aerospace Science and Technology*, *9*(7), 605–616. doi:10.1016/j.ast.2004.12.005

Cook, M. V. (2012). *Flight dynamics principles: a linear systems approach to aircraft stability and control*. Butterworth-Heinemann.

Deck, S., Gand, F., Brunet, V., & Ben Khelil, S. (2014). High-fidelity simulations of unsteady civil aircraft aerodynamics: Stakes and perspectives. Application of zonal detached eddy simulation. *Philosophical Transactions of the Royal Society A: Mathematical, Physical and Engineering Sciences*, *372*(2022), 20130325. doi:10.1098/rsta.2013.0325 PMID:25024411

Eberhardt, S. (2000). Airplanes for Everyone: A General Education Course for Non-Engineers. *Journal of Engineering Education*, *89*(1), 17–20. doi:10.1002/j.2168-9830.2000.tb00488.x

Goel, L. R., & Gupta, P. (1984). Analysis of a two-engine aeroplane model with two types of failure and preventive maintenance. *Microelectronics and Reliability*, *24*(4), 663–666. doi:10.1016/0026-2714(84)90213-0

Granichin, O., Khantuleva, T., & Amelina, N. (2017). Adaptation of aircraft's wings elements in turbulent flows by local voting protocol. *IFAC-PapersOnLine*, *50*(1), 1904–1909. doi:10.1016/j.ifacol.2017.08.263

Green, J. J. (2014). *Boeing rules out cyber sabotage connection to missing plane*. WTOP-FM.

He, Y. L., Wang, W. K., Zhang, Z. Q., & Xu, D. (2016). A hybrid positioning approach based on mechanical calculus model and grid simulation for drifting debris. *Ocean and Coastal Management*, *130*, 21–29. doi:10.1016/j.ocecoaman.2016.05.007

Kusunose, K., Matsushima, K., Goto, Y., Yamashita, H., Yonezawa, M., Maruyama, D., & Nakano, T. (2006, January). A fundamental study for the development of boomless supersonic transport aircraft. In *44th AIAA Aerospace Sciences Meeting and Exhibit* (p. 654). 10.2514/6.2006-654

Mamou, M., & Khalid, M. (2007). Steady and unsteady flow simulation of a combined jet flap and Coanda jet effects on a 2D airfoil aerodynamic performance. *Revue des Energies Renouvelables CER*, *7*, 55–60.

Mekki, K., Derigent, W., Rondeau, E., & Thomas, A. (2017). Wireless sensors networks as black-box recorder for fast flight data recovery during aircraft crash investigation. *IFAC-PapersOnLine*, *50*(1), 814–819. doi:10.1016/j.ifacol.2017.08.145

Mizobata, K., Minato, R., Higuchi, K., Ueba, M., Takagi, S., Nakata, D., & Tanatsugu, N. (2014). Development of a small-scale supersonic flight experiment vehicle as a flying test bed for future space transportation research. *Transactions of the Japan Society for Aeronautical and Space Sciences*, *12*(ists29).

Repperger, D. W., & Morris, A., Jr. (1989). *U.S. Patent Application No. 07/079,323*. US Patent Office.

Ringrose, R. (1997, April). Self-stabilizing running. In *Proceedings of International Conference on Robotics and Automation* (Vol. 1, pp. 487-493). IEEE. 10.1109/ROBOT.1997.620084

Wegener, P. P. (1997). *What makes airplanes fly?: history, science, and applications of aerodynamics*. Springer Science & Business Media. doi:10.1007/978-1-4612-2254-5

Chapter 3
Mathematical Theory of Flight MH370 Ditching

ABSTRACT

This chapter bridges the gap between the theory of flight ditching and the MH370 vanishing mechanism. In doing so, the Von Karman theory and volume-of-fluid (VOF) technique are used to deliver a logical understanding of MH370 crashing in the Indian Ocean. The most significant finding was that the fuselage would be broken down into some pieces, and wings and tails, for instance, would split away from the fuselage under air turbulence impacts and appear as floating debris in the crushing area. However, there is no recorded debris located at the crash site.

INTRODUCTION

Ditching is the keystone to investigate the MH370 dynamical vanishing in the Indian Ocean. It can be a new procedures to determine either MH370 was ditched into the Indian Ocean or not. In this regard, ditching entails landing on water, regularly the consequence of an emergency, such as engine failure. Nevertheless, they continuously land with the landing gear retracted, composing the bottom of the aircraft smoothly, and less probable to pitch the aircraft downward.

It is conceivable for an aircraft manoeuvring over water to be required to make an emergency water landing owing to a system failure, such as engine power malfunction as discussed in chapter 2. However, the Flight MH370 ditching did not occur, but rather it was plunged into the Indian Ocean. Though the exact fate of Flight MH370 endures hesitating, the offered data directs a crash into the Southern Indian Ocean. It is a confusing question, why did MH370 plunge into the Southern Indian Ocean? This chapter is devoted to reconstructing the scenarios of Flight MH370's entry into the body of water. The foremost aspect that need to be taken into account is how ditching mechanisms differ from an aeroplane crash into the water, which is also introduced here.

DOI: 10.4018/978-1-7998-1920-2.ch003

COMPUTATIONAL FLUID DYNAMIC

Up to date, Chen et al., (2015) are considered the only ones who have experienced the Flight MM370 crash from the point of view of the computational fluid dynamics (CFD). They used CFD, specifically, Von Karman theory and Volume-of-fluid (VOF) method to investigate the rationality explanations behind the Flight MH370 vanishing. They postulated that the Flight MH370 was ditched into two immiscible fluids.

Further, the Flight MH370 and fluid phases are separated by a deformable interface, which could go through topological changes such as destruction or coalescence (Erturk,2009). In fact, the VOF can easily deal with topological fluctuations of the interface. In this view, the CFD method can permit to simulate water entry for complex, common geometries rather than the abridged, ones, for instance, cones, cylinders, and wedges treated in the early era by encompassing nature into the two-phase fluid-structure interaction models (Chen et al., 2015).

Let assume that there are two phases with temperature T in the domain Ω satisfying. The density of them is, respectively $\rho_i : \Omega \text{ x } [0,\mathrm{T}_0] \rightarrow \mathbb{R}^+, i=1,2.$ Let $\sigma_i : \Omega \text{ x } [0,T_0] \rightarrow [0,1], i=1,2$ be the volume fraction of the phases [24]. Therefore, the velocity is $u:\Omega \text{ x}[0,T_0] \rightarrow \mathbb{R}^d$ and pressure $p:\Omega \text{ x}[0,T_0] \rightarrow \mathbb{R}$ (Gu, 2015). In this understanding, the mathematical expression of mass balance for each phase is given by

$$\nabla.(\alpha_i \rho_i u) + \frac{\partial(\alpha_i \rho_i)}{\partial t} = 0, \qquad i = 1,2,3,......N \tag{3.1}$$

where ρ is water density, α would be a step function across the interface for immiscible fluids, and u fluid flow through time t. Consequently, the energy balance for each phase is casted by (Gu, 2015):

$$\frac{\partial(\alpha_i \rho_i (e_i + k))}{\partial t} + \nabla.(\alpha_i \rho_i u(e_i + k)) - \nabla.(\kappa_i \nabla T)\frac{\alpha_i \rho_i}{\rho} + \nabla.(pu - \tau.u)\frac{\alpha_i \rho_i}{\rho} - \alpha_i \rho_i g.u = 0 \tag{3.2}$$

where κ is the thermal conductivity for phase i, τ is a deviatoric stress tensor, k is the kinetic energy per unit mass, and e_i is the internal energy per unit mass in phase i.

FLIGHT MH370 MECHANICS DITCHING

Ditching signifies that Flight MH370 or any plane has no alternative but to land on the water surface, for instance, sea or lake, given the safety of crews and passengers. Nonetheless, Flight MH370 was not ditched but crashed into the water. In the point of view of structural dynamics, a plane should be in extreme depression velocity when ditching. In this context, the distortion of a plane can be almost deliberated as elastic-plastic. Conversely, Flight MH370 was at high speed when it crashed into the Indian Ocean.

In ditching, a huge area of the aeroplane contacts with water, which is dissimilar a ground effect. The problem of water entry and effect can be investigated hypothetically, experimentally or mathematically. In this regard, the first theory developed was by Von Karman in 1929. This revolutionary work depletes the theory of added mass for the examination of influence loads on a seaplane during the ditching.

Von Karman Water Impact Theory

Von Karman's theory affirms that the momentum combined with the influence was conserved between the object and an 'allied' mass of water, which travels at the identical speed as the object. Let assume a body with a wedge-shaped under the water surface, which strikes a calm horizontal surface of the water (Figure 1). In other words, Von Karman concluded that the effective force on the body is associated with the instantaneous alteration of the entire momentum of the body with its own mass, but with an extra mass amplified by the "added mass" of the fluid around the plunged portion of the body.

Figure 1. Von Karman Concept

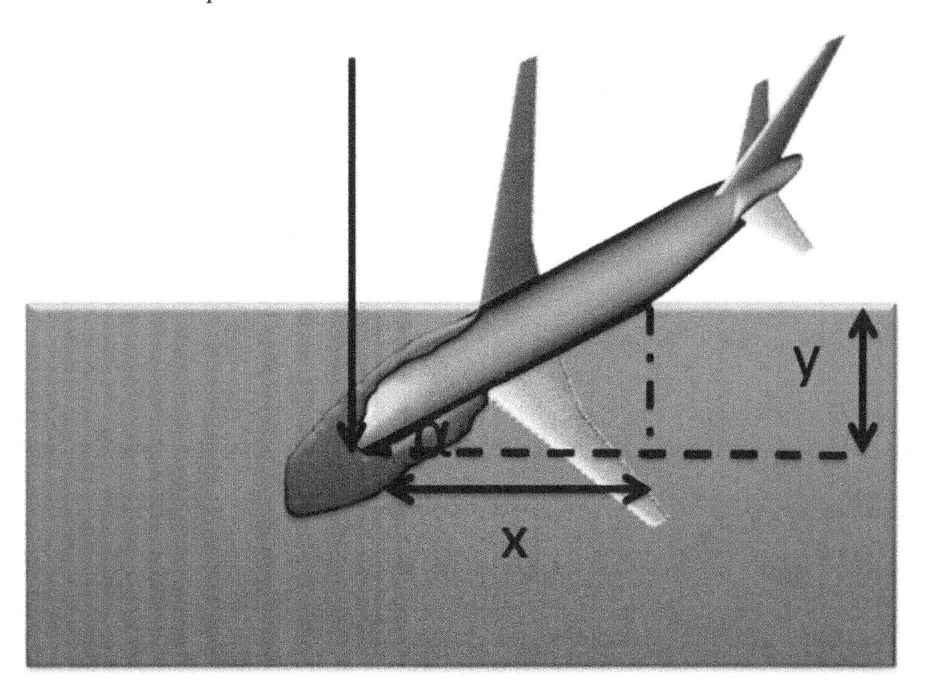

The boundary layer of mass flow ($\dot{m}$) can be mathematically expressed by

$$\dot{m} = \int_{0}^{\partial} \rho u x dy \tag{3.3}$$

where ρ is water density, x is the width of the width of the area for which the flow rate *u (x,y)* is being obtained. In this view, the added mass can cause wall shear stress τ_w, which is given by

$$\tau_w = \mu \left. \frac{du}{dy} \right]_{y=0} \tag{3.4}$$

Equation 3.2 demonstrates that the shear stress is caused by the added mass acceleration changes through the effective mixture viscosity μ.

At time *t*, the original momentum of the body is dispersed between the water and the Flight MH370. The amount of momentum already transferred from the MH370 at time *t* varies by *x*. This can be approximated by the increase in inertia of the MH370/water system. It is known that for a long plate of width $2x$ which is accelerated in a fluid, its inertia is increased by the amount $\rho x^2\pi$, where ρ is the density of the fluid. Therefore, Flight MH370 was initially being comprised half of water and half by air (Figure 1), owing to the varying domains above and below the ocean free-surface. The impact MH370 float would accelerate the water particles upfront about it, and sweep in the air behind it. Assume the effect of air is trivial contrasted to that of the water, only half of the apparent MH370 increase in mass is considered. In this understanding, the total momentum of the water/MH370 system post the added mass impact can then be mathematically expressed as:

$$M_g = F + \frac{d}{dt}\left[(M + m(t))d_p(t)\right] \tag{3.5}$$

where *M* is the mass of the projectile, *m(t)* is added mass, d_p is the depth of penetration into the Southern Indian Ocean. Therefore, *F* is the applied forces on the Flight MH370 which is computed as

$$F = \beta + C + D \tag{3.5.1}$$

where β, *C*, and *D* are buoyancy force, capillary force, and steady-state drag force, respectively. Let *U* and *V* are downward velocity components of the MH370 velocity v_0 at time *t* then are estimated by

$$U = v_0 \cot\alpha \left[1 + \frac{\rho\pi g x^2}{2W_i}\right]^{-1} \tag{3.6}$$

$$V = v_0 \left[1 + \frac{\rho\pi g x^2}{2W_i}\right]^{-1} \tag{3.7}$$

where *g* is the gravity acceleration, α is the deadrise angle i.e. angle of inclination of the sides of the Flight MH370 (Figure 1) with the horizontal and *W* is flight weight per unit depth. Then the Flight MH370 acceleration in the *y*-direction (Figure 1) or in the d_p depth of penetration can be obtained as:

$$\frac{dV}{dt} = v_0 \cot\alpha \left[1 + \frac{\rho\pi g x^2}{2W_i}\right]^{-3} \left[\frac{\rho\pi g x}{W_i}\right] \tag{3.8}$$

Based on Newton's second law of motion, the force between the Flight MH370 and the water can be mathematically expressed as:

$$F = v_0^{\,2}\cot\alpha\left[1+\frac{\rho\pi g x^2}{2W_i}\right]^{-3}\rho\pi x \tag{3.9}$$

The main force acts on the Flight MH370 plugged into the water is pressure (*P*). In two-dimensional (2-D), the pressure, *P* can be obtained as:

$$P = 0.5Fx^{-1} = 0.5\rho v_0^2\ \pi\cot\alpha\left[1+\frac{\gamma\pi x^2}{2W_i}\right]^{-3} \tag{3.10}$$

In equation 3.8, the term of $0.5\rho v_0^2$ presents the dynamic pressure corresponding to the initial impact velocity and π cot α is the theoretical factor of the pressure increments through the d_p depth of penetration. Further, the maximum pressure, arises when $x \rightarrow 0$. Hypothetically, this specifies that the maximum pressure is in the middle of the Flight MH370 at the first moment in contact with the water. In other words, the maximum pressure is implemented when the contact area between MH370 and the water body is minimal.

Furthermore, a sharp decline in the theoretical factor of increase of maximum pressure is a function of the increment of the deadrise angle. For instance, the deadrise angles less than about 10°, maximum pressure inclines to increase rapidly as the deadrise angle diminishes, resulting in a more 'crashing' effect of the Flight MH370 upon impact. Figure 2 indicates that the smallest acceleration of -15 m/sec^2 and deadrise angles of 0.35° leads to a fast time of 0.02 sec for water entry.

Wagner 's Theory

The von Karman theory is not able to model the non-linear water's free-surface. In other words, this theory has limitations to model the rise of the displaced ocean water, local flow velocity, splash, and pressure variations along the sides of the Flight MH370. Indeed, the offshore of the Southern Indian Ocean is not a calm surface but is dominated by a rough wave. In this understanding, the sea surface water must rise along the Flight MH370 as long contacted to the water surface (Figure 3).

Following Wagner 's theory for the free-surface elevation, the vertical distance between the highest wetted point on the side of the Flight MH370 and its apex to be a factor of 2π of the vertical penetration depth d_p as per the von Karman theory. In this regard, the maximum pressure was delivered by Wagner 's theory is given by:

$$P(x)_{\max} = 0.5\rho v^2\left[1+\frac{\pi^2}{4\tan\alpha^2}\right] \tag{3.11}$$

Equation 3.11 demonstrates that when the Flight MH370 entered the water, the water pile-up the free-surface of the water as the MH370 penetrates the water (Figure 4). In this circumstance, the highest pressure is acting just before the end of the wetted region along the side of MH370. Figure 4 depicts the maximum pressure acted on the Flight MH370 ranged between 1e+4 to 4e+4 $P_{\max}$/Pa. However, the

Figure 2. Acceleration variations with different aeroplane velocities and deadrise angles

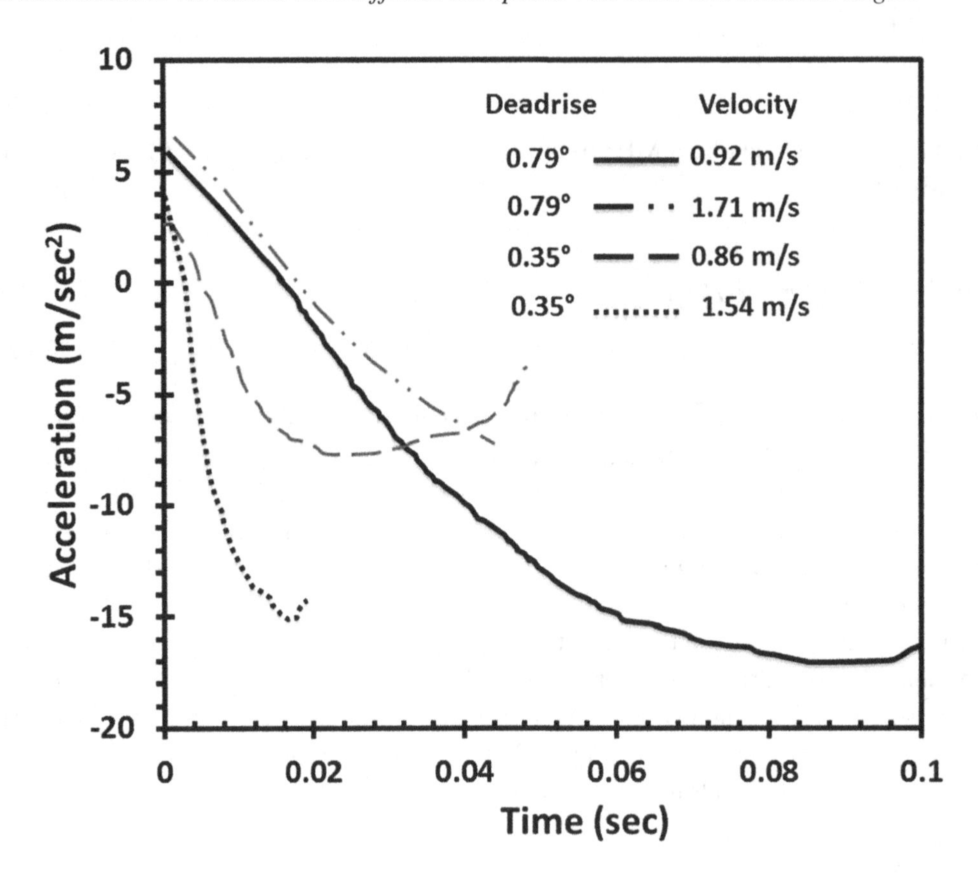

Figure 3. Pressure distribution along the side of MH370 through water entry

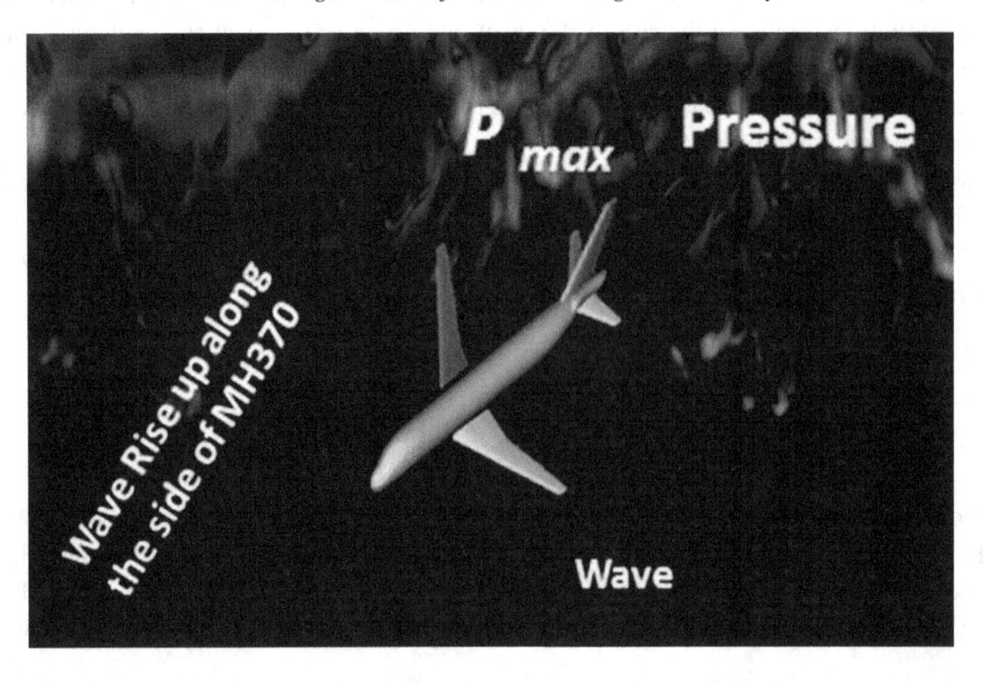

maximum pressure of 4e+4 P_{max}/Pa might act on the airframe. It is worth mentioning that, the maximum pressure increases along with the MH370 airframe as compared to the surrounding sea surface. In fact, the spatial distribution of the pressure along the surface of the Flight MH370 throughout impact is an important quantity during MH370 water entry. In most water impact cases, the event is very short and typically lasts for a few milliseconds.

Figure 4. Maximum pressure variations along MH370 on time contacted to the Indian Ocean surface

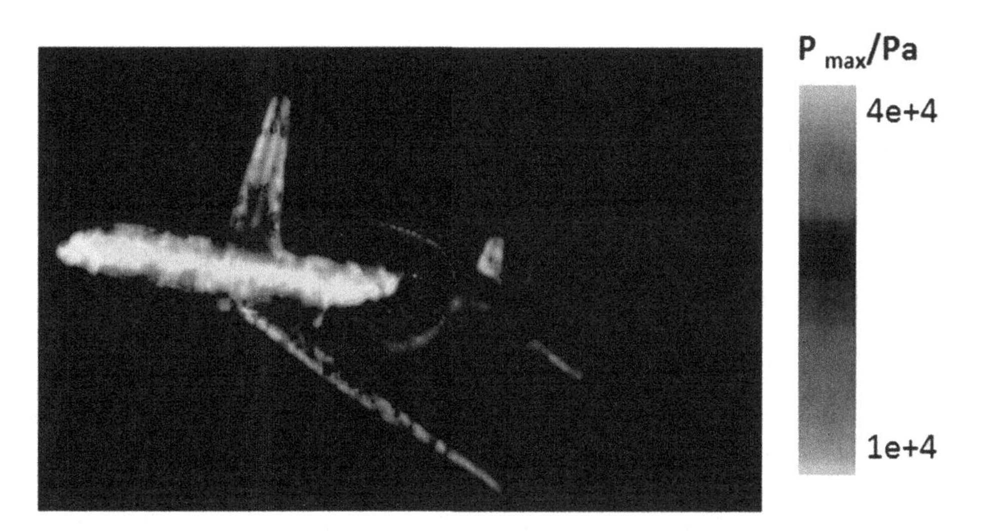

In general, the momentum of the Flight MH370 was transferred into the momentum of an associated mass of water as described by the Von Karman theory. Buoyancy and skin friction effects prevent the momentum from being completely conserved are considered. Nevertheless, viscosity, buoyancy, and gravitational effects were ignored as they are minor when compared to the inertia forces. Consequently, the Flight MH370 has two effects which are a function of impact angle: (i) it affected the growth of the associated mass of water; and (ii) it increased the amount of momentum in the wake.

The effect of rupture and structural disintegration is almost certain to ensue upon the water entry when the speed is extremely high. If the vertical component of the terminal impact velocity lies in the range of 62.5 m/sec and 80.5 m/sec, maximum decelerations could reach in the order of 100g to 150g (g is the gravitational acceleration constant) over a short period. In this view, the damage of the fuselage is possible upon the water entry.

VOLUME-OF-FLUID

The most common method used in CFD programs based on the finite volume method is the volume-of-fluid (VOF) model. Assume that each control volume contains just one phase (or the interface between phases). In this regard, solves one set of momentum equations for all fluids can be casted by:

$$\frac{\partial}{\partial t}(\rho u_i)+\frac{\partial}{\partial x_i}(\rho u_i u_j)=-\frac{\partial P}{\partial x_j}+\frac{\partial}{\partial x_i}\mu(\frac{\partial u_i}{\partial x_j}+\frac{\partial u_j}{\partial x_i})+\rho g_i+F_j \quad (3.12)$$

In VOF, surface tension and wall adhesion are modelled with an additional source term in the momentum equation. Tracking of the interface(s) between phases is accomplished by a solution of a volume fraction continuity equation for each phase:

$$\frac{\partial \varepsilon_k}{\partial t}+u_j\frac{\partial \varepsilon_k}{\partial x_i}=S_{\varepsilon k} \quad (3.13)$$

where ε is step function and equals to unity at any point occupied by fluid and zero elsewhere such that:

$$\varepsilon_k(cell)=\frac{\iiint_{cell}\varepsilon_k(x,y,z)dxdydz}{\iiint_{cell}dxdydz} \quad (3.13.1)$$

Equation 3.13.1 demonstrates for the volume fraction of kth fluid, three conditions are possible: (i) $\varepsilon_k = 0$ if the cell is empty (of the kth fluid);(ii) $\varepsilon_k = 1$ if the cell is full (of the kth fluid); and $0 < \varepsilon_k < 1$ if the cell contains the interface between the fluids. Moreover, mass transfer between phases can be modelled by using a user-defined subroutine to specify a nonzero value for $S_{\varepsilon k.}$

Consequently, the surface tension along an interface arises from the attractive forces between molecules in a fluid (cohesion). In this view, near the interface, the net force is radially inward. Surface contracts and pressure increases on the concave side. At equilibrium, however, the opposing pressure gradient and cohesive forces balance to form spherical bubbles or droplets.

One weakness of VOF, however, is that the free water surface denoted in the separate subdomain with tetrahedral cells is not properly smooth. The VOF technique involves regulated mesh along the flow direction to precisely replicate the water surface. Conversely, the other weakness is that the computational cost is exceptionally extraordinary owing to the mesh distortion in an external subdomain. In this regard, the free water surface acquired was inadequate owing to the existence of poor-quality cells during mesh distortion.

IMPACTS OF PITCH ANGLE AND ANGLE OF APPROACH ON MH370 DITCHING

Both angles can assist to understand the sensoria of the Flight MH370 entry water and also it's vanishing. In this regard, pitch angle in an aircraft is called "trim", however, in the plane, "trim" commonly refers to the equilibrium angle of attack, in preference to orientation (Figure 5). In other words, the pitch angle (attitude) is the angle between the longitudinal axis (where the aeroplane is pointed) and the horizon. However, commonplace utilization ignores this distinction between equilibrium and dynamic cases. Conversely, the approach angle is known as the flight path angle. In other words, it is an angle between the horizontal (or some other reference angle) and a tangent to the flight path at a point. It is also known as the flight-path slope (Figure 5).

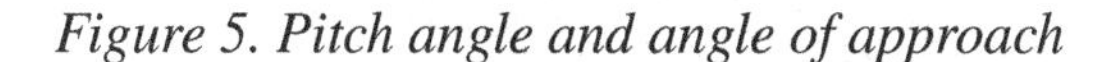

Figure 5. Pitch angle and angle of approach

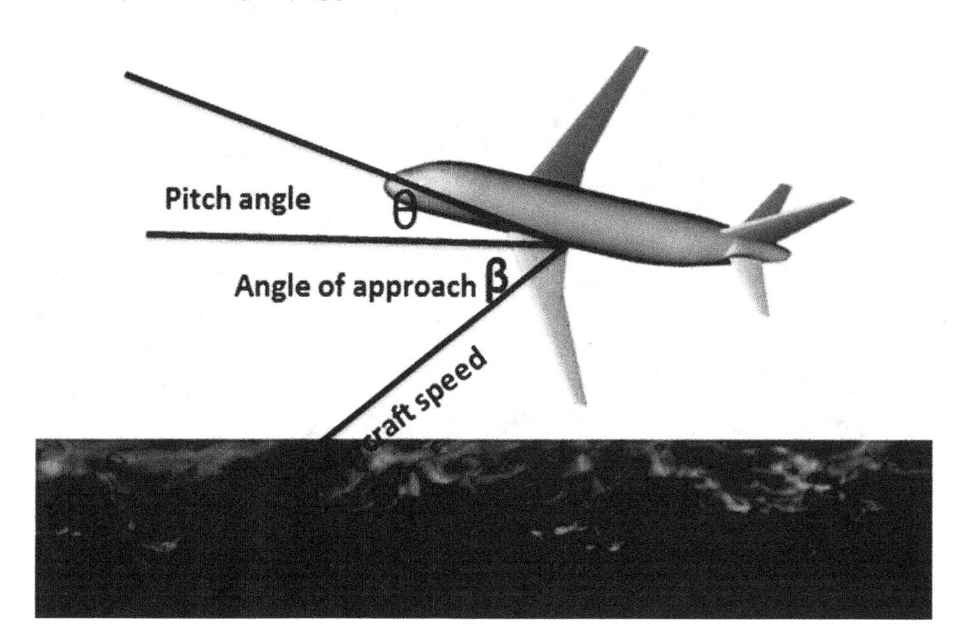

Let us assume the initial stage of the Flight MH370 water entry has an approach angle of 1° and a pitch angle of 9° which leads to glide ditching. In this view, the maximum pressure could be less than 4e+4 P_{max}/Pa and turbulent flow could be generated along with the two wings (Figure 6). In this circumstance, any of the structural failures must occur at weak joints, strut mounts, and stringer-stiffeners where the engine nacelles, wings, and fuselage are connected. Furthermore, the airliner's skin will be subject to surface pressure in the order of 6 MPa, which is considerably smaller than the yield stress 324 MPa of the 2024 T351 aluminum alloy's. In this circumstance, water flooding into the fuselage should be considered.

Figure 6 shows that the point of the peak pressure begins at the rear of MH370 then moves to the front. It can also be seen that the pressure achieves a greatest when this area crashes with water. Furthermore, the middle area of the fuselage is dominated by the maximum wave height of 10 m due to gliding water entry. This particular did not mention in the study of Chen et al., (2015) which is enough to split the two wings from the ditching fuselage. Additionally, as the wing joins the frame, there are stress concentrations in the middle of the Flight MH370, and an excessive vibration would occur in the middle point. Conversely, ditching the large Flight MH370 on the open Indian Ocean commonly would implicate waves of height more than a few which easily triggering fragmentation and the drip of debris.

On the contrary, MH370 would experience the high bending moment and surface pressure when the pitch angle is -3° and angle of approach is 3°. In this circumstance, the Flight MH370 would suffer huge bending moment, and collapsing and breakup can arise. The highest pressure experiences a middle and front cabinet of Flight MH370 of **$3x10^5$ and $5x10^5$** Pa, respectively. In this view, the maximum wave

Figure 6. MH370 ditching with an approach angle of 1° and pitch angle of 9°

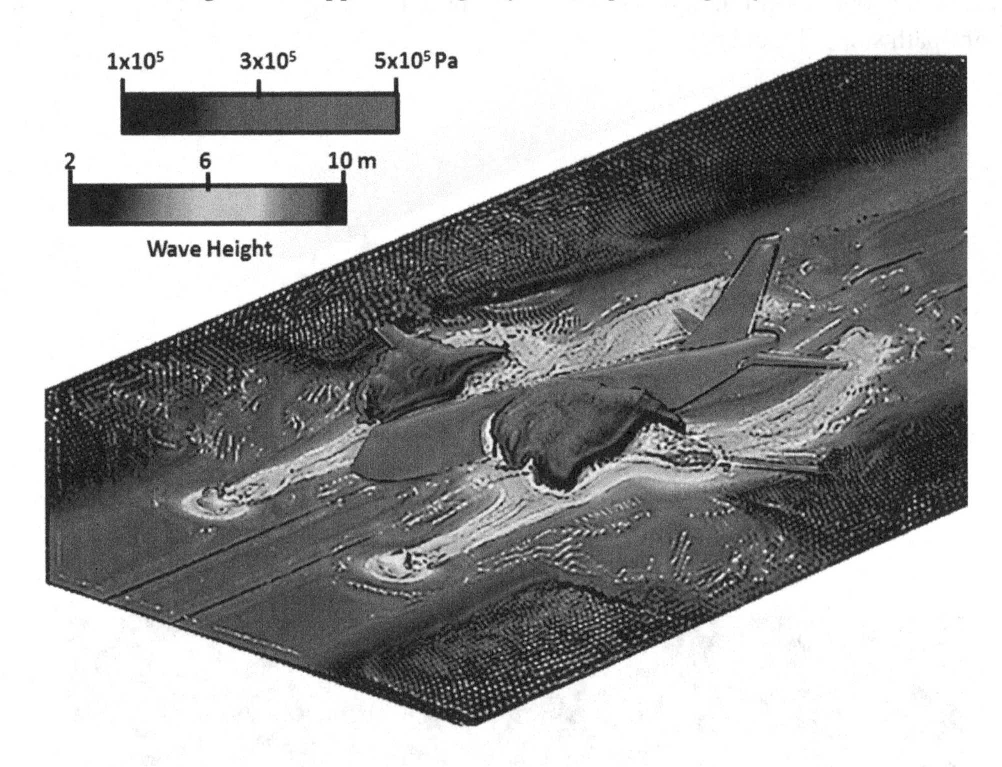

Figure 7. MH370 ditching with an approach angle of 3° and a pitch angle of -3°

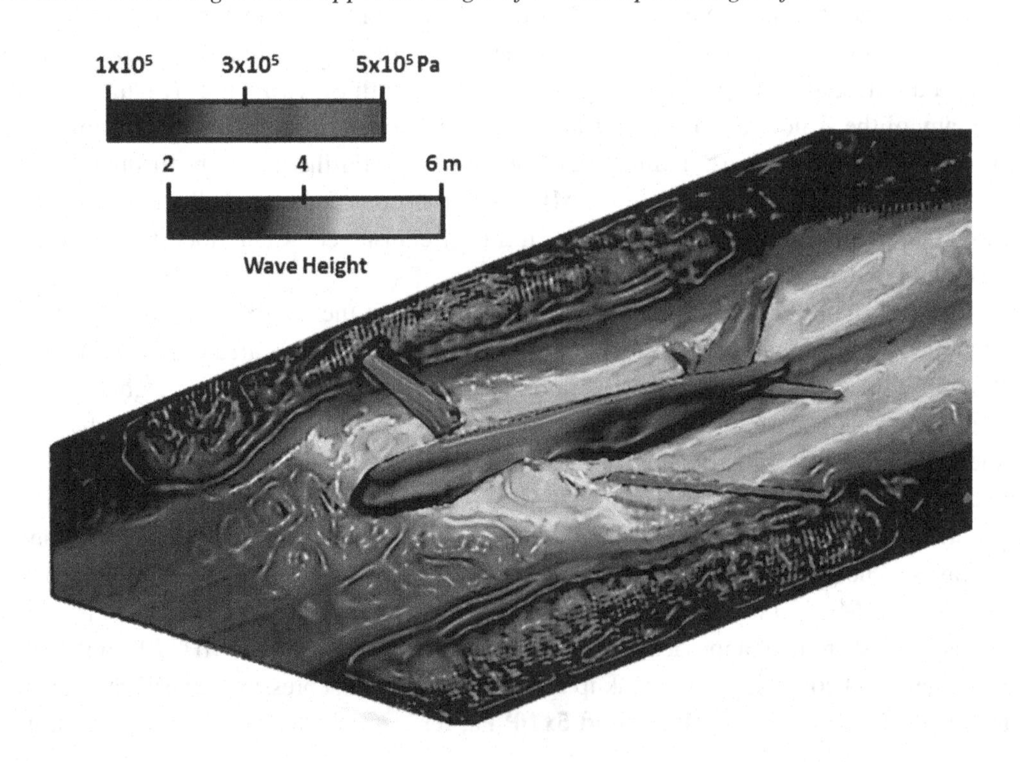

height occurred due to MH370 water entry is approximately 6 m. This occurs across the two wings and middle of the fuselage.

Figure 8 shows a great pressure of $4x10^5$ Pa occurred due to connecting with the ocean surface. This greater pressure could lead to a breakdown of the fuselage. Consequently, it must be much debris left behind the Flight MH370 water entry.

Figure 8. sub-fuselage pressure variations when MH370 was ditched

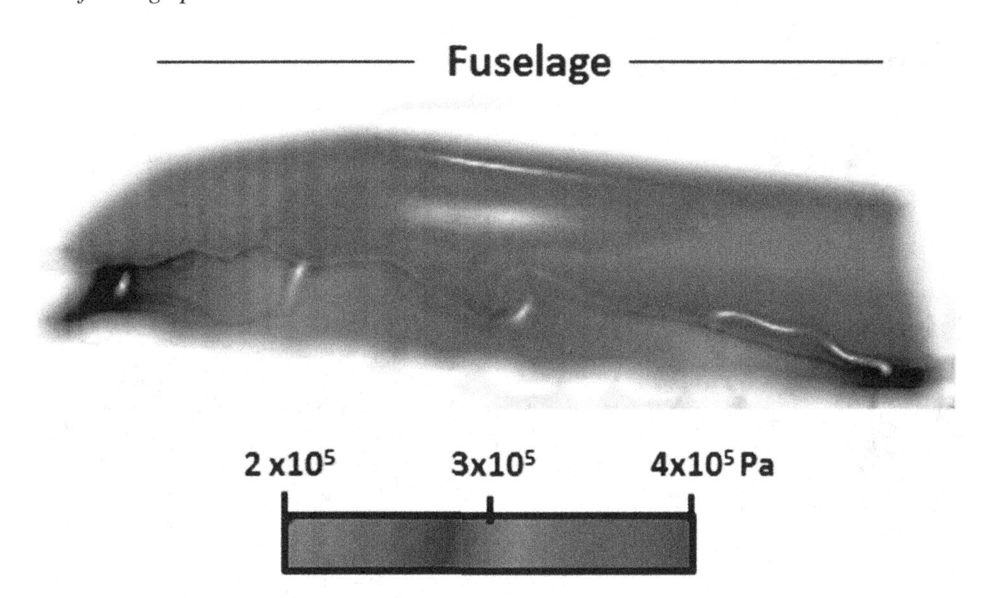

CAVITY FLOW

Flow past a sudden boundary opening of a finite extent is known as a cavity. During the ditching stage, the water flow throughout the rear fuselage has great velocity and squat pressure since the rear part of the fuselage middle has a usual convex-curved profile. The cavitation phenomenon will ensue when the local pressure is smaller than the saturated vapor pressure. In this regard, the complex cavitation model is not used; instead, the following simple method is employed (Batchelor 1956).

In general, the fluid is highly compressible in the two-phase flow areas (the Mach number can be as high as 4 or 5) and it is almost incompressible in the pure liquid areas. So the main difficulty consisted in managing these two different states of the fluid, without creating any spurious discontinuity in the flow field. Besides, the cavitation consists of a very sharp and very rapid process. It means that the density time fluctuations and the density repartition must be smoothed, to avoid numerical instabilities. In the strong density gradient areas such as in the Southern Indian Ocean, the cavitation sheet interface occurs sharply due to the Flight MH370 water entry (Shah, 2010).

As soon as the Flight MH370 wings enter the water, the leading edge of the wing also is exposed to an extreme pressure greater than 10^6 Pa. This scenario occurs when the pitch angle decreases to -30° and a large oblique angle of approach is approximately 30°. Consequently, the boundary outside was split into two parts: (i) the front pressure-inlet; and (ii) the pressure-outlet on the back, on which the pressures were defined in proportion to the depth below the interface in order to contain the water in the presence of gravity (Figure 9). The hydrostatic pressure in the air was neglected because of its small effect within the domain (Gu, 2015).

Figure 9. MH370 ditching with an approach angle of 30° and a pitch angle of -30°

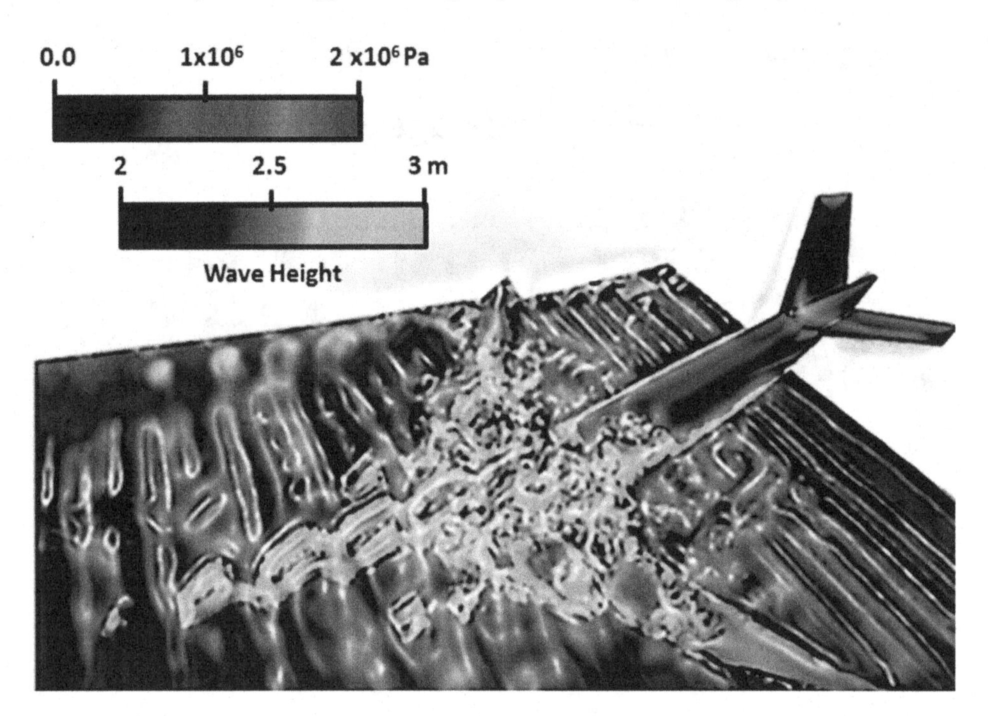

The sub-surface pressure increases along the submerged part of the fuselage to 2x 10^6 Pa (Figure 10). This is noticed on the nose and the middle part of the fuselage. It could be due to the cavity created along the submerged fuselage. In this circumstance, the fuselage must break up to numerous debris.

Flight MH370 's nose-dive situation would occur when the approach angle of 93° and a pitch angle of -90° (Figure 11). A strong current of 3m/s with large eddy can drift the Flight MH370 into the water column and make it sink down to the Southern Indian Ocean floor. In this circumstance, the wings and tail would be split away and the fuselage could dive depth of 30 m or 40 m within seconds, then descend without rising back to the Southern Indian Ocean surface. The pressure distributed vertically along the fuselage with more than 10^6 Pa. Conversely, wing pieces and other heavy debris would sink soon subsequently. This scenario creates a large eddy with a radius of 120 m and a wave height of 6 m. In a mathematical study, which consists of an inviscid core with uniform vorticity (Figure 11) coupled to boundary layer flows at the solid surface, to steady incompressible 2-D driven cavity flow at high Reynolds numbers. When the Reynolds number increases, thin boundary layers are developed along the solid walls and the core fluid moves as a solid body with uniform vorticity (Burggraf, 1966; Dennis and Chang, 1970;Tuann and Olson, 1978; Reboud et al., 1996;Tiesinga et al., 2002; Granichin, 2017 (Figure 11). Furthermore, the submerged part of the fuselage is exposed to the subsurface pressure of 2x10^6 Pa which occurred in the nose of Flight MH370 (Figure 12).

Figure 10. Pressure spatial distribution due to cavity

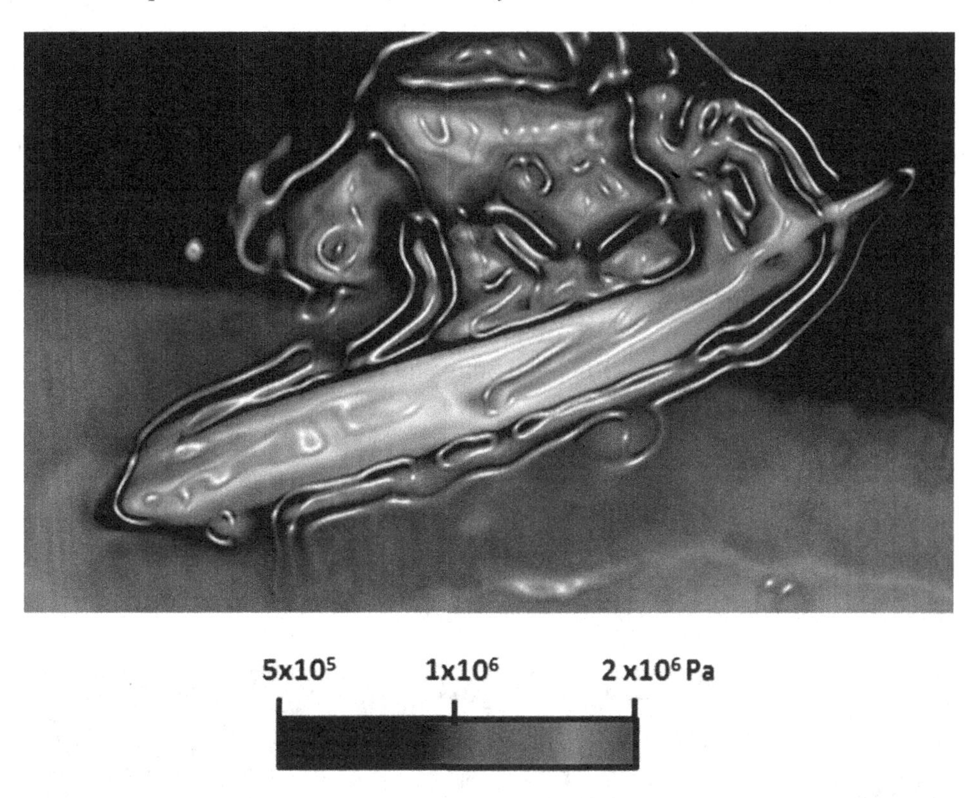

Figure 11. The approach angle of 93° and a pitch angle of -90° effects on Flight MH370

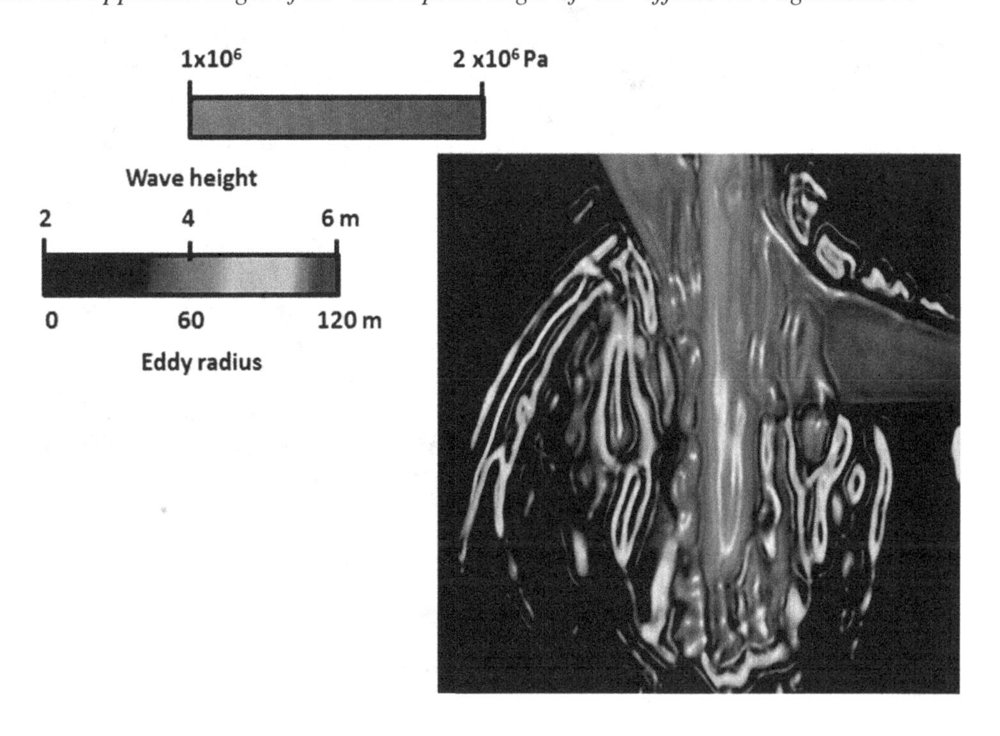

In general, it should be pointed out that almost any closure method will create a mixing layer springing from the upper left-hand lip to strike the opposite wall of the cavity (Figures 9 to 11). In this regard, a large critical flow will almost be created to fill the cavity and will cause the separation streamlines along the fuselage (Figure 11). Physically, the flow in a driven MH370cavity is neither two-dimensional nor steady (Figures 9 and 11), most probably, even at Re=1,000. Finally, in non- cavitating conditions, the boundary layer appears almost symmetric on both sides of the subsurface layer of the Southern Indian Ocean. In cavitating conditions, the important increase of the displacement thickness due to the presence of the cavity wake is visible all along the fuselage suction side. The interaction between that wake and the flow at the fuselage trailing edge leads to the great increase of the flow rate passing in the Flight MH370 nose in cavitation conditions, which modifies a turbulent flow circulation around the fuselage and the resulting sinking (Reboud et al., 1998 and Song and He 1998).

Figure 12. Pressure distribution along the submerged fuselage

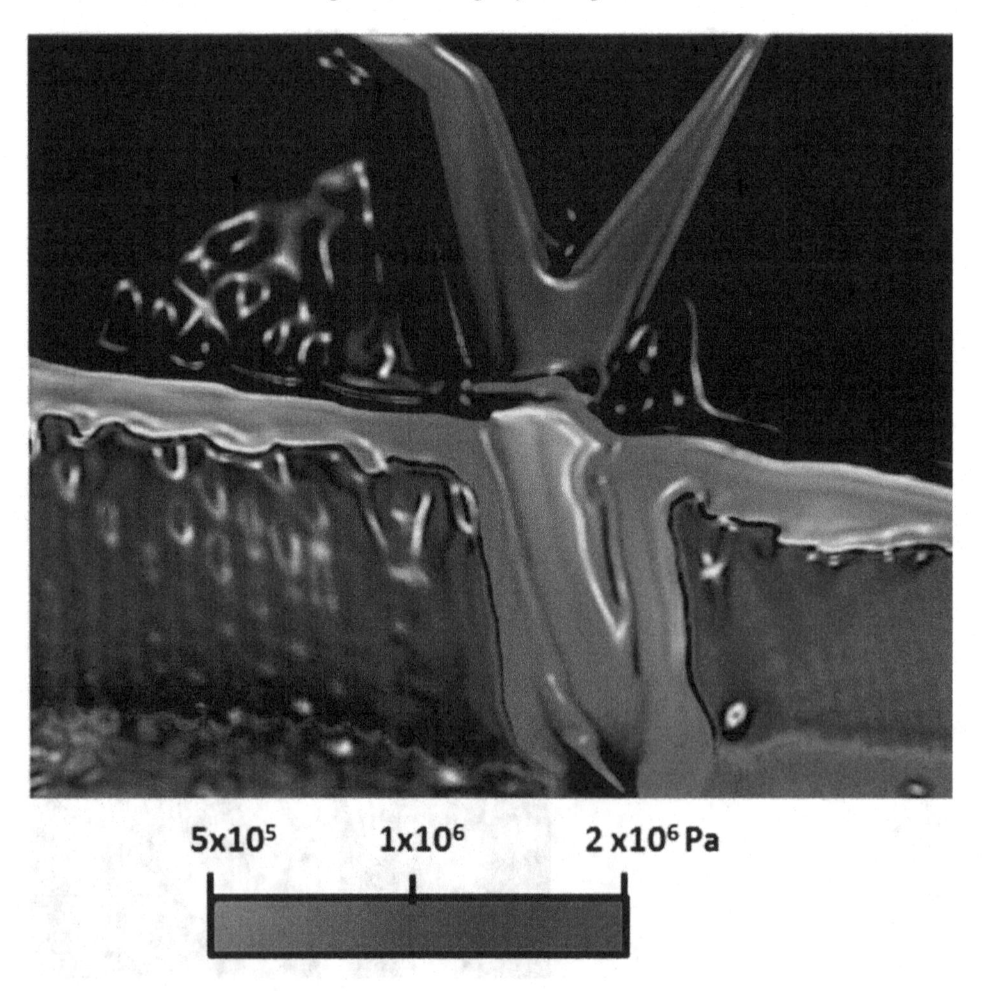

Debris Post Ditching

In general, if the aircraft hits the rough sea surface with a high density such as the Southern Indian Ocean intact the debris field will be relatively enormous. If an aircraft breaks up at elevation, the debris field will be large, and larger with more height and/or speed. Another factor is the angle of impact. If the aircraft hits the sea surface, if nose first and perpendicular, it will be very severely deformed and the debris field will be very contained. Nevertheless, if it hits at a very shallow angle, it will suffer less damage and the debris field will be larger.

Sea surface roughness such as wave breaking, high wave height, and strong current movements also affect damage and debris field. It would find things that have not decayed. The last things to go will be spars and other heavy parts. The human tissue will go fast, as will most textiles. For the metals, it depends on how quickly corrosion acts in the climate. It should be at least found things like spars, stringers, glass, weapons, belt buckles, metal skin components, wheel hubs, axles and such.

On an aircraft accident investigation, there are a lot of factors that are the keystones for aircraft accident investigations which are a function of debris. For instance, the tail debris is the keystone to understand how the flight did crash. First off, what is the position of the tail? If an aeroplane crashes nose down, then the tail will typically separate and end up ahead of the nose since it has more mass. If a plane crashes nose up, then the tail will be at the back of the wreck.

Another good clue is the state of the bodies. The bodies can provide a lot of information. If they still have clothes on, then the aeroplane probably did not break up in flight. However, if they were stripped, that meant the cabin disintegrated in the air. Also, if a water landing, is there water in the lungs? That can indicate if they were conscious of hitting the water or not.

Engines always provide a lot of information. You can look at how they were separated if at all. Also, looking at the damage can indicate if they were turning or not during impact which can provide information about engine failures.

Debris pattern also helps indicate the bearing of the aeroplane. If it was at a steep angle, then the debris field will be small and there will be very little recognizable debris. If the angle was less, then it will be more spread out and the debris items will be larger and more distinguishable.

IS A PLANE WATER CRASH WORSE THAN A TERRAIN CRASH? DO THE PASSENGERS USUALLY DROWN?

It is practically awkward to evaluate one crash to another if a crash is defined as an uncontrolled impact. It would be impossible to say if striking the ground would be worse or better than hitting the water. In an uncontrolled crash, the aeroplane is expected to fragment, and various or all passengers are expected to die to owe to trauma whether it is land or water.

Nonetheless, if any, passengers did persist an uncontrolled impact, they would probably be able to attain survival equipment or something else to hang on. Then it is a matter of environment and attainment rescued. There are not numerous crashes someplace surviving passengers have been saved whether they were in the water or in a strict environment on land.

In either case, it is commonly a difficulty of being initiated and rescued rapidly. In ditching event, a hostile environment usually far from rescue. Consequently, those survivors that do not drown will be expected freeze or worse prior they are located. Not to sound horrible, nonetheless, that is a sort of the approach it exists. It is easy, therefore, to ignore how far the crashes are from civilization when they are occurring over the middle of the ocean.

In case of a controlled emergency landing or ditching, then there are advantages and disadvantages to both. However, the universal perception is that land is worthier. In free of obstacle situation, an excellent control of the aeroplane is a very smooth landing at the slowest practical airspeed would be superior to a ditching. In either event, there are expected to have a lot of survivors, if an aeroplane can make a controlled crash landing or ditching.

Compared with an off-airport controlled landing in an airliner, it would rather be in a ditching. In fact, there are tons of stuff on land for the plane to hit, which is dangerous, for instance, trees, power lines, etc. This leads to the exploding of the flight. A properly executed ditching, consequently, can also be dangerous, particularly when the water is rough or aid is far away. Nevertheless, still, it sounds worse than "landing" in a forest.

IS THE DITCHING THEORY ACCEPTABLE IN CASE OF MH370?

Flight MH370 was uncontrolled when traveled down to the Southern Indian Ocean. The ditching theory is not acceptable in the case of MH370. If the flight failed down from high latitude it would spin clockwise and anticlockwise simultaneously before approaching the sea surface. In this view, the turbulent airflow around the Flight would crack its structure and split away some of the joint parts for instance wings and tails.

Further, whether floating debris from the passenger cabin—things as foam seat cushions, seatback tables, and plastic drinking water bottles—would float to the surface. Consequently, this would rely on whether the fuselage cracked on effect, and how was the critical damage. This scenario could be strongly acceptable only to answer why there is no much debris left after the event.

CONCLUSION

This chapter addresses the mathematical probable model description for MH370 ditching. In this view, the chapter also has addressed the differences between ditching and airplane crashing into the water. To this end, Von Karman theory and Volume-of-fluid (VOF) technique are implemented to investigate the rationality explanations behind the Flight MH370 vanishing. In this chapter, the most significant conclusion can be drawn is that If MH370 crashes nose down, then the tail will normally discrete and transpire ahead of the nose as it has heavy mass. If MH370 crashes nose up, then the tail would be at the back of the wreck. In conclusion, the ditching theory can be used in the case of MH370. As long as the MH370 was uncontrolled, then it must fall by a high speed and must be spinning in clockwise and anticlockwise motions simultaneously before hitting strongly due to gravity impact on the rough sea surface. In this understanding, the fuselage would be broken down into some pieces, and wings and tails, for instance, would split away from the fuselage under air turbulence impacts and appear as floating debris in the crushing area. However, there is no recorded debris that been located at the crash site.

REFERENCES

Batchelor, G. K. (1956). On steady laminar flow with closed streamlines at large Reynolds number. *Journal of Fluid Mechanics*, *1*(2), 177–190. doi:10.1017/S0022112056000123

Burggraf, O. R. (1966). Analytical and numerical studies of the structure of steady separated flows. *Journal of Fluid Mechanics*, *24*(1), 113–151. doi:10.1017/S0022112066000545

Chen, G., Gu, C., Morris, P. J., Paterson, E. G., Sergeev, A., Wang, Y. C., & Wierzbicki, T. (2015). Malaysia airlines flight mh370: Water entry of an airliner. *Notices of the American Mathematical Society*, *62*(4), 330–344. doi:10.1090/noti1236

Dennis, S. C. R., & Chang, G. Z. (1970). Numerical solutions for steady flow past a circular cylinder at Reynolds numbers up to 100. *Journal of Fluid Mechanics*, *42*(3), 471–489. doi:10.1017/S0022112070001428

Erturk, E. (2009). Discussions on driven cavity flow. *International Journal for Numerical Methods in Fluids*, *60*(3), 275–294. doi:10.1002/fld.1887

Granichin, O., Khantuleva, T., & Amelina, N. (2017). Adaptation of aircraft's wings elements in turbulent flows by local voting protocol. *IFAC-PapersOnLine*, *50*(1), 1904–1909. doi:10.1016/j.ifacol.2017.08.263

Gu, C. (2015). *Computational Mechanics for Aircraft Water Entry and Wind Energy* (Doctoral dissertation).

Reboud, J. L., Rebattet, C., & Morel, P. (1996). Effect of the leading edge design on sheet cavitation around a blade section. In *Hydraulic Machinery and Cavitation* (pp. 651–660). Springer. doi:10.1007/978-94-010-9385-9_66

Reboud, J. L., Stutz, B., & Coutier, O. (1998, April). Two-phase flow structure of cavitation: experiment and modelling of unsteady effects. In *3rd International Symposium on Cavitation CAV1998, Grenoble, France* (*Vol. 26*). Academic Press.

Shah, S. (2010). *Water impact investigations for aircraft ditching analysis (MSc thesis)*. RMIT University.

Song, C. C. S. (1998). *Numerical Simulation of Cavitating Flows by a Single-Phase Flow Approach*. In *Third International Symposium on Cavitation*, Grenoble, France.

Tiesinga, G., Wubs, F. W., & Veldman, A. E. P. (2002). Bifurcation analysis of incompressible flow in a driven cavity by the Newton–Picard method. *Journal of Computational and Applied Mathematics*, *140*(1-2), 751–772. doi:10.1016/S0377-0427(01)00515-5

Tuann, S. Y., & Olson, M. D. (1978). Numerical studies of the flow around a circular cylinder by a finite element method. *Computers & Fluids*, *6*(4), 219–240. doi:10.1016/0045-7930(78)90015-4

Chapter 4
Principles of Genetic Algorithms

ABSTRACT

This chapter aims to review the fundamentals of genetic algorithms. Consequently, the chapter correspondingly lectures on the dissimilarities between the alteration genetic and evolutionary algorithms. The decision mathematical rules based on the Pareto algorithm are similarly deliberated. The Pareto optimization rule can have a significant role in the examination of the precise position of MH370 vanishing. In this circumstance, the majority of multi-objective optimization algorithms exercise this principle to acquire the non-dominated set of solutions, as a result of the Pareto-front to investigate how MH370 ended up in the Indian Ocean.

INTRODUCTION

Although the Australian Transport Safety Bureau (ATSB) provided adequately technical reports concerning the Flight MH370 dynamic vanishing, it can deliver the dynamic scenario of this event and also fail to locate the site of the crash. Therefore, ATBS has documented the technical reports in one book named "Bayesian Methods in the Search for MH370". The book was interesting and considered as a first scientific book which deals with the Flight MH370 tragedy. Nevertheless, the Bayesian methods cannot detect any debris left beyond the Flight MH370 in the Indian Ocean. On the contrary, the Bayesian models just able to indicate the smallest geographic search area containing 99% of the probability distribution function (PDF). In this sense, the critical question is raised as: where is MH370 fuselage and its debris across that convinced search area?

In this understanding, Bureau (2015) reported that the features of the drifters are not precisely tallied to the flaperon, hence there is some vagueness in the relevant of the consequences. Nonetheless, it is understood to be the furthermost applicable data source accessible. In this context, most truthfully important sources of data on the expected of the flaperon's float are delivered from the Global Drifter Program. Indeed, the drifters are seated about 50 m below the sea surface which are arrayed with drogues. A significant variation in the drifter's mobility would notice as the drogue is separated. In this regard, Davey et al., (2016) stated that the buoyancy of the flaperon has not been considered in this phase. In fact, drogues by design instigate drifters to travel consistent with deeper currents. On the contrary, drogues

DOI: 10.4018/978-1-7998-1920-2.ch004

by design trigger drifters to move according to surface current not deeper currents. Besides, 15 m water depth below the sea surface is not considered as deep water but a subsurface layer.

This chapter is assigned a new technique to understand the real facts that accompany the Flight MH370 vanishing. This technique is based on a genetic algorithm. This chapter explains the GA's mechanisms which can be used in the next chapters to reconstruct the consequence of the Flight MH370 plug into the Southern Indian Ocean.

WHAT IS MEANT BY A GENETIC ALGORITHM?

Consistent with De (1975) and Holland, (1975), a genetic algorithm (GA) had been formerly delivered by John Holland within the 1970s due to investigations into the opportunity of computer applications experiencing a growth in the inner of the Darwinian rationality thought (De, 1975; Dhar and Stein, 1997; Fakhreddine and de Silva 2004).

GA is a ration of a larger soft computing model mentioned to as evolutionary calculation (Goldberg 1994). They endeavor to accomplish optimum clarifications over a technique disinterested similar to biological evolution (Kingdon,1997). This requires the perceptions of the persistence of the fitting and crossbreeding and mutation to create particular clarifications from a group of prevailing elucidations (Ackley, 1987).

Genetic algorithms were revealed to achieve solutions for a wide diversity of concerns for which no relevant algorithmic answers be existent. The GA practice is proper for optimization, a problem-solving technique in which an individual or a multi exceptional clarification is search for obtaining a space solution as well as an immensely an extensive variety of reasonable solutions (Medsker 2012). GA contracts the search cavity by continuously evaluating the existing initiation of applicant clarifications, removal those graded as ruthless, and producing an innovative group through crossbreeding and mutating the ones categorized as accurate. The ranking of aspirant clarifications has talented the convention of an insufficient pre-definite measure of integrity or appropriateness (Syswerda,1989).

WHAT IS THE DISSIMILARITY BETWEEN GENETIC AND EVOLUTIONARY ALGORITHMS?

The most significant kind of evolutionary algorithms is genetic algorithms. Conversely, they are the keystone of the evolutionary algorithms, which are continually articulated. In the 1970s, the restraint of an evolutionary algorithm was relatively advanced. Besides, its terminology is immobile a diminutive uncertain. The expression "Evolutionary Algorithms" integrates all stochastic algorithms created on "imitation" and "assortment" of contestant clarifications, which is insecurely mimicking how animal inhabitants cultivate in nature. Consequently, both genetic and evolutionary algorithms comprise in abundant prominent subgroups. The former is a Genetic Algorithm (GA), which is designated as a dualistic string that is accessible to an exemplification of the confident clarifications. Evolution Schemes (ES), which utilize vectors of real-valued quantity as a descriptor. One of the most well-known ES is the covariance matrix adaptation ES (CMA-ES).

The latter is Genetic Programming (GP) which is the most advanced technique (the 1990s), that utilizes binary trees. In other words, it depletes convoluted assemblies, for instance, fixed diagrams as an internal demo of applicant solutions. In this view, GP is extremely dominant, as it tolerates for developing virtually *everything*, from association programs to gene monitoring networks and from Bayesian Networks to complicated computing.

Additionally, evolutionary algorithms encompass quantum evolutionary algorithms. In this understanding, evolutionary algorithms are established by combining quantum superposition into genetic algorithm contexts, and genetic programming. Alternatively, they can develop assemblies similar to networks or computer programs.

HOW DOES GA WORK?

A genetic set of regulations is probabilistic which is searching for a technique that binary simulates the gadget of natural evolution. It mimics evolution in nature through over and completed renovating a population of applicant solutions up until the superlative solution is resolute. In this view, the GA evolutionary cycle commences advanced with accidentally determined initial inhabitants. Subsequently, the alterations to the population become about the procedures of optimal grounded absolutely on fitness, and modification exercising crossover and mutation (Figure 1). In this viewpoint, the software program of aspiration and amendment consequences in the inhabitants with the superlative percentage of advanced solutions. In this occurrence, the evolutionary cycle endures in anticipation of a fitting solution is established within the existing group of a population. Similarly, a bit operate parameters along the population creation are exceeded (Dhar and Stein 1997).

Accordingly, the tiniest constituent of a genetic algorithm is authorized as a *gsreene*, which signifies a portion of information in the predicament domain. The series of genes are called a *chromosome*, symbolizes individual promising explanation to the predicament. Certainly, an individual gene in the chromosome connotes one component of the revelation matrix (Goldberg, 1994;Dhar and Stein, 1997 ; Fakhreddine and de Silva 2004).

The furthermost common technique of representing a solution as a chromosome is the sequences of dualistic digits. Consequently, the discrete bit in this series is a gene. The practice of swapping the revelation from its inventive system into the sequences of dualistic digits is acknowledged as a *coding*. The specific coding structure discard is a practice resolved. The solution sequences of dualistic digits are decoded to simplify their consideration developing a fitness value (Michalewicz, 1996 and Fakhreddine and de Silva, 2004).

PRINCIPLES OF GENETIC ALGORITHM

Fitness Comparative Collection

In genetic evolution, furthermost operative the fittest remains to occur and their gene pond donates to the group of the succeeding time. The selection of a GA is similarly grunded on an analogous technique. In a conventional setting of the choice, stated to as the fitness comparative selection, each chromosome's probability of actuality resolved on as an exceptional one is relative to its fitness percentage. In the model, this fitness percentage is proportional as contrasting to absolute and is a castoff comparative evaluation of solutions in relationships of their adequacy.

Figure 1. The basic concept of a genetic algorithm

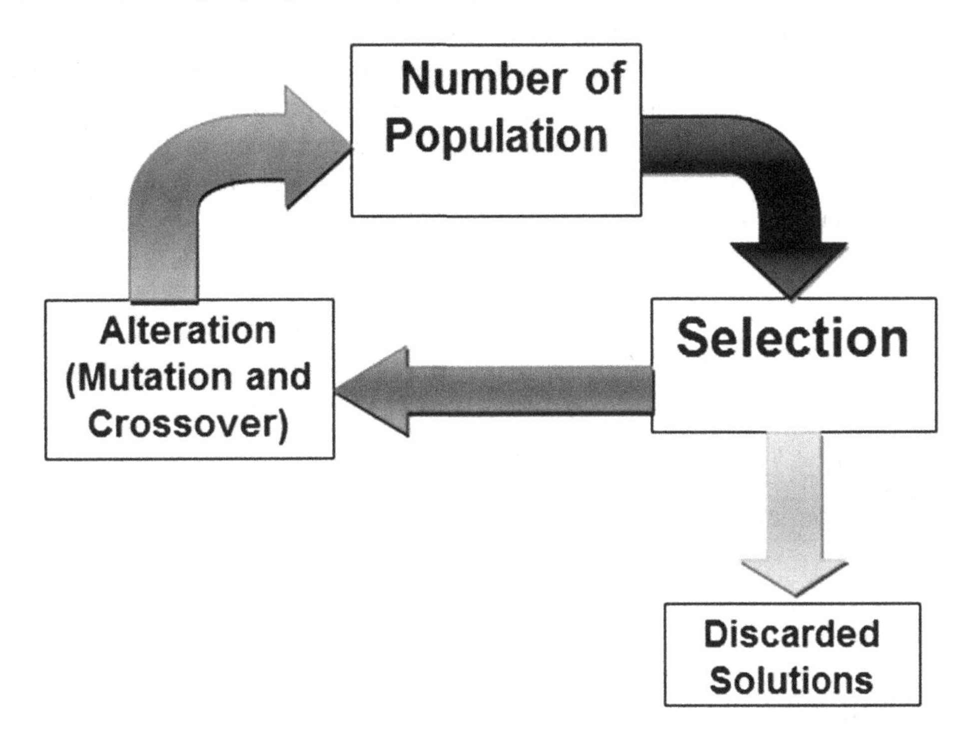

The most frequent fitness comparative desire technique is the roulette wheel selection. Every single chromosome has specified a slice of the spherical roulette wheel. This neighborhood is proportional to the fitness ratio, which is the ratio of the chromosome's fitness cost to the whole fitness of the population. To pick out a chromosome for the subsequent initiation, a random extent inside the interval [1,100] is generated, and the chromosome whose section spans the random extent is chosen, which is equivalent to spinning a roulette wheel for determining on a winner.

The fitness value of a precise answer is measured through a fitness characteristic, additionally referred to as an goal function. It takes a chromosome as a parameter and normally returns an actual-valued end result due to the fact of the chromosome's fitness value.

In a few instances, the preference of the fitness characteristic may additionally be incredibly simple, even as in special conditions, it can be expensive to be computed. The feature would possibly no longer be a mathematical solution and alternatively, be a dimension of the solution's overall performance in a simulation or actual-international investigation. The resolution operator selects high-quality solutions for comparable processing in particular based on their fitness values and discards the ultimate ones. The format of the fitness feature is one of the higher challenging phases in a GA.

Basic of Crossover

Crossover can appear like the artificial equivalent of genetic recombination through mating prevalent in nature. Within the crossover, chromosomes are signifying two singular options interior the current generation. In this fact, they are mixed to create the chromosome for the subsequent technological understanding of solutions. This is carried out through the use of splicing chromosomes from two tremendous alternatives at a crossover element and swapping the spliced aspects (Figure 2). In this vision, a few genes with the appropriate features from one chromosome can also add, as a consequence, combine with a few unique genes inner the one-of-a-compassionate chromosomes to generate a higher solution represented with the aid of the utilization of the producer new chromosome. Crossover is generally carried out with a possibility, with 0.7 typically producing fabulous consequences. Chromosomes are now no longer crossed over replicated with their precise copies which are added to the subsequent technology (Ackley 1987 and Kingdon, 1997).

Figure 2. Crossover point demonstration

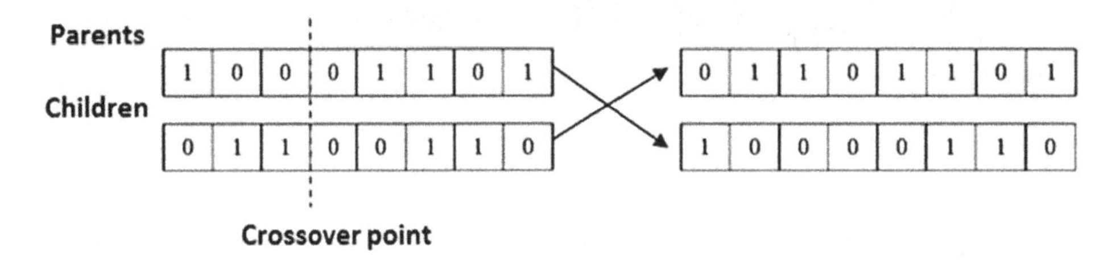

Principle of Mutation

The mutation is a random rectification through the genetic composition. In nature, mutations are illogical modifications in genetic elements which are ascribed to perilous chemical compounds or radiation. It is a long way advisable in GAs for acquainting with new qualities in an answer population – something now no longer performed via crossover alone. Crossover informal repositions the present features to deliver the new combinations. As an instance, if the foremost bit in every chromosome of a group takes vicinity to be a 1, any new chromosome created employing crossover will even have 1 due to the fact of the first bit.

The mutation operator selects a gene in a chromosome at random and changes its current rate to a special one. For bit string chromosomes, this change portions to flipping a zero bit to a 1 or vice versa.

Although recommended for introducing new dispositions within the solution pool, mutations may additionally be counterproductive and carried out the most advantageous occasionally and randomly (Figure 3). The likelihood of the mutation took place in nature is fairly minor. Consequently, in GA mutation chance commonly tiers between 0.001 and 0.01 (Michalewicz 1997).

Figure 3. Mutation points demonstration

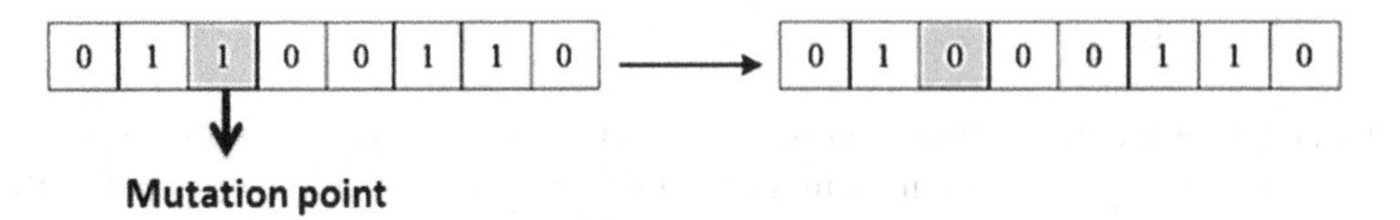

Interference of Mutation

Mutation interference occurs at a similar time as mutation values in a GA are tremendously disproportionate. Alternatively, solutions are recurrently or comprehensively mutated and as a consequence, the algorithm in no way accomplishes to determine any locality of the space systematically. In this concern, any exceptional solutions are determined to tend to be demolished rapidly. This difficulty is opposing to that of precocious computing convergence. A genetic algorithm is subjected to mutation interference perhaps by no converge incomes attributable to the fact its population is just excessively unstable. Ruling a mutation rate that authorizes the GA to converge, nonetheless, it tolerates an acceptable examination of the solution intermission which is critical for exceptional operation (Syswerda,1989).

CONVENTIONAL PHASES OF GENETIC ALGORITHM

De (1975); Dhar and Stein,(1997) and Kingdon (1997) stated that there are numerous stages tangled in the genetic algorithm procedures, which are designated as follows:

1. Signify the solution as a chromosome of the steady dimension.on the other hand, this stage contains the choice of population N, for crossover probability p_c and mutation probability p_m.
2. Allocate a fitness function f for formative the fitness of chromosomes.
3. Create an initial solution population arbitrarily of length N: $x_1, x_2, \ldots, x_N$
4. Training the fitness function f to weigh the fitness percentage of the discrete solution in a surviving generation: $f(x_1), f(x_2), \ldots, f(x_N)$
5. Establish "superlative" solutions as a function of a fitness level. The larger the fitness level of a solution, the greater the probability of it being chosen. Then the rest of the solution is banned.
6. If an individual or extra adequate solution is originated in a prevailing generation, or a maximum quantity of generations is exceeded, then discontinue.
7. Revise the solution population disbursing crossover and mutation to spawn a new population with length N of solutions.

To implement GA accurately, numerous inquiries request a precise answer.

- How is a discrete signified?
- How is discrete's fitness totaled?
- How is discrete chosen for breeding?
- How is discrete crossed-over?
- How is discrete mutated?
- What is the population dimension or length?
- What are the "execution circumstances"?

Along with the above perception, most of these explorations have a problem-specific retort. The former two, nonetheless, can be considered in a further mutual approach. With these concerns, the length of the population is tremendously asymmetrical. The length of the population must be as gargantuan as practicable. The pressuring factor, consequently, is deliberated the consuming phase of the

GA algorithm. In this perception, the larger population quantity which means the extra period crushing computation. Mainly, the accomplishment circumstances mean that there is no routine to dismiss the algorithm (Kingdon, 1997).

How is a Discrete Signified?

Let us adopt that $x,y,z,$ and w are the inconstant values in that are addressed on the succeeding equation:

$$f(x,y,z,w) = w^2 + x^2 - y^2 + 3wx + wz - xy + 4 \quad (4.1)$$

Equation 4.1 aims at locating the maximum value of the $f(x,y,z,w)$. In this perception, GA can be castoff to elucidate equation 4.1, nonetheless, it involves answering the above-listed inquiries. Further, the foremost inquiry is how to signify these four quantities. In this framework, a simple approach is to utilize an array of four tenets that comprise integers or floating-point numbers. In contrast, in GA, it is requisite to have bigger individual variations. To this end, the bit strings exploited to existing the individuals, which convey the greatest occurrence. In this outlook, it relies on the dimension selection of bits for each element. Then, it concatenates the four values composed into a distinct bit string (Michalewicz, 1996).

For instance, each element is chosen as a four-bit integer and producing a new entry individual of a 16-bit. Thus, an individual is prescribed as 1101 0110 0111 1100.

How is a Discrete's Fitness Calculation?

Predominantly, evaluation and fitness functions are dissimilar. In this concern, the evaluation function produces a perfect amount of the singular. Quite the reverse, the fitness function regulates an exact rate of the discrete bound by the rest of the population (Michalewicz, 1996).

Let us undertake 1010 1110 1000 0011,0110 1001 1111 0110,0111 0110 1110 1011, and 0001 0110 1000 0000, separately, as the 4 discrete of $x,y,z,$ and w. In this consideration, the fitness function can be designated from numerous decisions. For instance, the individuals could be organized from the nethermost of the uppermost evaluation function values, and an ordinal categorizing operated. Alternatively, the fitness function would be the discrete's evaluation value isolated by the mean evaluation rate (Negnevitsky, 2005). Hence, the greatest discrete has the utmost fitness. A discrete's fitness signifies the rate of the individual bound by the excess of population N dimension.

How Are Discretely Designated For Breeding?

The foundation of the selection technique is that it demands to be probabilistically weighted with the intention of improved fitness discrete has an enhanced probability of actuality decided on. Aside from these specifications, the procedure of selection is exposed to elucidation. The unique prospect is to smear the ordinal way to the fitness function, then estimate a probability of the selected genes, which are equivalent to the individual's fitness change divided by the full fitness of all of the individuals.

For instance, the first individual would deliver a 40% chance of actuality selected, the subsequent a 20% chance, the third a 30% chance, and the fourth a 10% chance. In this understanding, it delivers improved extra individual chances to be nominated (Michalewicz,1996;Medsker 2012; Negnevitsky, 2005).

How Does Crossed-Over Occur?

In a genetic algorithm computing algorithm, crossed-over designates bred. In this implication, from each fixed group of parents, probably, dual children are generated. In this event, binary sets within the individual are randomly designated, which describe corresponding substrings in every single individual. The role of substrings, consequently, is generated by twofold new-fangled children such as they performed between individual parents. In the one-location crossover, therefore, a location is termed as the crossover point, which is designated indiscriminate and the partitions to the right of this point are exchanged. Let undertake binary individual parents of p_1=101010 and p_2=010100, separately. Beyond, the crossover point is assumed to be between bits 4 and 5 (anywhere the bits are amount to from left to right beginning at 1). Formerly, binary children are c_1=101000 and c_2=010110, respectively (Figure 4).

Figure 4. Crossover result for initial and third individuals

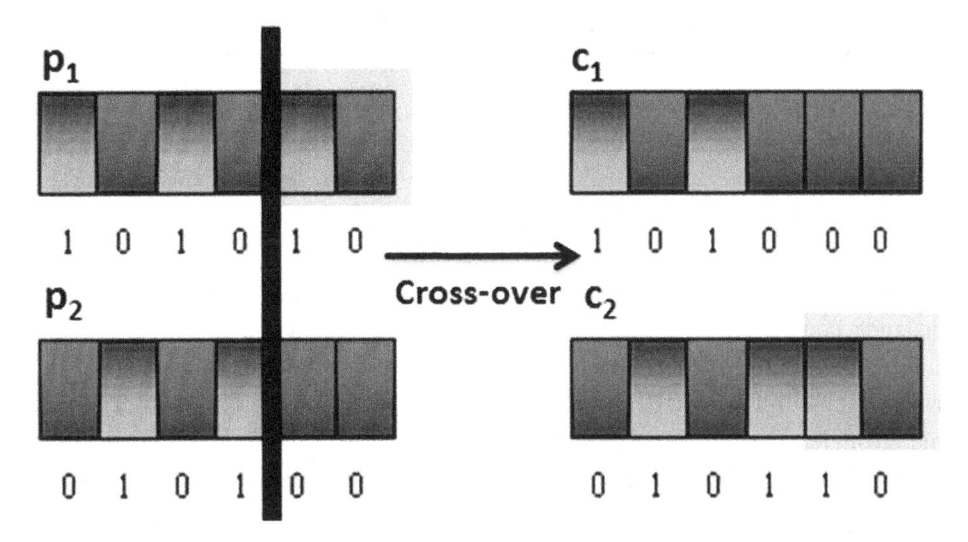

As stated by Michalewicz (1996), there are frequent varieties of crossover operatives, for instance, dual-point crossover, multi-point crossover, and identical crossover. In the dual-points crossover, as opposed to the linear strings, chromosomes are deliberated as ring-formed by way of a consequence of the association of the simultaneously ends. Subsequently, binary special points are designated at capricious, contravention the ring into binary segments. A segment from one ring is transacted with one from another ring to produce dual offspring. In this scenario, transaction a single segment, dual-point crossover accomplishes a comparable operation exploiting a one-point crossover (Fakhreddine, and de Silva 2004).

In contrast, the multi-point crossover is a growth of twofold-point crossover (Ackley, 1987). Analogous to twofold-point crossover, multi-point crossover covenants with the chromosome as a ring, which the crossover plugs to break into many segments (Fakhreddine, and de Silva 2004). In this framework, for each setting in the offspring, the constituent derivative from one or the other parent. Alternatively, it creates an indiscriminate crossover cover, which is termed as uniform crossover (Figure 5).

Figure 5. Uniform crossover

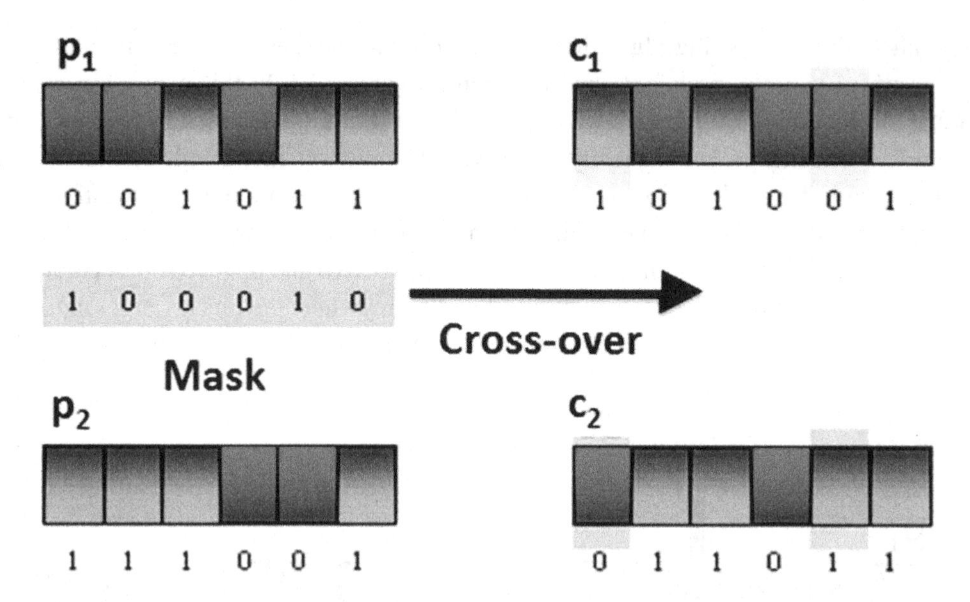

How do Individuals Mutated Achieved?

Once individuals are signified as bit strings, the mutation comprises receding a subjectively nominated bit. Let us, for instance, consider that the individuals are signified as binary strings. In a bit accompaniment, as soon as a bit is designated to mutate that bit can be inverted to be the counterpart of the distinctive bit rate. Furthest, let assume $p_1$is 101010 and guess that the mutational bit is 4 bits. Thenceforward, the bits are formed from left to right preparatory at 1 where the child c_1 is 101110 (Figure 6). Contrariwise, the mutation can involve the bit having "flip": 1 swap in 0 and 0, then shifts to 1. Another mutation is subjected by a noising mutation that smudges only to an unexperienced encoding gene. This mutation expands a white noise to the nominated gene from end to end of mutation.

Binary mutation signifies that the spinning of individual values, subsequently each individual, has merely dual stages. Irrelevance, the length of the mutation stage is continuously 1. A binary mutation for an individual with 11 variables, for instance, primes to a 4th mutable to be transmuted (Figure 7).

GENETIC ALGORITHM PROGRAMMING

Particularly, genetic programming is abundant prevailing than genetic algorithms. Certainly, the productivity of the genetic algorithm is a capacity, whilst the productivity of genetic programming is computer software. Intrinsically, genetic programming is the instigating of programming computing codes that

Figure 6. Concept of Mutation

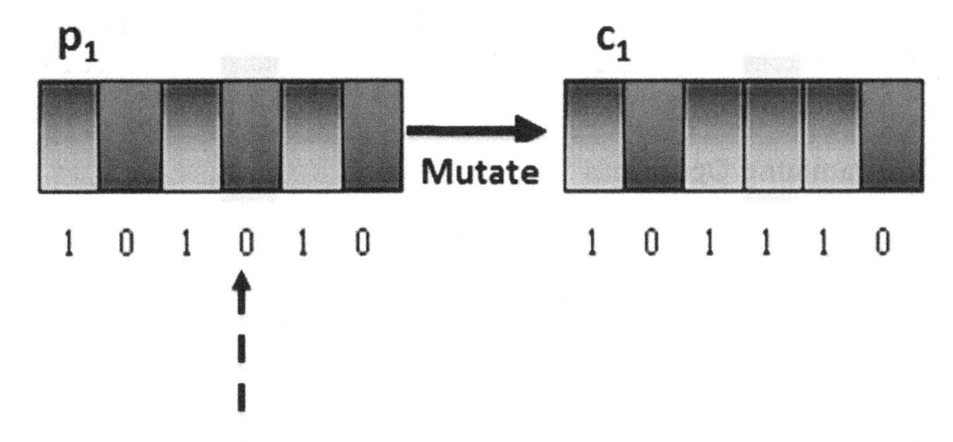

Figure 7. Individual prior and post binary mutation

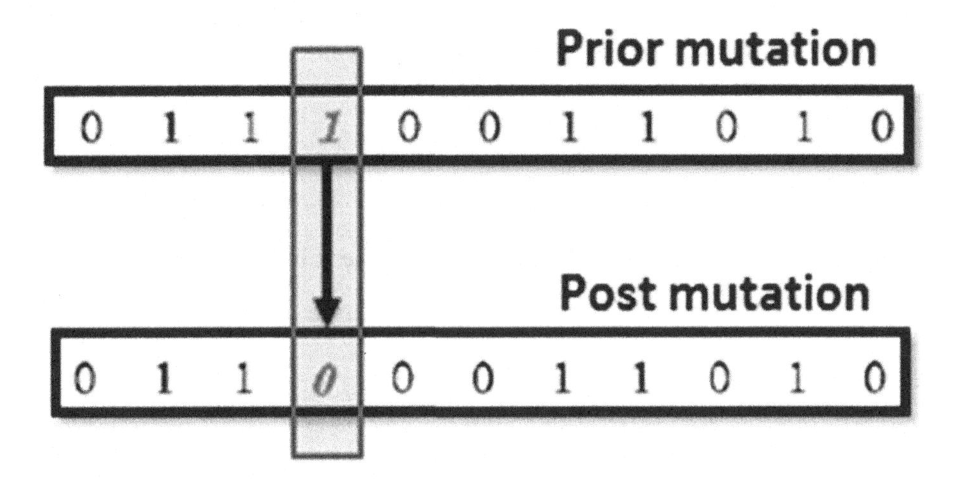

code themselves. Genetic programming, therefore, operates better for numerous sorts of issues. The primary sort is based on there is no impeccable solution. For instance, a scenario beyond the Flight MH370 vanishing.

Up to date, there is no one solution to mysterious of MH370. Therefore, a vanishing of MH370 consists of making compromises of Inmarsat's satellite telecommunication information about the flight location as well as many other variables such as debris, and the Southern Indian Ocean circulation. These will be discussed logically based on genetic programming in the following chapters. In this circumstance, genetic programming is employed to determine the best solution that challenges to concede and be the most effective solution to investigate the mystery of MH370.

The interpretation of the solution is the vital dissimilarity between genetic programming and genetic algorithms. In this exceptional incident, genetic programming creates processor tools in the construction of computer infrastructures as the revelation.

Genetic algorithms convey the categorization of amounts that continue to trendy the solution. Hypothetically, genetic programming uses numerous stages to resolve any problem. Once, a preliminary population of arbitrary constructions of the roles and sidings of the problem i.e., generating computer programs.

Genetic Programming Operators

Operator Crossover

Genetic programming involves dual keystone processes that exist for rereading the coding program. In this concern, the crossover is the furthermost vital information. The crossover algorithm is to create dual new offspring or solutions. Essentially, the fitness functions are implemented to select excellent-quality parents from the haphazard population. To this end, three approaches are implemented to select extraordinary solutions for running the crossover algorithm.

The notable technique exploits the probability created on the strength of the solution. Let us consider that f_{s_i} presents the fitness of the solution s_i that is chosen from the dimension of the population number n. The total (T) of the full population 's memberships can be obtained by:

$$T(f)_{s_i} = \sum_{i=1}^{n} f_{s_i} \tag{4.2}$$

The important parameter is required to be determined is the solution probability P that derivatives into the next generation as:

$$P_i(s_i) = f_{s_i}(T)_i^{-1} \tag{4.3}$$

The subsequent approach is competition selection, which is fixed precisely by determining a discrete population of chromosome variations in genetic programming.

In the larger event dimension, the imprecise chromosomes have an inferior probability to be selected. The stronger chromosomes have an advanced probability to be occurred along with a fragile one in a similar event. Consequently, the pseudo-code of the selected approach, perhaps be pronounced as (Miller et al., 1995):

Chose D (the event size) chromosomes commencing the arbitrary population number "n"
Chose the utmost chromosomes grounded on probability P
Chose the succeeding highest chromosomes over and done with probability P(1-P)
Chose the last optimum chromosomes expending probability $P((1-P)^2)$,
and so on

$$P_i(s_i) = \begin{cases} \frac{C_{n-1}^{D-1}}{[C_n^D]} & \text{if } i \in [1, n-D-1] \\ 0 & \text{if } i \in [n-D, N] \end{cases} \quad (4.4)$$

The third method is the rank selection which is recognized on the rank of chromosomes in place of their fitness. The rank (r) is reduced to the premium chromosomes, while the rank of the greatest gene equals 1. Irrelevance, the rank mathematical countenance is a utility of the probability is formulated as:

$$P_i(s_i) = \frac{r_i}{n(n-1)} \quad (4.5)$$

Equation 4.5 reveals that the choice is the purpose of the rankings. Otherwise, it is not grounded on the fitness mathematical principles of the population solutions. Additionally, equation 4.4 presents a linear selection rank. Nonetheless, the rank selection can be articulated in mathematical exponential grounded on the probability by a specified equation (4.6) (Jebari and Madiafi, 2013):

$$P(s_i) = 1.0e^{\frac{r_i}{C}} \quad (4.6)$$

Here C is obtained by:

$$C = \frac{[2n(n-1)]}{[6(n-1)+n)]} \quad (4.6.1)$$

In this method, only a persuaded probability P of the fittest chromosomes are chosen and simulated as P^{-1} times. Nonetheless, it is fewer trained in preparation than other events, exclusive of definite enormous population numbers.

Generally, the realization of offsprings of the crossover process is accomplished by demolishing the crossover ration of the initial parent. To this end, the subsequent parent is implanted by the crossover splinter. The subsequent offspring, consequently, is designed in the asymmetric procedure.

Mutation Machinist

The mutation is another energetic feature of genetic programming. Binary categories of mutations are feasible. In the former category, a function can simply switch a function. On the other hand, an incurable can swap a terminal. In the latter category, a seamless subtree can swap an alternate subtree (Figure 8).

In this considerate, genetic programming gears dual dissimilar categories of mutations. The topmost interprets tree is a single alternative. The bottom left, designate tree validates a mutation of a specific terminal (2) for the alternate separate terminal (a). Additionally, it consistently validates a mutation of a single function (-) for alternate single function (+).On the bottom right, the construe tree elucidates a swap of a subtree by a different subtree (Goldberg, 1994).

Figure 8. Mutation operation

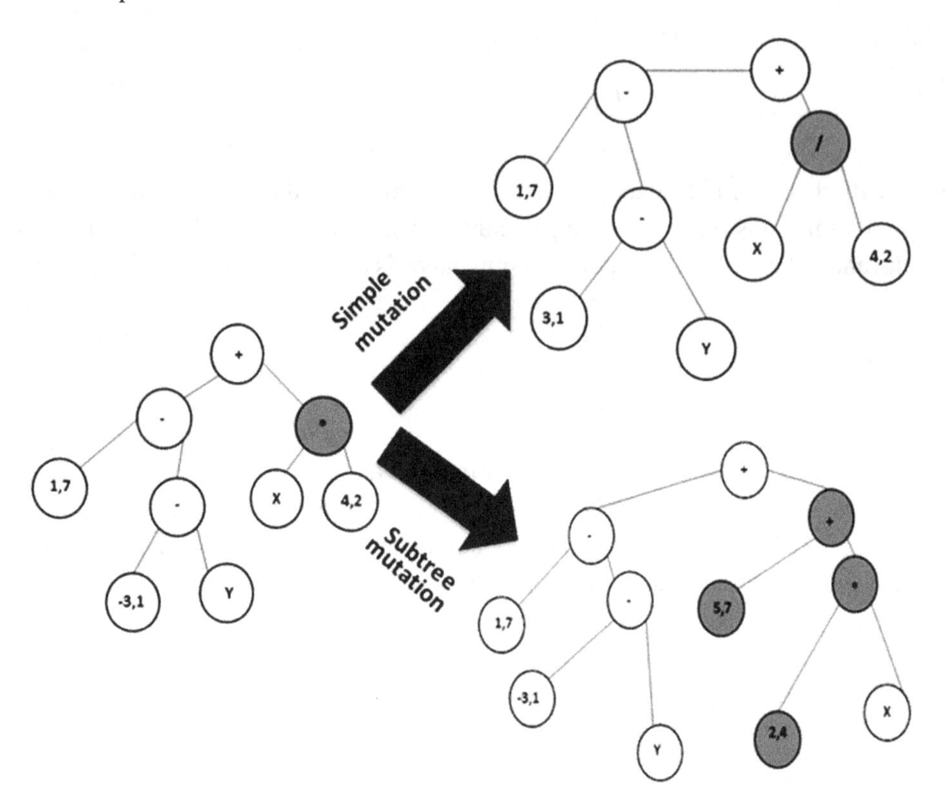

MULTI-OBJECTIVE GENETIC PROGRAMMING

In a multi-objective optimization (MOO) dilemma, one augments with esteem to several aims or fitness roles $f_1, f_2, \ldots, f_n$. The MOO procedure aims to attain optimal solutions. On another hand, the solutions would be at the slightest acceptable, and reliable with all the events concurrently. On the other hand, the difficulty of altered solutions is one of the furthermost stimulating subjects to be functioned in evolutionary schemes, such as, genetic programming.

Generally, the foremost contest in the multiobjective algorithm is to regulate the optimal solutions to problems of the formula

$$P_i = 1.0e^{r_i C^{-1}}$$

$$\text{Maximize } f(x) = f_1(x), f_2(x), \ldots, f_m(x) \tag{4.7}$$

Subject to $x \in X$,

Equation 4.7 indicates a search space or *decision space* that presents as x Therefore, X presents the potential solutions or *decision alternatives*. In this perspective, there is frequently no additional restraint on the province of the decision alternatives. Let us consider that $Y=f(X)$ presents the *objective space* as the component $y \in Y$presents an *objective vector* and its constituent's *objective values*. Consequently, the objective function $f : X \mapsto \mathbb{R}^m, m \in \mathbb{N}$ is well-known as the practical association between the decision alternatives and the decision criteria. To entirely comprehend multi-objective optimization problems (MOOP), algorithms and principles particular definitions must be explained.

Decision Adjustable and Objective Space

The variable constraints of an optimization problem bound distinctly decision constraints to a lower and upper bound. Alternatively, it presents a space recognized as the decision variable space. In multi-objective optimization, consequently, the number of element functions creates a multi-dimensional space recognized as an objective space. In this concern, the individual decision element in element space is similar to a target in impartial space.

Feasible and Infeasible Solutions

Multi-objective algorithms comprise dual categories of solutions that are feasible and infeasible. A feasible solution fulfills all linear and non-linear constraints. Nonetheless, infeasible elucidation presents a solution that would fulfill all the constraints and dissimilarity element constraints and equivalence. Conversely, the infeasible solution contains a solution which does not influence all element constraints and bounds. On the other hand, a specific solution perhaps is infeasible that does not involve infeasible problems. Nevertheless, infeasible problems do happen.

Ideal Objective Vector

Let us assume that v^{*i} is a vector of variables that optimizes maximize or minimize the ith objective which is accessing a multi-objective optimization problem. This can be expressed mathematically by:

$$\exists v^{*(i)} \in \Omega,\ v^{*(i)} = [v_1^{*(i)}, v_2^{*(i)}, v_3^{*(i)}, \ldots\ldots, v_N^{*(i)}]\ : f_i(v^{*(i)}) = OPTf_i(v) \tag{4.8}$$

Then, the vector V can be given by:

$$V^* = f^* = [f_1^*, f_2^*, f_3^*, \ldots\ldots, f_N^*]^T \tag{4.9}$$

being f_N^* is the optimum of Nth the objective function which is ultimate for a problem of multi-objective optimization. Moreover, the perfect solution of this vector can be resolute by the object fit in $\Re^N$. Basically, the ultimate objective vector is attained if a vector has slighter constituents than that of a perfect objective vector. On the other hand, the ultimate objective vector is described as:

$$\forall i = 1, 2, 3, 4, \ldots\ldots, n\quad : V_i^{**} = V_i^* - \varepsilon_i, \qquad \varepsilon_i > 0. \tag{4.10}$$

Domination

Contradictory objects are the foremost concern in the utmost real-world requests. In this understanding, optimizing a solution approximately one object cannot be a consequence, of a perfect solution regarding the other objects.

Let undertake that $\triangleleft$ is the operator between binary solutions s_i and s_j i.e. $s_i \triangleleft s_j$ for N objective MOP. Conversely, it clarifies that a solution s_i is added precise than the solution s_j to a confident objective. In this viewpoint, the description of domination for both minimization and maximization MOP can be specified as: a feasible solution $s^{(1)}$ is predictable to govern alternate feasible solution $s^{(2)}$ as, $s^{(1)} \leq s^{(2)}$, in the following circumstances:

(i) (i) The solution $s^{(1)}$ is no poorer than $s^{(2)}$ with style to all objectives rate. On the other hand, mathematical manifestation can be specified by:

$$f_j(s^{(1)}) \triangleright f_j(s^{(2)}) \quad \text{for all } j = 1,2,3,....,N \tag{4.10}$$

(ii) (ii) The solution $s^{(1)}$ is decisively consistent than $s^{(2)}$ within the tiniest one objective value. This can precisely be formulated as:

$$f_i(s^{(1)}) \triangleleft f_i(s^{(2)}) \text{ for at least one } i \in \{1,2,3,\ldots,n\} \tag{4.11}$$

In these statuses, the solution $s^{(1)}$ governs the solution $s^{(2)}$. Nonetheless, the solution $s^{(1)}$ is non-governed by the solution $s^{(2)}$. Alternatively, the solution $s^{(2)}$ is governed by the solution $s^{(1)}$.

Pareto-Optimal Set (Non-Dominated Set) and Pareto Front

In the decision mutable space, uncertainty the solution is not governed by any other solution is termed as the Pareto-optimal solution. In this framework, the Pareto-optimal is the well-known (optimal) solution about entire objectives. Additionally, it cannot be finalized in some objective deprived of fading the alternative objective. In this considerate, the group of completely feasible solutions that are non-governed by somewhat extra solution is recognized as the Pareto-optimal or non-dominated set. Contrariwise, the overall Pareto-optimal set is predictable when the non-dominated set is delimited by the precise feasible analysis space. The Pareto-optimal (P_O) is accurately formulated by:

$$P_o = \{s \in \Omega \neg \exists s' \in \Omega \quad f(s') \leq f(s)\}. \tag{4.12}$$

The objective rate functions which are associated with each clarification of a Pareto-optimal set in objective space is identified as the Pareto-front (Figure 9). In this concern, a symbolic Pareto-front of a binary objective fading category optimization obstacle in impartial space.

On the other hand, Pareto-front (P_F) is mathematically expressed as:

Figure 9. Pareto-front

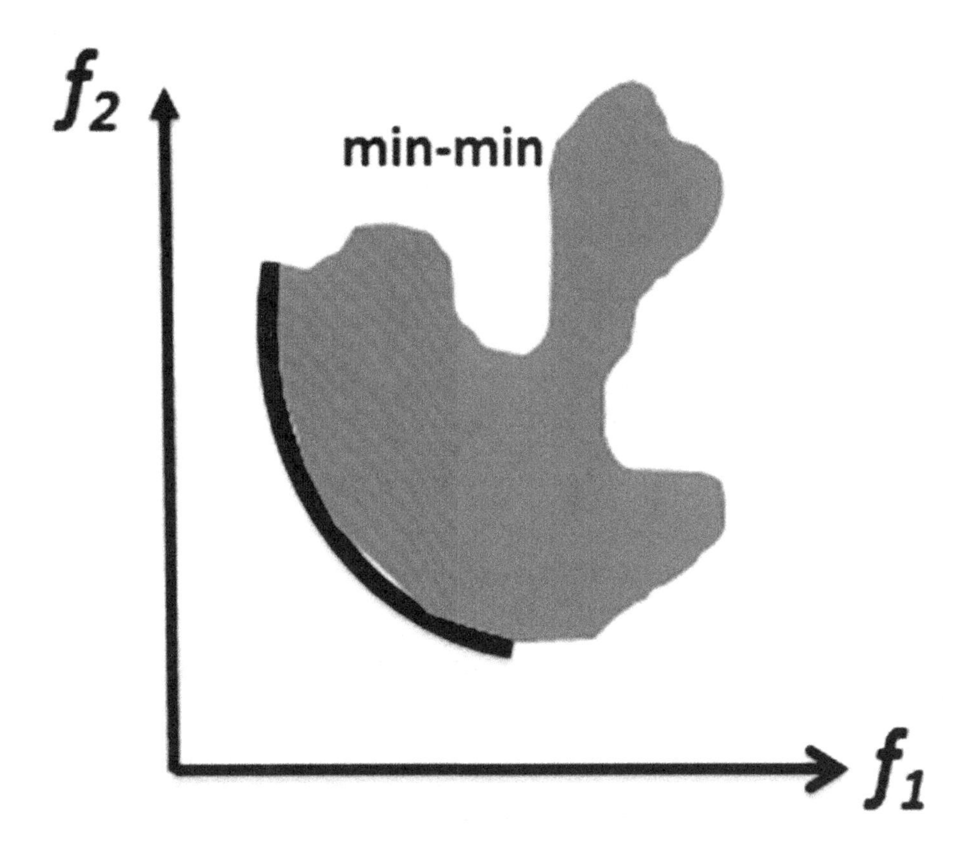

$$P_F := \{f(s) \mid s \in P_o\}. \tag{4.12}$$

Equation 4.12 demonstrates that as long the idea of domination permits evaluation of the solutions regarding multi-objective. In this circumstance, the majority of multi-objective optimization algorithms exercise this principle to acquire the non-dominated set of solutions, as a result of the Pareto-front.

CONCLUSION

In a literature review, only one approach which is based on the Bayesian models was utilized to investigate the search zone of the MH370 and also to locate any debris belonging to MH370. However, this method is not able up to date to locate the fuselage and exact search zone of MH370 vanishing. In this regard, this chapter introduces a possible approach based on a theoretical mathematical description of the genetic algorithm. Therefore, the chapter also addresses the differences between the difference between genetic and Evolutionary Algorithms. Moreover, the decision mathematical rules based on the Pareto algorithm are also discussed. In fact, the Pareto optimization rule can have a significant rule in the investigation the accurate location of MH370 vanishing. In this understanding, the next chapters will implement the MOP based GP broadly to comprehend the contrivance of the Flight MH370 vanishing and to answer critical questions, for instance, did MH370 end up in the Southern Indian Ocean? Then, where are its fuselage and debris? Therefore, the next chapters will deal intensely with the applications of MOP and GP to determine the exact scenario of the Flight MH370 disappearing.

REFERENCES

Ackley, D. H. (1987). *A Connectionist Machine for Genetic Hill climbing*. doi:10.1007/978-1-4613-1997-9

Bureau, A. T. S. (2015). *MH370 search area definition update.* ATSB transport safety report, External Aviation Investigation AE-2014-054.

Davey, S., Gordon, N., Holland, I., Rutten, M., & Williams, J. (2016). *Bayesian Methods in the Search for MH370*. Springer Singapore. doi:10.1007/978-981-10-0379-0

De Jong, K. *a.(1975). An analysis of the behavior of a class of genetic adaptive systems* (Doctoral dissertation). University of Michigan.

Dhar, V., & Stein, R. (1997). *Seven methods for transforming corporate data into business intelligence*. Prentice-Hall, Inc.

Goldberg, D. E. (1994). Genetic and evolutionary algorithms come of age. *Communications of the ACM*, *37*(3), 113–120. doi:10.1145/175247.175259

Holland, J. H. (1975). Adaptation in natural and artificial systems. Univ. of Mich. Press.

Jebari, K., & Madiafi, M. (2013). Selection methods for genetic algorithms. *International Journal of Emerging Sciences*, *3*(4), 333–344.

Karray, F., Karray, F. O., & De Silva, C. W. (2004). *Soft computing and intelligent systems design: theory, tools, and applications*. Pearson Education.

Kingdon, J. (1997). From Learning Systems to Financial Modelling. In *Intelligent Systems and Financial Forecasting* (pp. 1–17). Springer.

Medsker, L. R. (2012). *Hybrid intelligent systems*. Springer Science & Business Media.

Michael, N. (2005). *Artificial intelligence a guide to intelligent systems*. Academic Press.

Michalewicz, Z. (1996). GAs: What are they? In *Genetic algorithms+ data structures= evolution programs* (pp. 13–31). Springer. doi:10.1007/978-3-662-03315-9_2

Miller, B. L., & Goldberg, D. E. (1995). Genetic algorithms, tournament selection, and the effects of noise. *Complex Systems*, *9*(3), 193–212.

Syswerda, G. (1989, June). Uniform crossover in genetic algorithms. In *Proceedings of the 3rd international conference on genetic algorithms* (pp. 2-9). Academic Press.

Chapter 5
Simulation of MH370 Actual Route Using Multiobjective Algorithms

ABSTRACT

This chapter delivers the mathematical model to retrieve the definite route of MH370 and its debris, which is based on a multi-objective evolutionary algorithm. The chapter shows that the appropriate short route for Captian Zaharie to murder-suicide is the Gulf of Thailand, not in the Southern Indian Ocean, which is specified by 1000 iterations and 100 fitness. Needless to say that the MH370 path reclaimed from Inmarsat 3-F1 satellite data was not delivering the real scenario of MH370's vanishing, which is proving the multiobjective genetic algorithm.

INTRODUCTION

The Flight MH370 ended in the Southern Indian Ocean as based on the measurements made by the British company Inmarsat. Nevertheless, Zweck (2016) argued in support of Inmarsat by using Doppler frequency shift, trigonometry, time and location of pings. In other words, Inmarsat's radar tracking approach and data analysis have not yet assured each one that they are ironclad (Finkleman, 2014 and Chen et al., 2015). Conversely, during flight, the MH370 exploited satellite communications contacts to trade information with ground stations through the Aircraft Communications Addressing and Reporting System (ACARS) (Wikipedia, (2018).

Despite the fact most of the functionality of ACARS was immobilized earlier in the MH370, for six hours, post the last radar communicated with the Flight Mh370 and ground station traded a sequence of dumpy memos, which are well known as *pings*. Indeed, the Inmarsat 3-F1 satellite transmitted these pings to a ground station in Perth, Australia. In this sense, the Inmarsat 3-F1 satellite was in a geosynchronous orbit in excess of the equator at longitude 64.5°E (Zweck, 2016). In this regard, the British satellite company Inmarsat engineers created mathematical approaches to define flight paths that greatest fit the Burst Timing Offset (BTO), and the Burst Frequency Offset (BFO). In this view, they are able to

DOI: 10.4018/978-1-7998-1920-2.ch005

identify the Flight MH370 paths in addition to the search area in the Southern Indian Ocean. However, these efforts are challenged by changing the search area numerous times between 9th March and 26th June 2014 (Zweck, 2016). In other words, there are ambiguities in determining the particular location of the missing flight. On March 17th, 2014, the search was refocused to a region in the Southern Indian Ocean, about a 3000 km southwest of Perth, Australia. The search area, however, relocated an 800 km farther southwest of Perth, on October 8th, 2014. On the other hand, the exploration was postponed until January 17th, 2017. In October 2017, the final search area is believed to be at, approximately, 35.6°S 92.8°E. In January 2018, the search area based on these geographical references was continued by a private company of Ocean Infinity.

The important questions arise up: why was the search area changed several times? Did the Inmarsat 3-F1 satellite data regarding the Flight MH370 involve uncertainties? If the Inmarsat 3-F1 satellite data are accurate why the search teams cannot detect the flight 's fuselage and its debris? In fact, all the models reconstructed the Flight MH370 path and its debris trajectory movements are based on the information delivered from Inmarsat 3-F1 satellite data. Can the multi-objective evolutionary algorithm based on a genetic programming answer the above-mentioned questions?

Inmarsat 3-F1 Satellite Data

Inmarsat is the brief of " International Maritime Satellite Organization Inmarsat" which was founded on July 16th, 1979, at the initiative of the International Maritime Organization (IMO) and primarily had a standing of a federal organization. Herewith, the foremost goal of the launch of Inmarsat was to offer the marine vessels with steadfast broadcastings, predominantly for the development of vessel securities, which involved the communication of SOS signals. These signals are transmitted between other vessels and shore assistances, contact between crew followers and passengers with the shore service. Further, the main objective of Inmarsat is to guarantee the operation of maritime satellite communication system in harmony with the requests of the Global Maritime Distress and Safety System (GMDSS) (Venkatesan,2013; Santamarta,2014; Abdelsalam et al., 2017; Swanwick, 2017).

Inmarsat was initiated as a maritime telecommunications machinist. It is foremost exclusively-possessed constellations which are the Inmarsat-2 (I-2) and Inmarsat-3 (I-3) series. They were situated directly above the world's ocean-lanes to develop four ocean districts: (i) West Atlantic Ocean Region, at 54° W;(ii) East Atlantic Ocean Region, at 15.5° E;(iii) East Indian Ocean Region, at 64°E; and (iv) Pacific Ocean Region, at 178° E (Figure 1).

Satellite Constellation Inmarsat contains 14 geostationary satellites. A unique aspect of geostationary satellites is its immovability above a directed position over the Equator. Herewith, this superficial immovability is accomplished via satellite spin sideways globular orbit corresponding to the equatorial plane, with an angular velocity equivalent to the momentum of the Earth revolving speed. In this regard, the orbital altitude makes approximately 35,786 km.

Inmarsat Satellite Frequencies

The Inmarsat system conducts in frequency arrays assigned by the International Telecommunications Union for mobile satellite service. Herewith, L-frequency is implemented for Telecommunications which is involving 1626.5-1660.5 MHz bands, which are directed from Earth to satellite. On the contrary, the frequency rates of 1525.0-1559.0 MHz bands which are directed between satellite and Earth. Conversely,

Figure 1. Inmarsat Spot beams

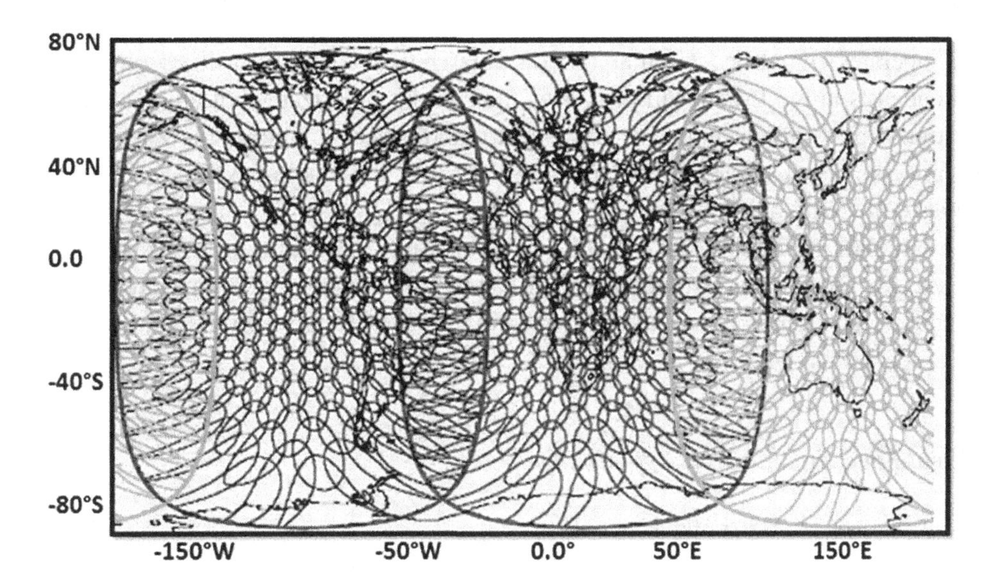

Feeder lines conduct in the C-range, which are involved 6425-6450 MHz bands which their direction is between Earth and satellite. On the other hand, the frequency of 3600-3623 (3600-3630) MHz bands conduct between the satellite and Earth (Swan et al., 2012).

Conversely, the satellite does not have predictable transponders. Truthfully, there is no immobile frequency relationship between the L-band frequencies and the extended C-band feeder connection frequencies. In fact, the connectivity between the feeder linkage C-band spectrum and L-band spectrum. In both bands, the forward and return routes, have been designated as presenting all obtainable feeder link spectrum lashed to the presented L-band spectrum (Iwai et al., 1991 and Spiridonov, 1994).

Inmarsat-4 F3

Inmarsat-4 F3 is a telecommunications satellite activated by the British satellite operative Inmarsat. It was launched into a geosynchronous orbit at 22:43 GMT on August 18th, 2008 by a Proton-M/Briz-M Enhanced carrier rocket. It is presently positioned at 97.65° West longitude. Herewith it is providing coverage of the Americas (Swan et al., 2012 and Swanwick, 2017).

Inmarsat-4 F3 was created by EADS Astrium similar the former satellites of Inmarsat-4 F1 and F2 satellites. Further, it is using a Eurostar E3000 bus with a mass of 5,960 kilograms. Moreover, it is expected to activate for 13 years. It was formerly scheduled for launch by means of an Atlas V 531. However, it was transferred to Proton owing to a huge excess of Atlas launches (Spiridonov,1994;Swan et al., 2012 ; Swanwick, 2017).

In the United States, Inmarsat ground stations are certified to activate at frequency ranges of 1525-1559 MHz and 1626.5-1660.5 MHz, respectively. Therefore, the frequency ranges of 1544-1545 MHz and 1645.5-1646.5 MHz bands, respectively, are retained for security and distress telecommunications (Spiridonov,1994 and Swan et al., 2012).

Polarization

Inmarsat-F3 operates right-hand circular (RHC) polarization which is utilized on both uplink and downlink broadcasts in the L-band. In this view, the circular polarization of an electromagnetic wave is a polarization express in which, in a separate position, the electric field of the wave has a continuous magnitude, nonetheless, its bearing revolves with the cycle at a stable frequency in a plane which is perpendicular to the route of the wave.

Hence, a circularly polarized wave involves two possible statuses: (i) right circular polarization wherein the electric field vector revolves in a right-hand view in regard to the wave propagation direction (Figure 2); and (ii) left circular polarization in wherein the vector revolves in a left-hand view (Figure 3).

Figure 2. Right-hand circular (RHC) polarization

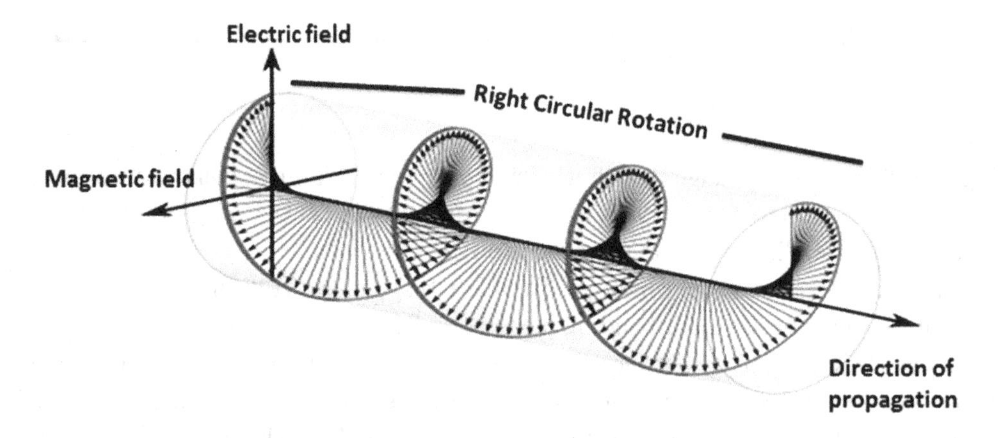

Figure 3. Left-hand circular (RHC) polarization

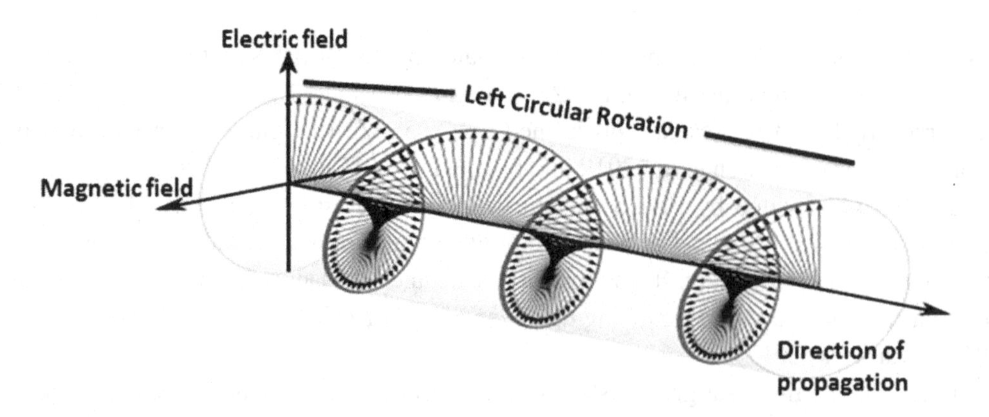

Satellite Transmit Capability

Satellite operates within both Feeder downlink and global beams. The Inmarsat-3 F1 satellite delivers dual C-band worldwide downlink beams, one in RHCP and the other one in LHCP. Both transmitted beams are technically indistinguishable in individual polarization. Further, both beams envelop entirely the spots in the field of view of Inmarsat-3 F1. This operates with a peak gain of 20 dB, which is delivering a greatest of equal to 30.5 dB of downlink effective isotropic radiated power (EIRP) on every polarization. In this understanding, EIRP is known as the Equivalent Isotropic Radiated Power. In antenna quantities, the calculated emitted energy in a single direction. In other words, for an antenna emission configuration, quantity, if a particular value of EIRP is specified, this will be the maximum rate of the EIRP inclusive restrained angles (Spiridonov, 1994 and Abdelsalam et al., 2017).

On the contrary, the Inmarsat-3 satellites specify a pulse in L-band global beam which is in RHCP. This global beam overlays entirely the positions in the field of view satellite. This beam has a maximum peak gain of 19.5 dB. In this view, it is delivering a maximum of capable of 41.5 dB of downlink EIRP (Iwai et al., 1991).

Satellite Receive Capabilities

Satellite receive operations include Feeder Uplink and service uplink or global beam. In these regards, the Inmarsat-3 F2 satellite occupies dual C-band global uplink beams: primarily operates in RHCP and the latter activates in LHCP. Both polarization beams are supposedly matching in separately polarization. The beams, therefore, have the greatest peak gain of 20.5 dB and an entire device thermal noise of about 891 K. The greatest G/T of the C-band uplink global beams is -9.0 dB/K. The cross-polarization isolation of the beams, therefore, is 30 dB diagonally the examination area (Iwai et al., 1991; Spiridonov, 1994; Swan et al., 2012; Venkatesan et al., 2013; Swanwick, 2017).

Nevertheless, global uplink beam in L-band with RHCP has an antenna gain peak of 18.5 dB. In this context, the thermal noise in the receiver is about 562 K which is involving antenna losses. Then, the beam peak ratio of antenna gain to thermal noise is approximate 9.0 dB/K.

Communications Payload

The forward route obtains signals from immobile Earth positions at C-band in the frequency of 6.4 GHz. Then, it relays these signals to portable Earth positions at L-band in the frequency of 1.5 GHz. Conversely, the reoccurrence route obtains signals from mobile positions at L-band in the frequency of 1.6 GHz. Subsequently, it transmits them to permanent ground positions at C-band in the frequency of 3.6 GHz.

On the accelerative route, uplink conductions are sensed in a global beam. In fact, the antenna is designed as dual circularly polarized. In this manner, the LHCP antenna output port is linked straight to the C-band receiver, while the RHCP port is associated with a C-band Diplexer. Therefore, one output of this Diplexer delivers for the telecommunication signals which are received in the RHCP uplink. The latter output specifies for the RHCP examination uplink to be associated with the C-band receiver (Swan et al., 2012; Venkatesan et al., 2013; Swanwick, 2017).

The C-band receiver down transmits the frequency of 34 MHz response spectrum to L-band, whereas, creating the accelerative connection of gain beam/transmitted beam. Conversely, the forward beamformer offers an exclusive established of amplitude and phase weightings for a piece of the 8 beams of the onward transponder. The beam is formed by eight inputs, seven spots and one global), and 22 outputs. Every input beam sustains a splitter board which splits the input signal into 22 routes and creates the required amplitude and phase arrays for every beam (Spiridonov, 1994; Swan et al., 2012; Venkatesan et al., 2013; Swanwick, 2017).

Subsequently, the L-band obtains module of the return transponder which is considered in a corresponding routine. In this context, fully completed low noise amplifiers are linked to all of the 22 receive antenna elements. Then, the 22 L-band signals are supported by the arrival combiner which modifies the L-band spot beam and global beam signals. Finally, the input signals are down-transformed and sustained to the channelling surface acoustic wave (SAW) filters which achieve a multiplexing utility post the filter-to-beam adjustment arrays (Venkatesan et al., 2013 and Swanwick, 2017),

The forward link and the return link have the maximum "transponder" gain of 137dB and 127dB respectively. These occur between the output of the receiving antenna and the input of the transmitting antenna on both links. Consequently, the forward path and return path have the gain of transmission value of 2dB over a 24 dB and 23 dB, respectively (Iwai, et al.,1991 and Kimura, et al., 1996).

HOW DID INMARSAT 3-F1 DETECT THE PATH OF MH370

Doppler Effect is the keystone theory beyond the tracking of the Flight MH370 paths. The general theory of the Doppler effect delivers information about any target's location in the space-dimension and its direction of propagation with respect to the Doppler effect. In this view, let us assume that an object which emits a signal, the signal's frequency relies on the velocity and direction the object is moving beside the device 's speed and direction of measuring the signal. In this understanding, as the MH370 moved with respect to the Inmarsat 3-F1 satellite, the frequency of every signal would change. In this sense, if the original frequency is known at which the Flight MH370 emitted the pings (Figure 4), then the MH370 is calculated with respect to satellite as a function of modification frequency.

Figure 4. Pings emitted by MH370

In the other words, the Doppler effect is a property of an electromagnetic signal that is due to the relative motion between a source (the aircraft) and a receiver (the satellite). If the distance between the aircraft and the satellite is decreasing, then the frequency of the received signal will be higher than that of the transmitted signal, and if the distance is increasing, the received frequency will be lower. This change in frequency is called the Doppler shift. The Doppler shift is proportional to the component of the relative velocity vector of the two moving objects that are in the direction of the displacement vector between them.

Specifically, the keystone to deal with the Doppler frequency is Brust Frequency Offset (BFO). The difference between the expected and actual frequencies of the signal received from the aircraft is known as BFO. In the other words, the BFO is primarily caused by the Doppler shift—a shift in frequency caused by the relative movement of the aircraft, satellite, and ground station—along with several other factors which can be calculated and removed, allowing the Doppler shift between the aircraft and satellite to be isolated. The Doppler shift between the aircraft and satellite indicates the relative motion of the aircraft relative to the satellite, although multiple combinations of aircraft speed and heading exist that match a given Doppler shift value. Moreover, a ping arc angle, which is the angle between the aircraft and the satellite, as measured from the centre of the earth.

Pings do not enumerate the Flight MH370 's position or the path travelling, but they deliver two types of clues. The former is the period it acquires for the pulse to travel between Inmarsat 3-F1 satellite and Flight MH370. In this sense, the distance between satellite and aircraft was calculated. The latter is the radio frequency upon which the retort is sensed by the Inmarsat 3-F1 satellite (the pitch of its voice), from which can be calculated whether it was moving towards or away from the satellite when it was transmitted, the so-called Doppler effect, which we commonly experience as the sound of a train approaching and leaving a platform.

On the contrary, velocity alone cannot determine any facts regarding the Flight MH370 site. In this circumstance, the Inmarsat is a reference position in space with an identified location which permits, finding out the position of Flight MH370 using sending and return signals from satellite to flight (Figure 5).

If the satellite and the Flight MH370 were both at the same height exceeding the Earth's surface and the aircraft was hovering in a straight route away from the Inmarsat 3-F1 satellite. This simplified description disruptions the problem down into one measurement. In this circumstance, If the position and velocity of Inmarsat 3-F1 satellite and the velocity of the plane are known, it could easily detect the plane's paths over time fluctuations.

In the sense of a realistic scenario, where the Inmarsat 3-F1 satellite is located higher than the plane by thousands of kilometres, which is required three-dimensional of some trigonometry. Nonetheless, the solution of the Flight MH370 turned into complicated due to kind different velocities;(i) satellite movement in orbit as well as (ii) the movement of the plane. In this context, the single Inmarsat satellite, which sensed the plane's pings determined the frequency of only eight pings in total.

In this view, engineers conspired with the MH370's probable path diagonally the Indian Ocean. In doing so, a combination of the Doppler effect to compute the velocity of MH370 with respect to the Inmarsat 3-F1 and trigonometry. Consequently, then map the plane's flight path and finally determine where it expected to crash afterwards discharging its fuel tanks.

The orbit of the INMARSAT satellite is inclined at 1.66° to the equator, which directs to a location -reliant on Doppler shift with a 24 hour period. In this understanding, the Flight MH370 velocity and orientation, and the short-term constancy of the carrier frequency from the plane's transmitter, which involved the possible variations in the power supply voltage. In this regard, it will be a band of velocity-

Figure 5. Sending a signal from the satellite and a return signal from MH370

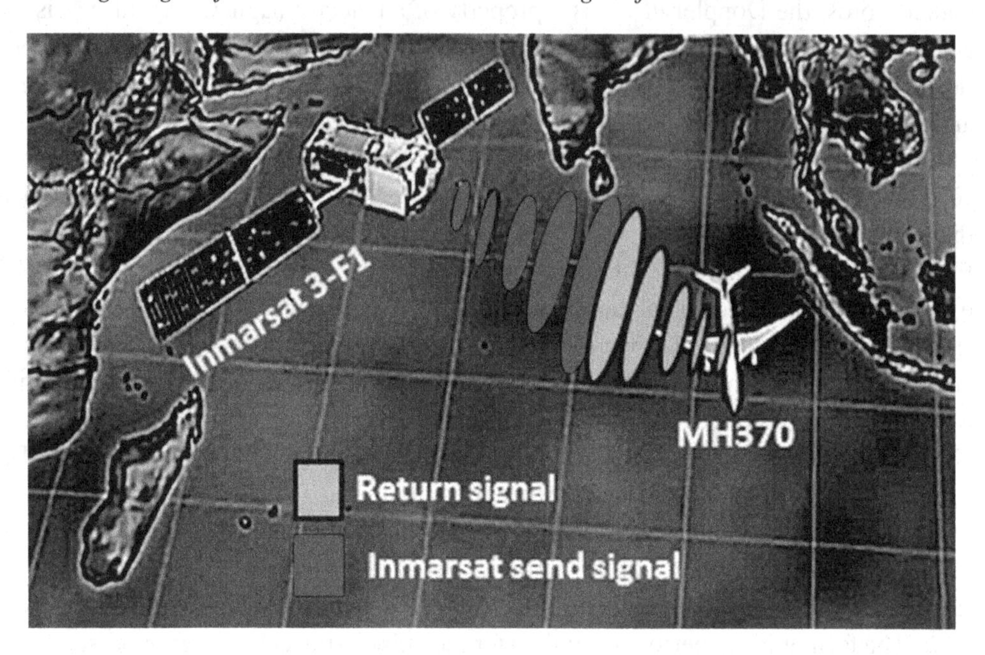

relies on positional solutions for the individual ping. This can be gauged for an exclusive steady solution for a flight path if one obtains. However, If the transmitted frequency differs at the source and/or if the MH370's path and speed varied unreliably, then there is no distinctive solution.

However, the Inmarsat-3 satellites have L-band transponders (1.6Ghz) so, the calculated Doppler shift is approximately 1.2Hz as based on "shift = (speed of plane/speed of light) x 1.6GHz". This also suggests that the plane was shifting exactly away from the satellite.

More importantly, the delusion of the Inmarsat 3f-1 satellite standard frequency use, as Inmarsat has specified "Signal Handshakes are usually accomplished. If the receiving or transmitting device is motorized on and an assembly ping is accomplished on an hourly basis, on average, to create and revive the signal propagation for device use if they are needed which is basically wrong.

Therefore, Inmarsat 3F-1must have a signal in bounce mode to complete the handshake operation. In this regard, Inmarsat 3F-1 is Burst Mode or Microburst Mode. In fact, Inmarsat 3F-1 directs incessant beam burst and receive an assembly when a device is spreading beams outward.

Consequently, The Flight MH370 would have to be sent a transmission beam outward, in which accumulating a burst signal for the Handshake to rebound and return the signal. Then, burst mode sends out signal cells endlessly in anticipation of a transmission is endeavoured or assembly accomplished by a device itself. Consequently, the Flight MH370 continuously attempted these transmission signals in emergency mode. However, the calls were terminated and continuous retreat the data stream undecoded in the routine with complete data. These data are being discarded into cloud computing as the emergency network. However, this procedure is never obtained by being premeditated by Inmarsat to adopt such an occurrence.

MULTI-OBJECTIVE OPTIMISATION

It seems that there are contradictory with Inmarsat 3-F1 satellite records. Inmarsat 3-F1 satellite did not record the departure of the MH370 from KLIA. In fact, the Inmarsat 3-F1 satellite "pinged" the MH370 7 times over a 6 hour period since the last radar contact. This is considered a great conflict. Indeed, the MH370 supposed to be pinged by satellite since it being taken-off in KLIA.

Another conflict is symmetric about the great ping circle enclosing the satellite and starting location since the last radar record. In this view, this conflict can lead to different a priori unknown speed of MH370. Further, the satellite includes position error of less than 6 km and velocity error less than 0.5 km/hr. The most significant challenges are the uncertainty of both MH370-Independent frequency shift and dependent bias. In this context, the dependent bias is approximately 150 $\pm$5 Hz. In this understanding, these conflicts and uncertainties are considered the multi-objectives which are required an optimal solution. The trade-off solution begins owing to the actuality that the diverse objectives are contradictory to one another.

Let us assume that the initial position of MH370 overpass the Andaman Sea is $r_0(t_0)$ and the final position is $r_n(t_n)$ where MH370 plugged into the Southern Indian Ocean. In this context, the search space is defined by spherical coordinates (Figure 6). The latitude and longitude of the starting and destination

Figure 6. Spherical coordinates for the MH370 path

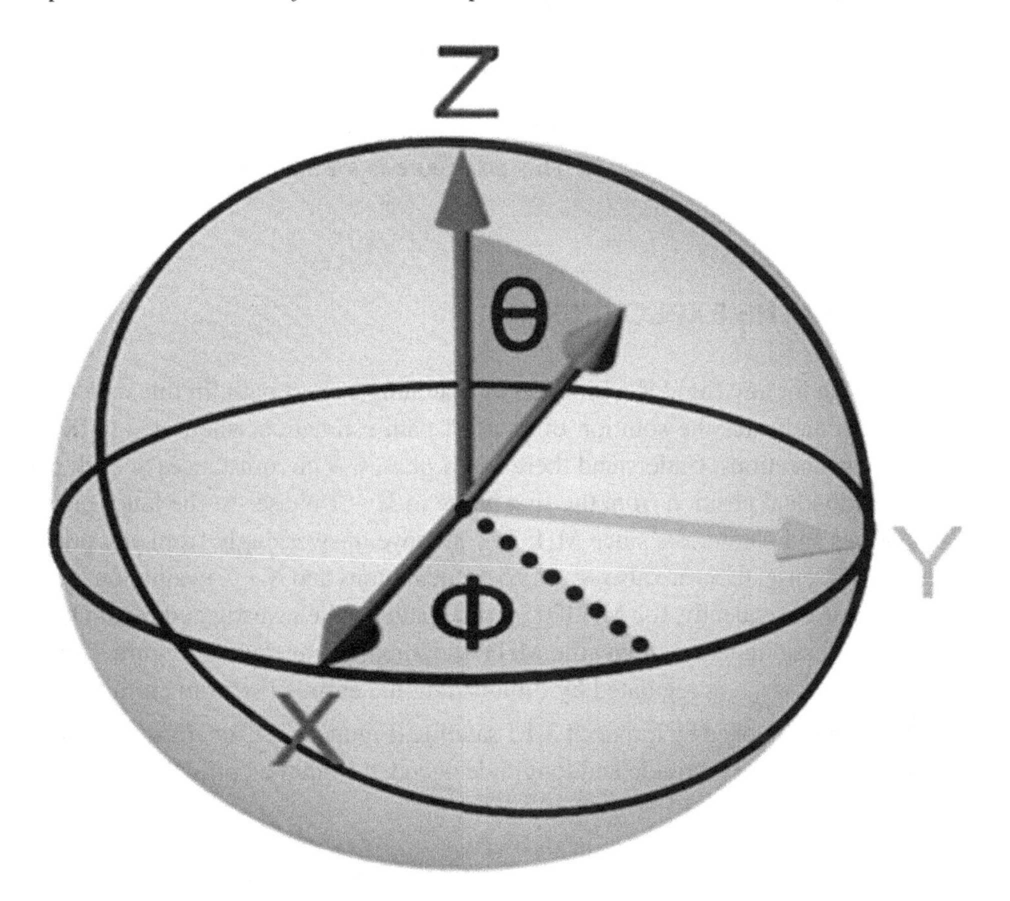

points are transformed into the spherical coordinates. When the size of the search space and the dimension of the grids are specified, the total number of admissible states and stages can be determined. Then the flight path fluctuations can be quantified by (Ng, et al., 2009):

$$r_{\phi,\Theta+1}(t_{\phi,\Theta+1}) = r_{\phi,\Theta}(t_{\phi,\Theta}) + d_{\phi,\Theta}(t_{\phi,\Theta}) \tag{5.1}$$

Here, $t_{\phi+1,\Theta+1}$ is the vanishing time of MH370 on the spherical grid off ϕ and Θ i.e. latitude and longitude, respectively. Further, $d_{\phi,\Theta}$ is the decision variable identified in a set of acceptable influences $D_{\phi,\Theta}$? In any circumstance, the minimum cost-to-go function $\mathrm{H}((r_{\phi,\Theta}(t_{\phi,\Theta}))$, in dynamic programming can be expressed mathematically by:

$$\mathrm{H}((r_{\phi,\Theta}(t_{\phi,\Theta})) = \min_{\phi,\Theta}[J_{\phi,\Theta,t_{\phi,\Theta}}^{\phi',\Theta+1,t_{\phi,\Theta+1}} + w_{\phi,\Theta,t_{\phi,\Theta}}^{\phi',\Theta+1,t_{\phi,\Theta+1}} + C_{\phi,\Theta,t_{\phi,\Theta}}^{\phi',\Theta+1,t_{\phi,\Theta+1}} + \mathrm{H}_{\phi,\Theta,t_{\phi,\Theta}}^{\phi',\Theta+1,t_{\phi,\Theta+1}} + BFO_{\phi,\Theta,t_{\phi,\Theta}}^{\phi',\Theta+1,t_{\phi,\Theta+1}} \tag{5.3}$$

From $r_{\phi,\Theta}(t_{\phi,\Theta})$ to $r_{\phi,\Theta+1}(t_{\phi,\Theta+1})$, $J_{\phi,\Theta,t_{\phi,\Theta}}^{\phi',\Theta+1,t_{\phi,\Theta+1}} = \left\| d_{\phi,\Theta}(t_{\phi,\Theta} \right\| s_{fuel}^{-1}$ is the expected fuel cost, s_{fuel} is a consumer quantified conversion persistent. $w_{\phi,\Theta,t_{\phi,\Theta}}^{\phi',\Theta+1,t_{\phi,\Theta+1}}$ is the component of the weather dynamical fluctuations along the MH370 path. $C_{\phi,\Theta,t_{\phi,\Theta}}^{\phi',\Theta+1,t_{\phi,\Theta+1}}$ is the element of the cost related to overpass an obstructed zone. $BFO_{\phi,\Theta,t_{\phi,\Theta}}^{\phi',\Theta+1,t_{\phi,\Theta+1}}$ is the burst frequency offset which is the difference between the expected and actual frequencies of the signal received from the MH370. The BFO is mainly triggered by the Doppler shift in which a shift in frequency instigated by the absolute movement of the MH370 and Inmarsat 3-F1 satellite. The optimal MH370 path can be resolved by diminishing the entire flying cost from initial to the vanishing location of MH370 (Figure 7). This procedure is a function of obtaining the minimum cost-to-go function.

OPTIMAL MH370 PATHS EXPLORING

The optimal path search for any flight is a function of searching a short path. In this understanding, let us assume that an *O*(*e*)indicates the solution of shortest path estimation where *e* signifies the entire quantity of tolerable connections. Understand there are *M* phases, *N* circumstances at each phase and *L* connections at the individual position from the first phase to *M* – 1 phase. At the latter phase *M*, there are *N* cost calculations and 0 contrasts since MH370 can move unswervingly from any position of the vanishing point. In this sense, there are also *N* × *L* cost calculations and *N* × *L* evaluations from the first phase to *M* – 1 phase. Consequently, the MH370 optimal path can be investigated by in *O*(*M* × *N* × *L*) = *O*(*e*) operations. In these understandings, the MH370 optimal flight route I (Figure 7), which is the chain of optimal controls $d_{\phi,\Theta}$, is regulated by diminishing the entire cost from an initial point to the vanishing points which are tracked by Inmarsat 3-F1 satellite (Figure 8). Thenceforth, the MH370 flight route is quantified by an array of latitude and longitude decoded from the optimal controls.

For exploration space, delineate an array of grids $r_{\phi,\Theta}$.

Figure 7. A grid of MH370 path

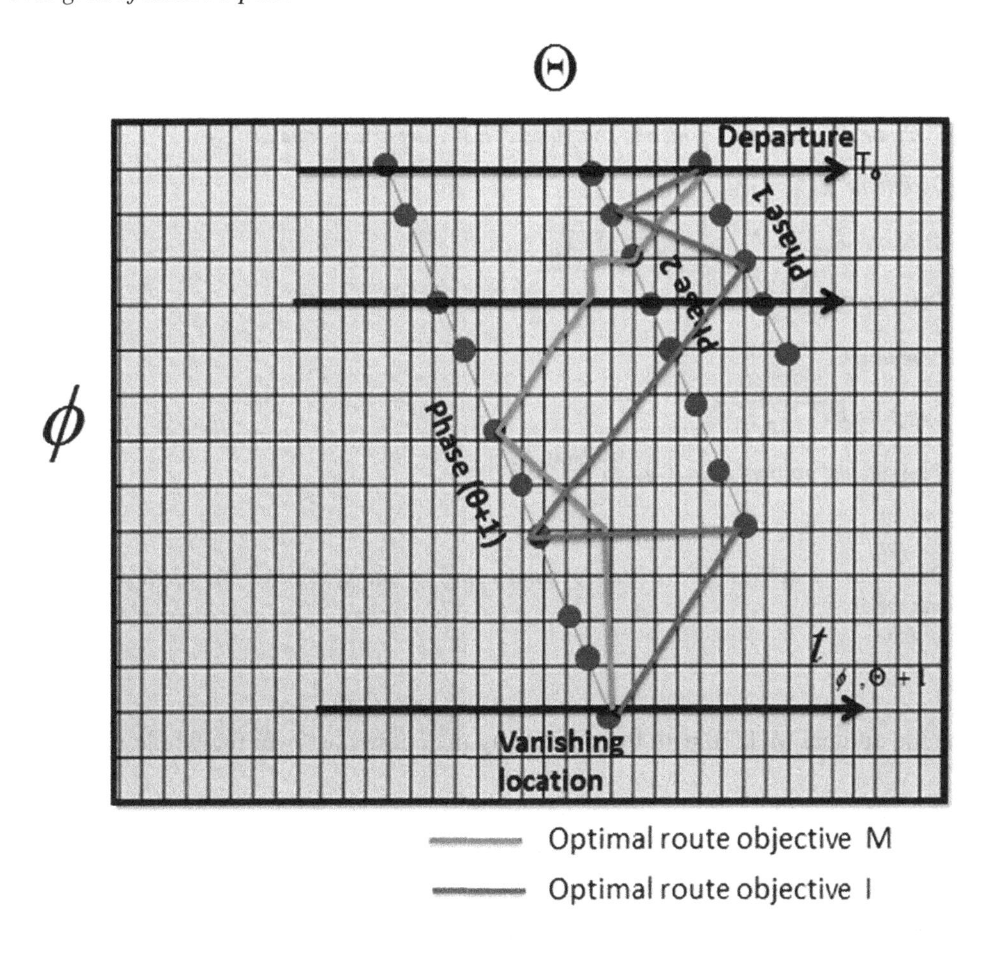

for every hesitant departure period,

Evaluate $t_{\phi,\Theta}^{\min}$ and $t_{\phi,\Theta}^{\max}$ which is computed MH370 departure time at $r_{\phi,\Theta}$

Determine $d_{\phi,\Theta}$,

Estimate $J_{\phi,\Theta,t_{\phi,\Theta}}^{\phi',\Theta+1,t_{\phi,\Theta+1}} = \left\| d_{\phi,\Theta}(t_{\phi,\Theta} \right\| s_{fuel}^{-1}$,

Compute $w_{\phi,\Theta,t_{\phi,\Theta}}^{\phi',\Theta+1,t_{\phi,\Theta+1}}$,

Compute $C_{\phi,\Theta,t_{\phi,\Theta}}^{\phi',\Theta+1,t_{\phi,\Theta+1}}$,

Compute $BFO_{\phi,\Theta,t_{\phi,\Theta}}^{\phi',\Theta+1,t_{\phi,\Theta+1}}$,

Compute the optimal cost-to-go $\mathrm{H}_{\phi,\Theta,t_{\phi,\Theta}}^{\phi',\Theta+1,t_{\phi,\Theta+1}}$,

end for

end for

Figure 8. The pseudo code of searching MH370 route

For exploration space, delineate an array of grids $r_{\phi,\Theta}$.
for every hesitant departure period,
Evaluate $t^{min}_{\phi,\Theta}$ *and* $t^{max}_{\phi,\Theta}$ *which is computed MH370 departure time at* $r_{\phi,\Theta}$
Determine $d_{\phi,\Theta}$,
Estimate $J^{\phi',\Theta+1,t_{\phi,\Theta+1}}_{\phi,\Theta,t_{\phi,\Theta}} = \left\| d_{\phi,\Theta}(t_{\phi,\Theta} \right\| s^{-1}_{fuel}$,
Compute $w^{\phi',\Theta+1,t_{\phi,\Theta+1}}_{\phi,\Theta,t_{\phi,\Theta}}$,
Compute $C^{\phi',\Theta+1,t_{\phi,\Theta+1}}_{\phi,\Theta,t_{\phi,\Theta}}$,
Compute $BFO^{\phi',\Theta+1,t_{\phi,\Theta+1}}_{\phi,\Theta,t_{\phi,\Theta}}$,
Compute the optimal cost-to-go $H^{\phi',\Theta+1,t_{\phi,\Theta+1}}_{\phi,\Theta,t_{\phi,\Theta}}$,
end for
end for
Determine the optimal MH370 path by minimizing the entire cost over the whole phases.
end for
end for

Determine the optimal MH370 path by minimizing the entire cost over the whole phases.

end for
end for

The Flight MH370 path with a speed of 400 knots/hr was delivered by Inmarsat 3-F1 satellite (Figure 9) is agreed with the one reported by Zweck [14]. In this view, the Flight MH370 path was simulated without considering other parameters such as weather conditions and fuel consuming. On the contrary, there is a great uncertainty between Flight MH370 path with a velocity of 350 knots/hr, 400 knots/hr and 450 knots/hr.

Consistent with Zweck (2014), the slight motion of the Inmarsat 3-F1 satellite in the orbit breaks the symmetry between the paths. Furthermore, the Inmarsat communications system is not able to directly record the aircraft satellite Doppler shift. Figure 10 shows that the Pareto front is not able to determine the best solution for multi-objectives of both BFO and MH370 speed. This indicates that Inmarsat 3-F1 satellite data included a great ambiguity. These ambiguities lead to the disastrous search area. Consequently, the MH370 fuselage and related debris are difficult to be tracked since March 8th 2014.

PROBABILITY OF FLAPERON AND WEATHER EFFECTS OF THE MH370 PATH DEVIATION

In rehearsal, there are numerous causes of ambiguity that can trigger the Flight MH370 to deviate from its route was tracked by Inmarsat 3-F1 satellite. In this chapter, two caused are considered: (i) dropping of flaperon; and (ii) weather for flight deviation. However, these occur at a certain set of the period. To

Figure 9. MH370 paths with different speed

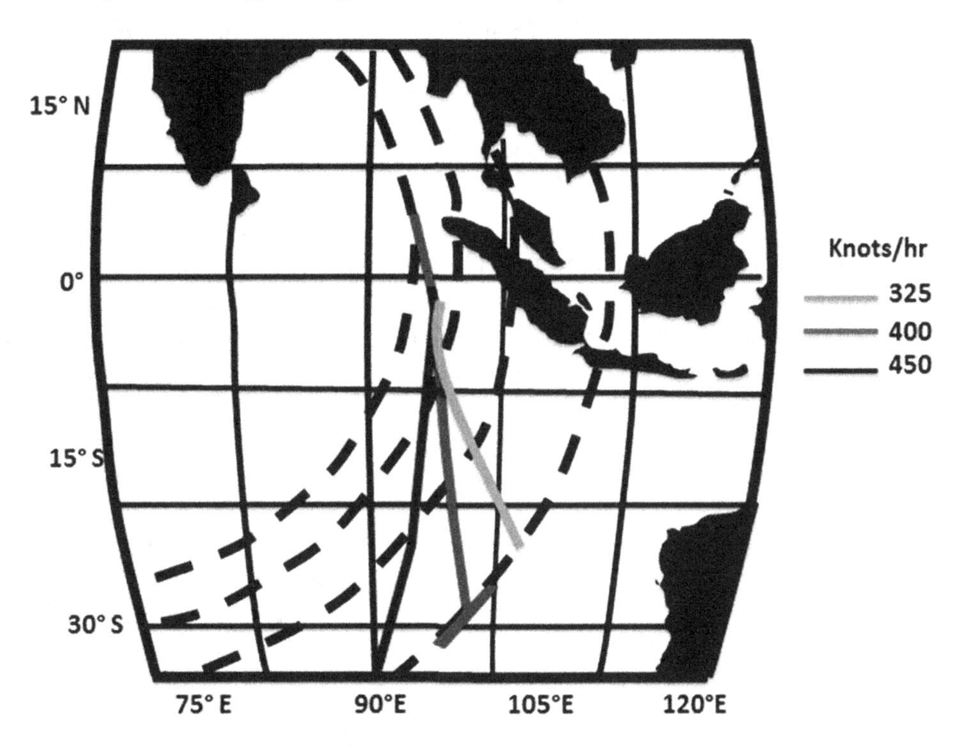

Figure 10. Pareto front to determine the best solution for tracking MH370

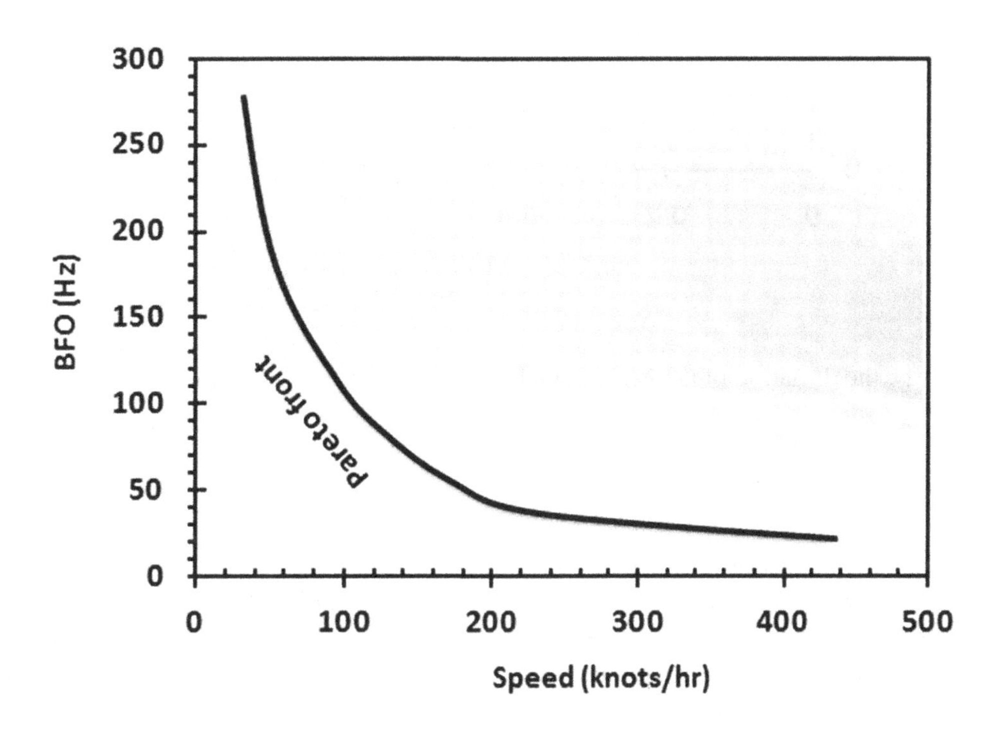

that end, let us assume that E_{Div} is the deviation event which occurs along the trajectory of the path between an initial time t_0 to vanishing time $t_{\phi+1,\Theta+1}$. Then, the probability of deviation $P(E_{Div})$ for these causes is casted as:

$$P(E_{Div})_{\phi,\Theta} = P(\bigcup_{t_k} E_{Div})_{flaperon} \tag{5.4}$$

Equation 5.4 demonstrates the deviation of MH370 due to flaperon separation prior to the flight vanishing in the Southern Indian Ocean. The splitting of flaperon away from the Flight MH370 would cause instability and fluctuation of the flight around its axis (Figure 11). This is indicated by the highest maximum probability value of 0.9 along the Pareto front with a great normalized instability value of 0.8.

Figure 11. Pareto front of the Flight MH370 instability due to the splitting of flaperon

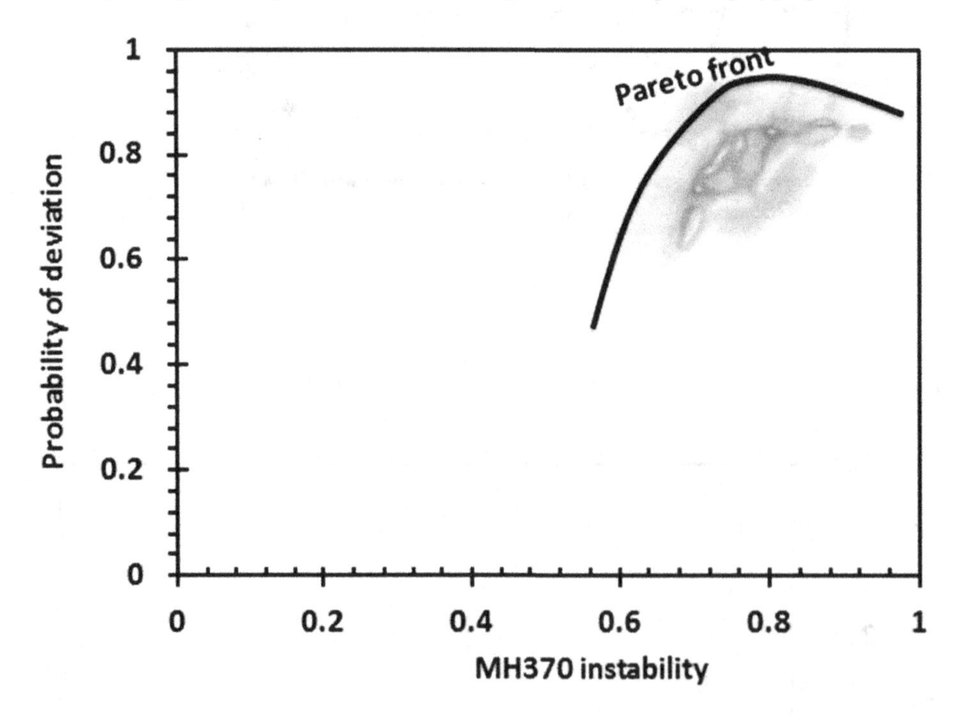

Consequently, the deviation of the MH370 path due to the weather impact is defined as:

$$P(E_{Div})_{\phi,\Theta} = 1 - \prod_{t_k} (1 - P(\bigcup_{t_k} E_{Div}))_{weather} \tag{5.5}$$

Equation 5.5 presents the predicted probability of deviation for MH370 as a function of forecasted convective weather events from the Convective Weather Avoidance Model CWAM. The weather data have been obtained from http://www.weathergraphics.com/malaysia/contrail.shtml. The wind speed was about 96 km/hr which was located at a latitude of -15°S. The zone between Indonesia and north of Australia dominated by strong wind propagation towards the west-north (Figure 12). In this understanding, the Flight MH370 would be deviated either toward the south-west or north-west due to wind impacts (Figure 13). This could be due to the impact of the autopilot mechanism as long the pilot and copilot are not able to control the flight path.

Figure 12. Wind pattern during 40° INMARSAT 3-F1 arc along the coastal waters of Western Australia

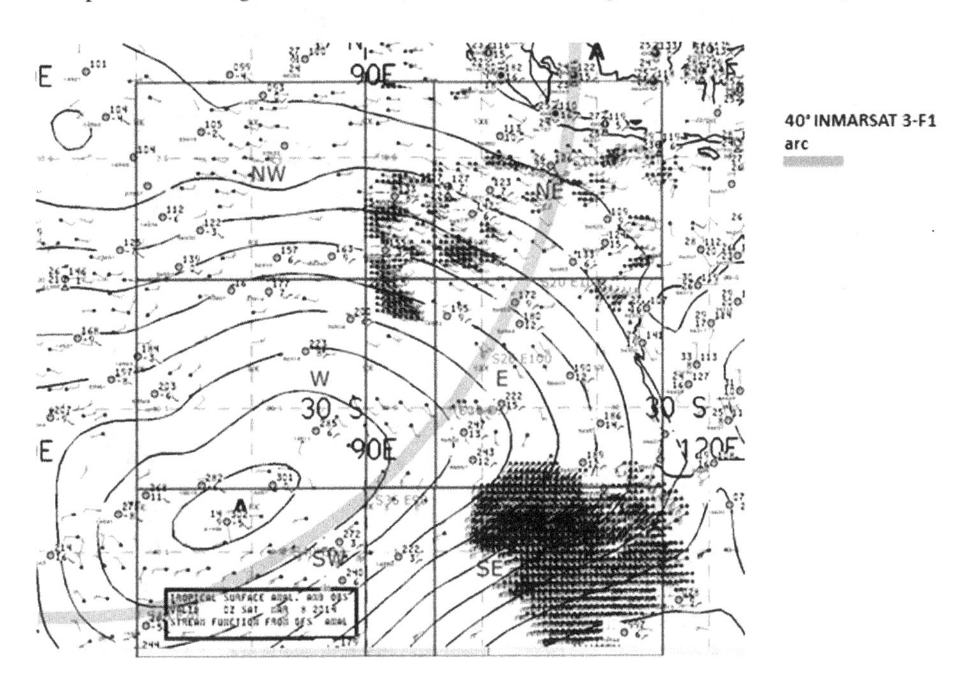

The GFS 8/0000 UTC analysis for 250 mb showed flight level winds across this area of 230° true at 110 knots/hour. Assuming this contrail is representative of the Flight MH370 track of 149° true (magnetic 196°) the aircraft would have had to assume a heading of 210° magnetic to fly a track of 196° magnetic, which would produce this contrail. Going from the hypothesis that the Boeing 777 autopilot was flown in heading mode of 210° with a normal heading reference, without any change (possibly from pilot incapacitation or an intention to ditch the plane over open ocean), the flight track was backtracked in time. This backtrack fully compensates for magnetic declination and wind field changes, producing a curved and slightly uneven path. In this understanding, the Pareto simulation MH370 path would be dissimilar to the one simulated by Inmarsat3-F1 satellite data. It is clear the wind impact considerations can cause a great gap between Inmarsat3-F1 satellite results and the Pareto optimal simulation. In this view, the track spectacles a distinct curve to the left farther the end was determined by Inmarsat3-F1 satellite. Furthermore, the effects of rapidly increasing magnetic declination (as much as 47°W at the contrail point) and drift due to a jet stream at 45°S. This proves the uncertainties have found in the Inmarsat3-F1 satellite investigations and reported by Zweck (2014).

Figure 13. Pareto optimal simulation of the Flight MH370 path due to wind impacts

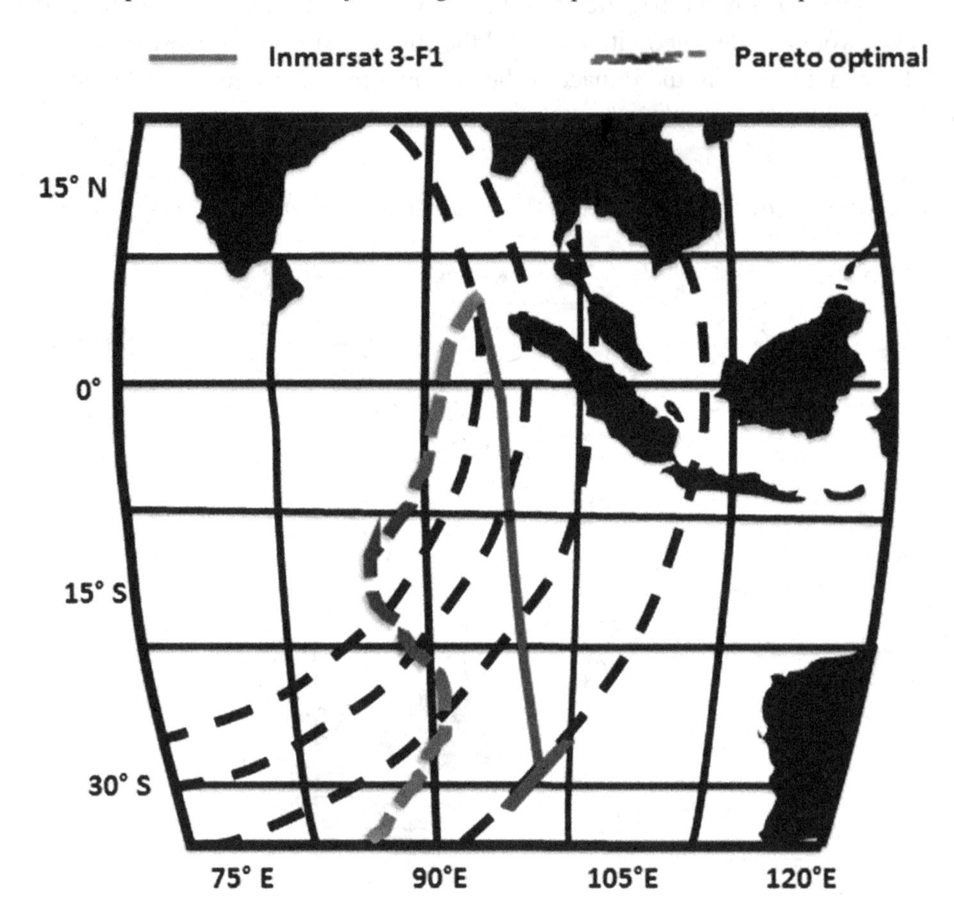

OPTIONAL FILTERING OF ROUTES

In practice, the uncertainties of Inmarsat 3-F1 tracking of MH370 trajectory movements which are determined by Zweck (2014 and 2016) and must be filtered to ensure the precise flight path. In this sense, a possible filtering to only retain a maximum quantity of path per Pareto front is comprised. Subsequently, it is anticipated to have non-assembled clarifications on the Pareto front. The filtering routine eradicates the answers that are the most assembled. In this view, the solutions must be compared with a consequential approach. In doing so, first, the objective functions are scaled.

$$\bar{f}(H)_{\phi,\Theta} = \frac{f_{\phi,\Theta}(H) - G^{I}_{\phi,\Theta}}{G^{N}_{\phi,\Theta} - G^{I}_{\phi,\Theta}} \tag{5.6}$$

here, $\bar{f}(H)_{\phi,\Theta}$ and $f_{\phi,\Theta}(H)$ present the scaled objective function and the objective function of minimum cost-to-go function $\mathrm{H}((r_{\phi,\Theta}(t_{\phi,\Theta}))$, respectively. Figure 14 shows the impact of the scaled function on the Flight MH370 path. The equation 5.6 produces an irregular pattern of the route due to the impacts of wind, fuel consumption, and missing of flaperon. In this context, the irregular pattern of paths could be due to the falling of the flaperon from MH370.

Conversely, Figure 14 confirms the results of Figure 13 where there is a great gap between Inmarsat 3-F1satellite data and Pareto optimal solution. In fact, Inmarsat 3-F1satellite simulation coincides with neither Pareto optimal solution nor scaled objective function. In fact, both Pareto optimal solution and scaled objective function indicate that the vanishing zone is farther southern west of one simulated by Inmarsat 3-F1satellite.

Figure 14. MH370 paths simulated by objective scaled function and Inmarsat 3-F1satellite data

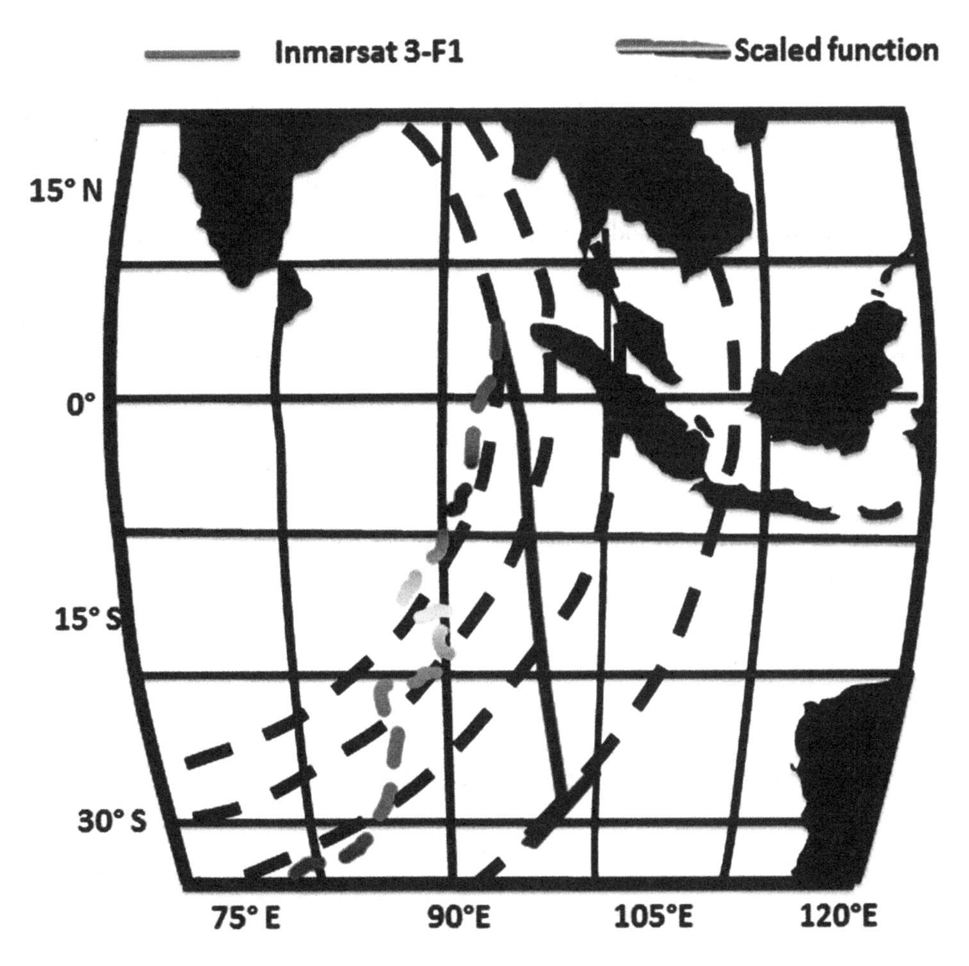

In addition, $G^{I}_{\phi,\Theta}$ and $G^{N}_{\phi,\Theta}$ are ideal point and nadir points for best and worst possible value of every objective function, respectively. In these regards, an ideal point is not a part of the feasible set. On the contrary, the nadir point is as a replacement for the individual worst quantity of every objective function when is appraised on the Pareto front. In practice, the algorithm foremost locates the binary points with the minimum Euclidean space of the scaled quantities of the objective functions. Subsequently, the one with the subsequent smallest space beyond the duality is eliminated. This iteration is done pending there are single as numerous points as the preferred maximum lift.

Following Osyczka and Kundu (1995), the distance-based Pareto genetic algorithm (DPGA) is used to predict the accurate MH370 path which based on equation 5.3.

The DPGA practices a discrete elite set with an infinite size. In the problem of MH370, there is a huge quantity of stages of the Pareto front. In consequence, the elite sets rapidly develop incredibly greatly. Since the computational period of the algorithm relies on the number of elite set members, an upper frontier is set on the number of elites as an accumulation to the innovative DPGA. Equation 5.3, also used to filter the elite set in every generated phase in order to attain a permanent maximum quantity of individuals.in the other words, the solutions on the Pareto front are speckled by filtering which addresses on equation 5.6.

Then, the elite population is modified to the multi-objective optimal solution for a scaled of equation 5.3. In the standard population, every individual is equated to every one of the elites with a sense of the consequences of the objective functions. Fitness, therefore, is allocated both be contingent on exactly how talented the quantities of the objective functions are, and on the space $R^{m}_{\phi,\Theta}$ for an individual at every point in spherical coordinates ϕ and Θ, respectively to the elite member m in the objective space. The distance $R^{m}_{\phi,\Theta}$ is calculated using:

$$R^{m}_{\phi,\Theta} = \sqrt{\sum_{m=1}^{M}\left(\frac{e^{m} - f^{\phi,\Theta}_{m}}{e^{m}}\right)^{2}} \tag{5.7}$$

where e and f denote the objective function quantities of the elite and normal population respectively, and M is the number of objective functions. In the case of the dominant individual, the fitness of the individual is determined from:

$$F_{\phi,\Theta} = \max[0, F(e^{m_{\phi,\Theta}}) - \min_{m=1}^{E_t} R^{m}_{\phi,\Theta}] \tag{5.8}$$

On the contrary, the fitness of a non-dominant individual is casted as:

$$F_{\phi,\Theta} = F(e^{m^{*}_{\phi,\Theta}}) + \min_{m=1}^{E_t} R^{m}_{\phi,\Theta} \tag{5.9}$$

where E_t represents the elite set and $m^{*}_{\phi,\Theta} = \{m : R^{m}_{\phi,\Theta} = R^{\min}_{\phi,\Theta}\}$. Figure 15 proves that the Flight MH370 due to its flaperon splitting away would be spun, rotated, and fluctuated upside down and experienced instability. This leads to a crash of MH370 in the air prior to plunging into the water. In addition, the vanishing zone must be dissimilar with one simulated by Inmarsat 3-F1satellite data.

In this circumstance, the input the fitness function acquires from the objective function is whether or not the point is on the Pareto front. In this understanding, the fitness can be a positive scalar irrespective of when the measured problem is a minimum or a maximization problem. Hence, of Equation (5.8) and Equation (5.9) a point on the Pareto front is extremely evaluated when it is isolated from other points on

Figure 15. MH370 path simulated by the fitness of non-dominant

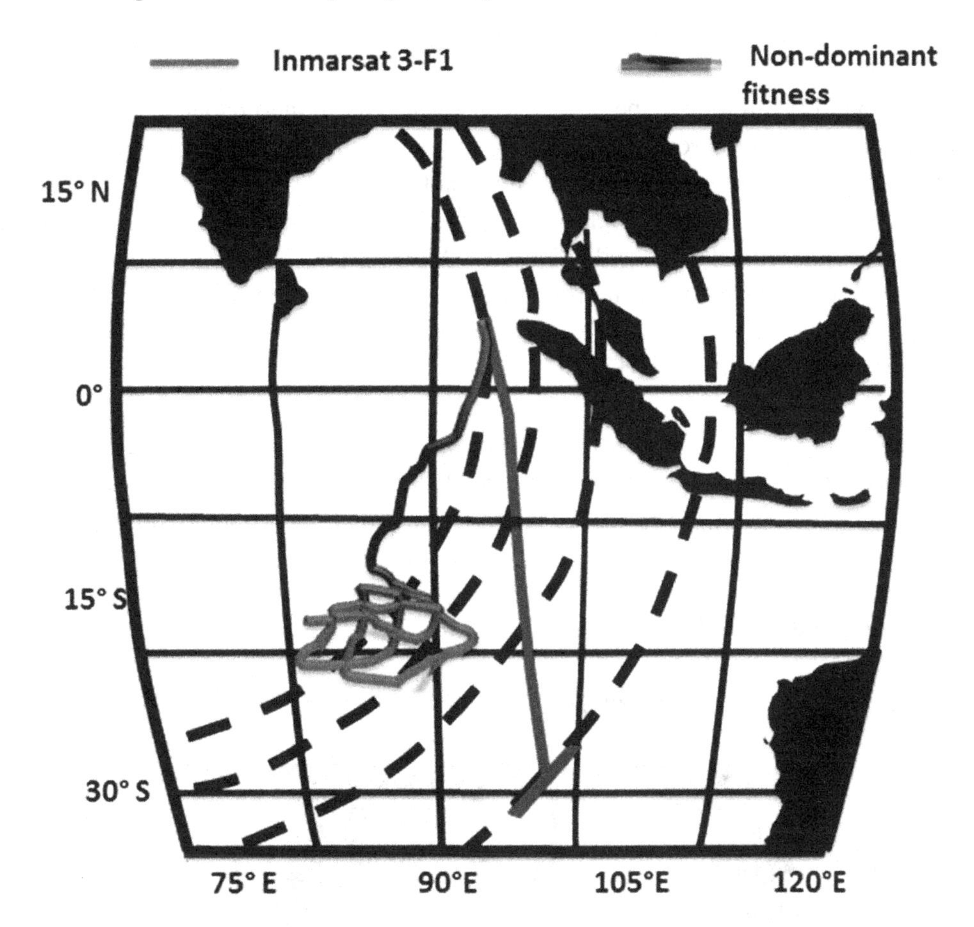

the Pareto front. Although, a point which is not on the Pareto front is extremely estimated when a point is close to the Pareto boundary. This assists in routing the population concerning a Pareto boundary with utmost scattered. Conversely, any former elites subjugated by the innovative elite fellow are eradicated from the elite array. When the fitness of all individuals has been calculated, the fitness of entire elites is an array to the maximum fitness of the elites. Subsequently, any point on the Pareto boundary should not be evaluated greater than any other point on the Pareto boundary (Andersson, 2015).

MUTATION OPERATOR

Irwin{Hall distribution} used to implement the creep mutation. In this manner, a 12-segment eleventh-order polynomial guesstimate to the ordinary distribution is implemented (Hall, 1027). A creep mutation, therefore, is a small modification of the recent elucidation which is retained neighbouring to the recent solution by exploiting approximately probability variation with the existing amount as the average quantity (Irwin 1927).

If the latitude of a phase in the route is nominated for the mutation the latitude changes to its new quantity post-examination that the new MH370 path is a function of equation 5.3. If the new latitude results in a route that is a function of weather effects, separation of flaperon, and fuel consumption are acquired by incrementally fluctuating between increasing and decreasing latitudes and longitudes, respectively (Figure 15).

So that sustains stability throughout the MM370 route which it is required that the period of the phase prior to the mutation phase. On the other hand, the period in the phase post the mutation phase are both retained the identical prior and post the mutation. This denotes that the rate both to and from the mutation phase hypothetically requirements to swap in to accomplish the request period. Subsequently, the latest latitude and velocity to and from the mutated phase of the MH370 route have been resolved, the objective functions to and from the mutated phase are revised exploiting the contemporary latitude and speeds.

What is the Precise Route of MH370?

The significant question can rise up: what is the precise MH370 path? The distance-based Pareto genetic algorithm (DPGA) demonstrates that the Flight MH370 as a function of the autopilot must fly on a slight curve directly towards Perth. On the contrary, the MH370 particular path must be closer to

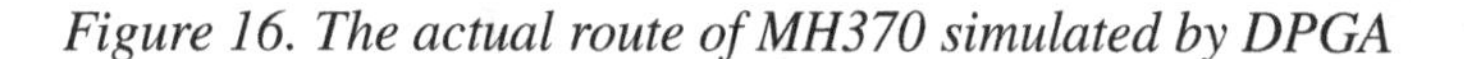

Figure 16. The actual route of MH370 simulated by DPGA

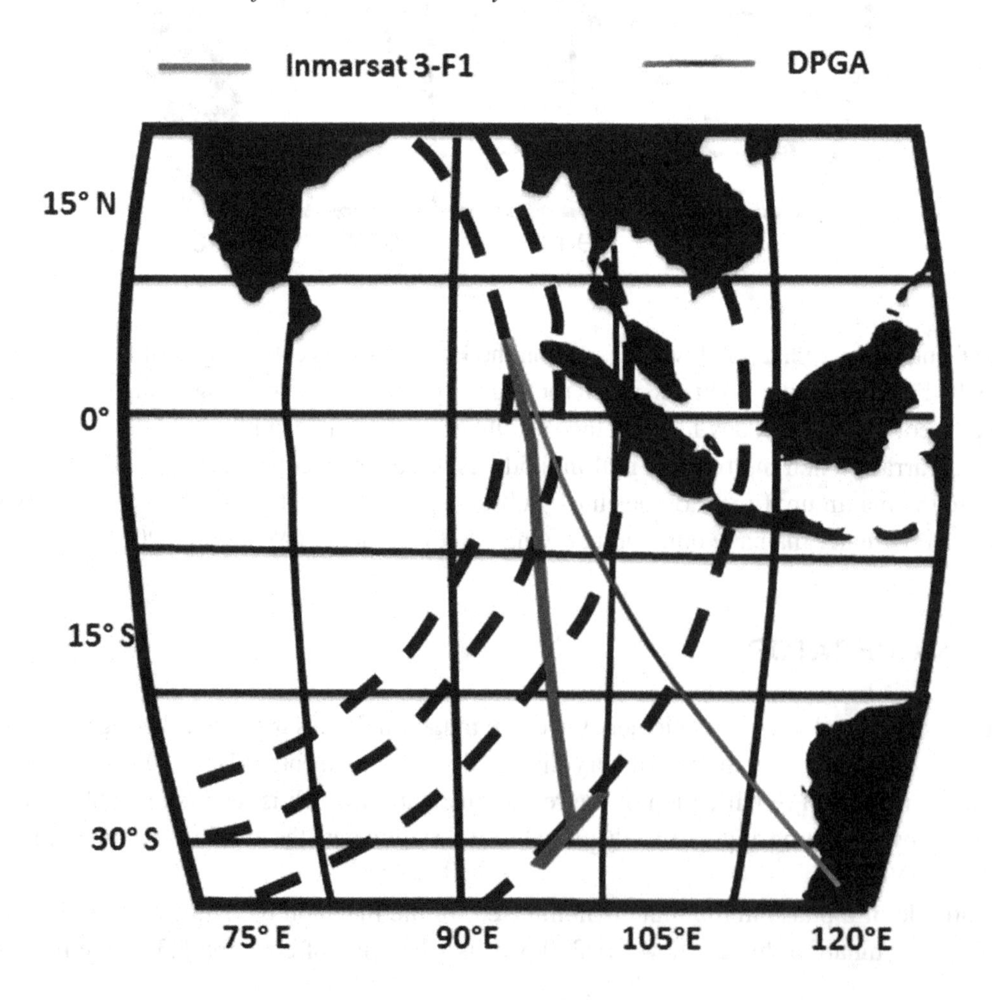

the land rather the open ocean. In this sense, the simulated MH370 route was determined by Inmarsat 3-F1satellite data is not completely logical. Figure 16 demonstrates that the route simulated by DPGA does not coincide with the one is retrieved by Inmarsat 3-F1satellite data.

The main theory supports the Inmarsat 3-F1satellite data, which is the case of murder-suicide. On the contrary, if Zaharie, the plane's pilot was planned for the murder-suicide, he must fly in short paths as shown in Figure 17. In fact, any person is planning to murder-suicide must achieve it in a short period, not a long period. DPGA agrees that the appropriate short path for Zaharie to murder-suicide is the Gulf of Thailand not in the southern Indian Ocean which indicates by 1000 iterations and 100 fitness. These results are not agreed with Inmarsat 3-F1satellite data and the report was delivered by (Bureau 2014).

Figure 17. The actual route of MH370 due to murder-suicide simulated by DPGA

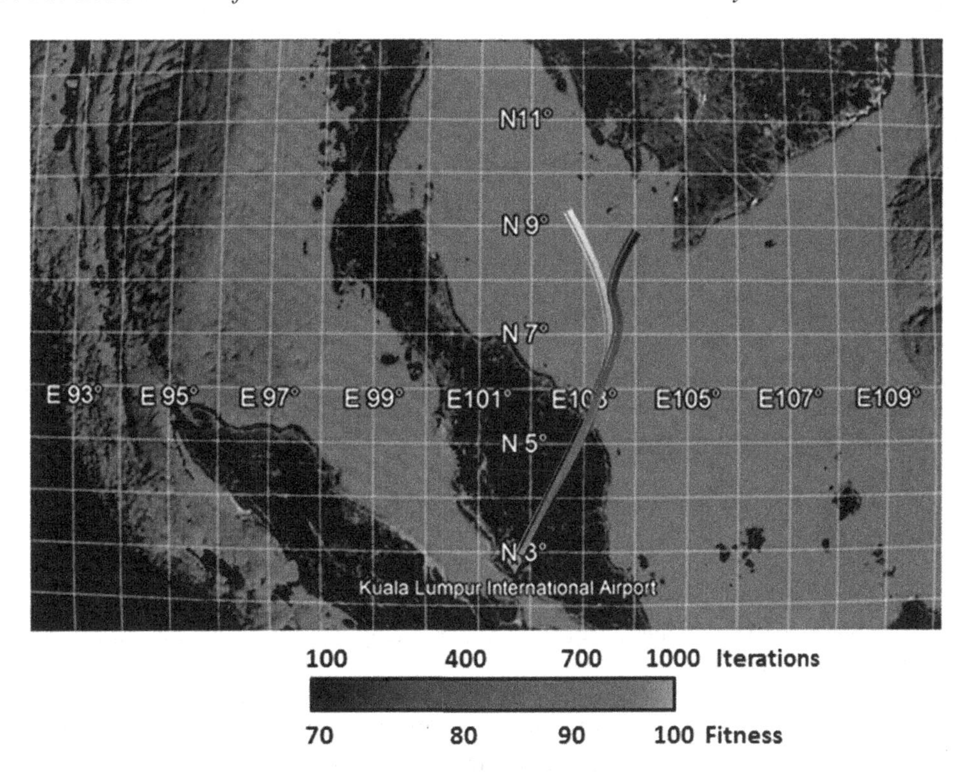

According to the above perspective, Zaharie must ditch vertically with extremely high speed in the Gulf of Thailand to accelerate the vanishing of the Flight. In fact, the vertical ditching under high speed will accelerate the sinking of the flight as the strong stream water will flow through the fuselage and pull it down.

CONCLUSION

This chapter has demonstrated the mathematical model to retrieve the actual path of MH370 and its debris which is based on a multi-objective evolutionary algorithm. The most significant result is that the appropriate short path for Zaharie to murder-suicide is the Gulf of Thailand not in the southern Indian Ocean which indicates by 1000 iterations and 100 fitness.

It can be said that the MH370 route was retrieved from Inmarsat 3-F1satellite data is far away from the actual reality. This proves, by comparison with one simulated by the multiobjective genetic algorithm. Besides, the international search teams cannot find the precise location since the Flight MH370 vanished. The next chapters perhaps can discuss further the possible vanishing scenarios.

REFERENCES

Abdelsalam, A., Caragata, D., Luglio, M., Roseti, C., & Zampognaro, F. (2017). Robust security framework for DVB-RCS satellite networks (RSSN). *International Journal of Satellite Communications and Networking*, *35*(1), 17–43. doi:10.1002at.1154

Andersson, A. (2015). *Multi-objective optimisation of ship routes* (Master's thesis). ABB Corporate Research/Chalmers University of Technology, Göteborg, Sweden.

Bureau, A. T. S. (2014). *MH370: Flight Path Analysis Update*. Australian Transport Safety Bureau.

Chen, G., Gu, C., Morris, P. J., Paterson, E. G., Sergeev, A., Wang, Y. C., & Wierzbicki, T. (2015). Malaysia airlines flight mh370: Water entry of an airliner. *Notices of the American Mathematical Society*, *62*(4), 330–344. doi:10.1090/noti1236

Finkleman, D. (2014). *A mathematical and engineering approach to the search of MH370*. Powerpoint Presentation. (unpublished)

Hall, P. (1927). The distribution of means for samples of size n drawn from a population in which the variate takes values between 0 and 1, all such values being equally probable. *Biometrika*, *19*(3/4), 240–245. doi:10.2307/2331961

Irwin, J. O. (1927). On the frequency distribution of the means of samples from a population having any law of frequency with finite moments, with special reference to Pearson's Type II. *Biometrika*, *19*(3-4), 225–239. doi:10.1093/biomet/19.3-4.225

Iwai, H., Yasunaga, M., & Karasawa, Y. (1991). A fading reduction technique using interleave-aided open loop space diversity for digital maritime-satellite communications. *IEICE Transactions on Communications*, *74*(10), 3286–3294.

Kimura, K., Morikawa, E., Kozono, S., Obara, N., & Wakana, H. (1996). Communication and radio determination system using two geostationary satellites. II. Analysis of positioning accuracy. *IEEE Transactions on Aerospace and Electronic Systems*, *32*(1), 314–325. doi:10.1109/7.481271

Ng, H. K., Grabbe, S., & Mukherjee, A. (2009, August). Design and evaluation of a dynamic programming flight routing algorithm using the convective weather avoidance model. In AIAA guidance, navigation, and control conference (p. 5862). doi:10.2514/6.2009-5862

Osyczka, A., & Kundu, S. (1995). A new method to solve generalized multicriteria optimization problems using the simple genetic algorithm. *Structural Optimization*, *10*(2), 94–99. doi:10.1007/BF01743536

Santamarta, R. (2014). *A wake-up call for satcom security.* Technical White Paper.

Spiridonov, V. V. (1994). Inmarsat systems and services. In *Satellite Communications, 1994. ICSC'94., Proceedings of International Conference on*, (*vol. 1*, pp. 45-52). IEEE.

Swan, P. A., & Devieux, C. L. Jr., (Eds.). (2012). *Global Mobile Satellite Systems: A Systems Overview*. Springer Science & Business Media.

Swanwick, J. (2017). *What are The Future Prospects for Data Communications within Space Exploration?* (Doctoral dissertation). Cardiff Metropolitan University.

Venkatesan, R., Arul Muthiah, M., Ramesh, K., Ramasundaram, S., Sundar, R., & Atmanand, M. A. (2013). Satellite communication systems for ocean observational platforms: Societal importance and challenges. *Journal of Ocean Technology*, *8*(3).

Wikipedia. (2018). *Aircraft Communications Addressing and Reporting System*. https://en. wikipedia.org/w/index.php?title=AircraftCommunicationsAddressing and Reporting System&oldid=672562871

Zweck, J. (2016). Analysis of Methods Used to Reconstruct the Flight Path of Malaysia Airlines Flight 370. *SIAM Review*, *58*(3), 555–574. doi:10.1137/140991996

Zweck, J. (2014, May 1). How did Inmarsat deduce possible flight paths for MH370? *SIAM News*.

Chapter 6
Potential of Optical Remote Sensing Sensors for Tracking MH370 Debris

ABSTRACT

The main question is about how optical remote sensing can be implemented to investigate the HH370 debris. The perfect understanding of the principles of remote sensing and optical satellite data can assist to answer this question. This chapter aims at reviewing the fundamental of optical remote sensing satellite data. From the point view of the electromagnetic spectrum to physical characteristics of optical satellite sensors with high and low resolution, the MH370 debris can be recognized in satellite images. In this understanding, the chapter carries a novel explanation of remote sensing technology of MH370 as a specific and unique case. This clarification is deliberated with particular debris imagined in satellite images as quantum information, which is presented somewhere in the Indian Ocean.

INTRODUCTION

Remote sensing technologies play an incredible task in comprehending the mechanisms and characteristics of the Flight MH370 route and debris tracking. This was accomplished through the exploitation of both optical and microwave remote sensing technologies. The main aim of this chapter is to clarify how remote sensing technologies are used to investigate the MH370 debris. Consequently, both techniques are mechanically connected with techniques that determine and reckon electromagnetic energy which has interacted with tsunami wave propagation, the atmosphere and materials that are made through debris influences.

Understanding of electromagnetic (EM) spectrum and EM radiation, of that light, radar, and radio waves are based totally on the photoelectrical theory.

DOI: 10.4018/978-1-7998-1920-2.ch006

ELECTROMAGNETIC SPECTRUM

The electromagnetic spectrum definitely refers to the wavelengths of light. In this regard, the electromagnetic spectrum is the time period used to describe the all-inclusive range of light that is existent. Conversely, most of the light in a universe from radio waves to gamma rays is vague to us!

Light is a wave of irregular electric and magnetic fields. The transmission of light is not much exceptional than wave propagation in an ocean. A vital descriptive characteristic of a waveform is its wavelength or distance between succeeding peaks or troughs. In remote sensing, the wavelength is most frequently measured in micrometres, each of which equals one-millionth of a meter. The variation in the wavelength of electromagnetic radiation is so significant that it is usually proven on a logarithmic scale (Figure 1) (Lillesand et al., 2007 and Bakshi and Godse, 2009).

Figure 1. Electromagnetic wave spectra

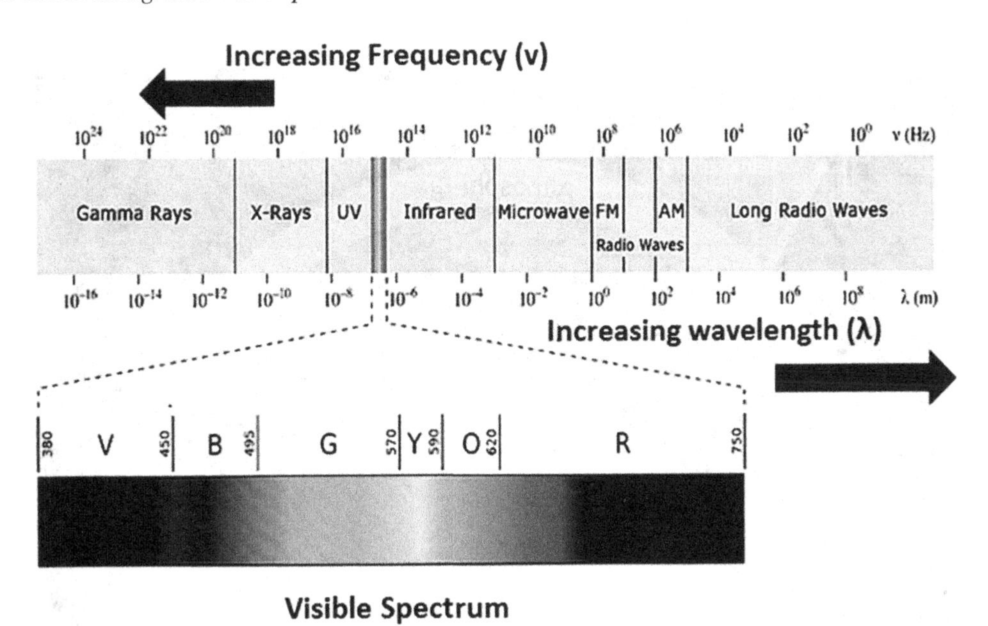

Physically, the Earth is brightened through electromagnetic radiation from the Sun. The peak solar power is in the wavelength variation of visible light (between 0.4 and 0.7 μm) (Figure 1). Additional large components of incoming solar energy are in the configuration of invisible ultraviolet and infrared radiation. Merely tiny portions of solar radiation encompass the microwave spectrum. Imaging radar systems used in remote sensing create and transmit microwaves, and then measure the component of the signal that has backscattered to the antenna from the Earth's surface (Lillesand et al., 2007).

The electromagnetic spectrum ordinarily breaks up into seven regions, in order of decreasing wavelength and growing energy and frequency: radio waves, microwaves, infrared, visible light, ultraviolet, X-rays and gamma rays.

Radio Waves

Radio waves are at the lower range of the EM spectrum, with frequencies of up to about 30 billion hertz, or 30 gigahertz (GHz). These correspond to the wavelengths as low as 30 cm and as high as 1000 m. Radio waves are used specifically for communications, including voice, data and leisure media (See Chapter 5).

For instance, a radio programme receiver does not require to be absolutely in the outlook of the transmitter to obtain the signals. Diffraction, however, allows low-frequency radio waves to be received behind the hills, in spite of repeater stations are often used to improve the quality of the signals. The lowest frequency radio waves are also reflected from an electrically charged layer of the upper atmosphere, called the ionosphere. This means that they can still reach receivers that are not in the line of sight because of the curvature of the Earth's surface (Figure 2).

Figure 2. Radio wave reflection from the ionosphere

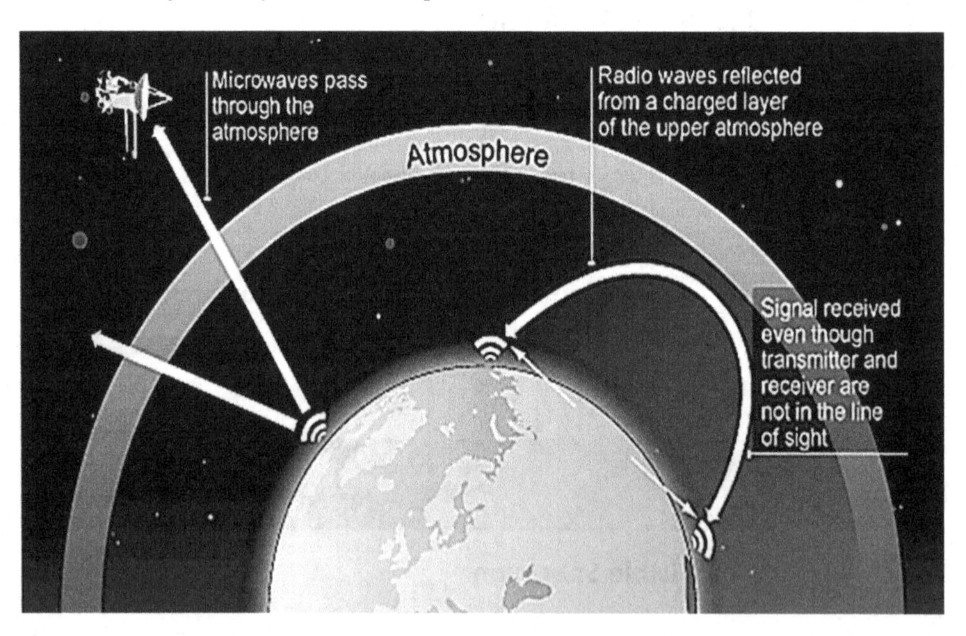

Microwaves

Microwaves fall in the range of the EM spectrum between radio and IR. They have frequencies from about 3 GHz up to about 30 trillion hertz, or 30 terahertz (THz), and wavelengths of about 10 mm (0.4 inches) to 100 micrometres (μm), or 0.004 inches (Figure 3). Microwaves are used for high-bandwidth communications, radar and as a source of warmth for microwave ovens and industrial applications.

Figure 3. Microwave spectrum

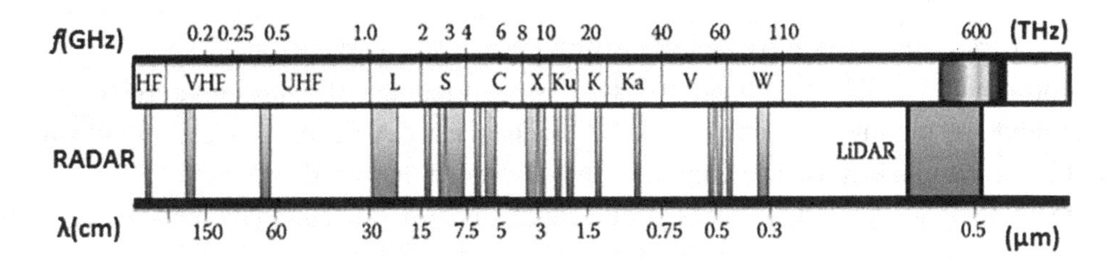

Infrared

Infrared is in the range of the EM spectrum between microwaves and visible light. IR has frequencies from about 30 THz up to about 400 THz and wavelengths of about 100 μm (0.004 inches) to 740 nanometers (nm) (Figure 4), or 0.00003 inches. IR light is invisible to human eyes, but we can feel it as heat if the intensity is sufficient.

Figure 4. Infrared spectrum

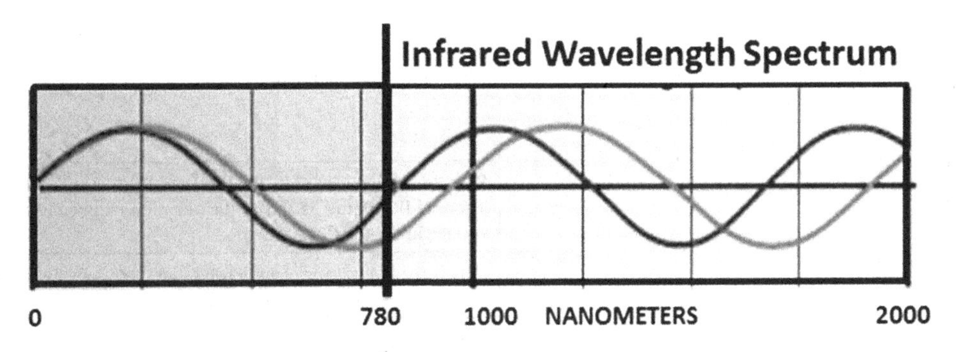

Visible Light

Visible light is found in the middle of the EM spectrum, between the IR and UV. It has frequencies of about 400 THz to 800 THz and wavelengths of about 740 nm (0.00003 inches) to 380 nm (0.000015 inches). More generally, visible light is defined as the wavelengths that are visible to most human eyes.

Ultraviolet

Ultraviolet light is in the range of the EM spectrum between visible light and X-rays. It has frequencies of about 8×10^{14} to 3×10^{16} Hz and wavelengths of about 380 nm (0.000015 inches) to about 10 nm (0.0000004 inches). UV light is a component of sunlight; however, it is invisible to the human eye. It has numerous medical and industrial applications, but it can damage living tissue.

X-Beams

X-beams are generally arranged into two sorts: soft X-beams and hard X-beams. The soft X-beams have frequencies of around 3×10^{16} to around 10^{18} Hz and wavelengths of approximately 10 nm (4×10^{-7} inches) to around 100 picometers (pm), or 4×10^{-8} inches. Hard X-beams possess an indistinguishable area of the EM of gamma beams. The only difference between them is their source: X-beams are delivered with the guide of quickening electrons, while gamma beams are created by nuclear cores (Table 1).

Table 1. Summary of Principal of Electromagnetic Spectrum

Spectra Wavelength	Description and Usages
Gamma rays	Gamma rays
X-rays	X-rays
Ultraviolet (UV) region 0.30 μm - 0.38 μm (1μm = 10^{-6} m)	This band is the violet portion of the visible wavelength, and hence its name. Approximately earth's surface material primarily rocks and minerals emanate visible UV emission. Nevertheless, UV emission is generally scattered by the earth's atmosphere and hence not used in the field of remote sensing.
Visible Spectrum 0.4 μm - 0.7 μm Violet 0.4 μm -0.446 μm Blue 0.446 μm -0.5 μm Green 0.5 μm - 0.578 μm Yellow 0.578 μm - 0.592μm Orange 0.592 μm - 0.62 μm Red 0.62 μm -0.7 μm	This is merely the portion of the spectrum that can be correlated with the concept of color. The color of an object is identified through the color of the reflected light.
Infrared (IR) Spectrum 0.7 μm – 100 μm	Remote sensing is used by the reflected IR (0.7 μm - 3.0 μm). Thermal IR (3 μm - 35 μm) is the radiation radiated from the Earth 's surface in the form of heat.
Microwave Region 1 mm - 1 m	This is the longest wavelength used in remote sensing, which has the potential to penetrate through clouds
Radio Waves (>1 m)	This is the longest portion of the spectrum regularly used for meteorology and commercial broadcast

Gamma-Rays

Gamma-rays are in the range of the spectrum above soft X-rays. Gamma-rays have frequencies greater than about 10^{18} Hz and wavelengths of less than 100 pm (4×10^{-9} inches). Gamma radiation causes damage to living tissue, which makes it useful for killing cancer cells when applied in carefully measured doses to small regions. Uncontrolled exposure, though, is extremely dangerous to humans (Table 6.1). Instruments aboard high-altitude balloons and satellites like the Compton Observatory provide our only view of the gamma-ray sky.

Generally, the electromagnetic spectrum is composed of the low frequencies used for contemporary radio communication to gamma radiation at the short-wavelength (high-frequency) end, thereby covering wavelengths from thousands of kilometres down to a fraction of the dimension of an atom. Visible light lies towards the shorter end, with wavelengths from 400 to 700 nm. The restriction for long wavelengths is the dimension of the universe itself, while it is understood that the short wavelength constraint is in the vicinity of the Planck length which is a unit of length and equal to $1.616229(38)\times10^{-35}$ meters. Until the middle of the 20th century, it used to be believed by most physicists that this spectrum was countless and continuous (Marghany, 2018).

Most waves are both longitudinal and transverse. For instance, sound waves are longitudinal. Nonetheless, all electromagnetic waves are transverse. Moreover, EM is created through the movement of electrically charged particles. Consequently, EM can travel in a "vacuum" (they do NOT want a medium) at the speed of light, i.e., 300,000 km/sec in space (] Campbell et al., 2011 and Marghany 2018).

ENERGY IN ELECTROMAGNETIC WAVES

The wave theory of EM radiation is adopted based on the concept of particle waves. Consequently, EM can behave like a wave or like a particle whereas a "particle" of light is called a photon. In other words, the insufficiency of the wave theory preceded a revival of the idea that EM radiation might better be thought of as particles, dubbed photons. The energy of a photon can mathematically be written as:

$$E = h * v \tag{6.1}$$

where E is energy, ν is the frequency for electromagnetic radiation, h is constant Planck's quantum of energy, and equals:

$$\begin{pmatrix} 6.626 \text{ x } 10^{-34} & \text{joule seconds} \\ 4.136 \text{ x } 10^{-15} & \text{eV seconds} \end{pmatrix}.$$

The electron volt (eV) is a convenient unit of energy related to the standard metric unit (joules, or J) by the relation 1 eV= $1.602 \text{ x } 10^{-19}$ J. Photon energy E is determined by the frequency of EM radiation: the higher the frequency, the higher the energy. Even though the photons propagate at the speed of light, they have zero rest mass, so the rules of special relativity are not ruined.

The relationship between the wavelength (λ) and frequency (ν) for electromagnetic radiation is formulated as:

$$\lambda * v = c \tag{6.2}$$

From these relationships, we can define the relationship between energy and wavelength as:

$$E = h * \frac{c}{\lambda} \tag{6.3}$$

or, rearranging as:

$$\lambda = h * \frac{c}{E} \tag{6.4}$$

The relationship between the wavelength (λ) and momentum (m*v) for DeBroglie's "particle-wave" is determined from:

$$\lambda = \frac{h}{\mathrm{m} * \mathrm{v}} \tag{6.5}$$

From the above relationships, we can calculate the relationship between energy (*E*) and momentum (m*v) as:

$$\frac{h}{\mathrm{m} * \mathrm{v}} = h * \frac{c}{E} \tag{6.6}$$

Simplify, and solve for *E*:

$$E = m * v * c \tag{6.7}$$

The highest velocity (v) attainable by matter is the speed of light (*c*), therefore, the maximum energy would seem to be:

$$E = m * c * c \tag{6.8}$$

or

$$E = m * c^2 \tag{6.9}$$

Equation 6.5 is the De Broglie equation where photons behave like matter particles and extend it to all particles (Figure 5). Consequently, Equation 6.7 relates the wave-like the behaviour of matter to its momentum. Figure 5 summarizes the concepts of EM energy spectra. For instance, visible-light wavelengths correspond to a wavelength range from 0.38-0.75 μm of 2-3 eV.

PHOTOELECTRIC EFFECT

The concept of energy in photons is very important for remote sensing technology. The photoelectric effect was the first example of a quantum phenomenon to be seen at the end of the 19th Century. Light or other forms of electromagnetic radiation shone onto metals release electrons; the energy supplied by the radiation frees the electrons from the metal. The way in which the numbers and energies of electrons released changes when the frequency and intensity of the radiation changes cannot be explained using the classical wave model of light. Increasing the intensity of the radiation does not increase the energy of the electrons, but releases more of them per second. Increasing the frequency of the light increases

Figure 5. De Broglie concept

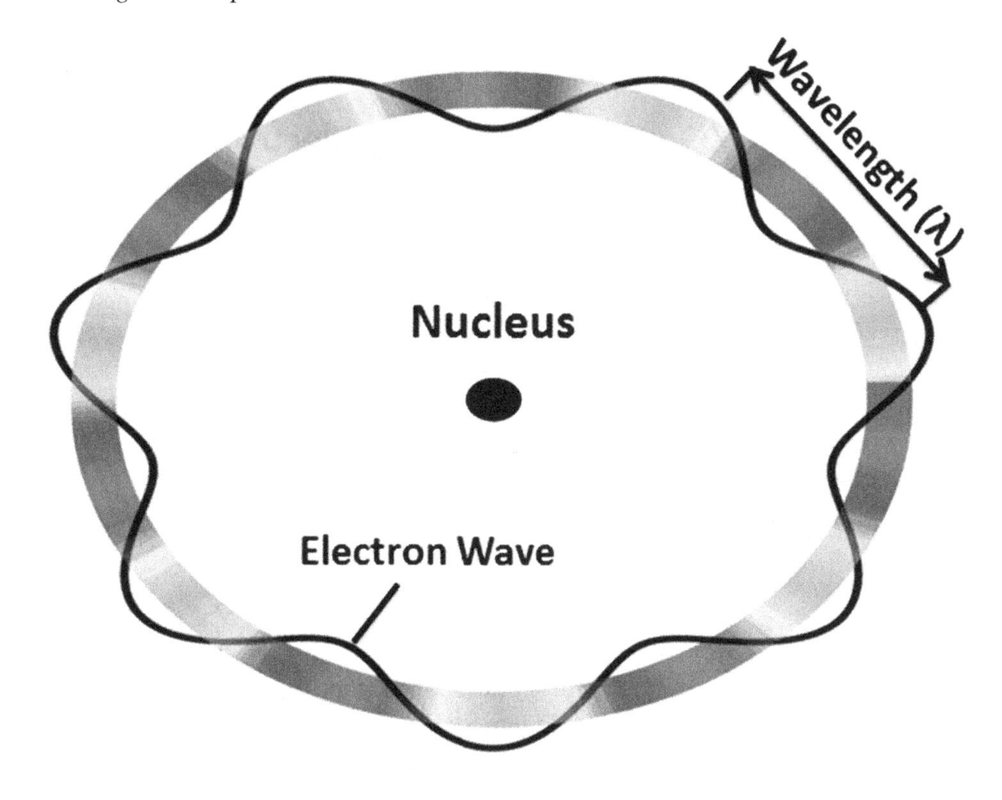

Figure 6. Summary of electromagnetic spectra

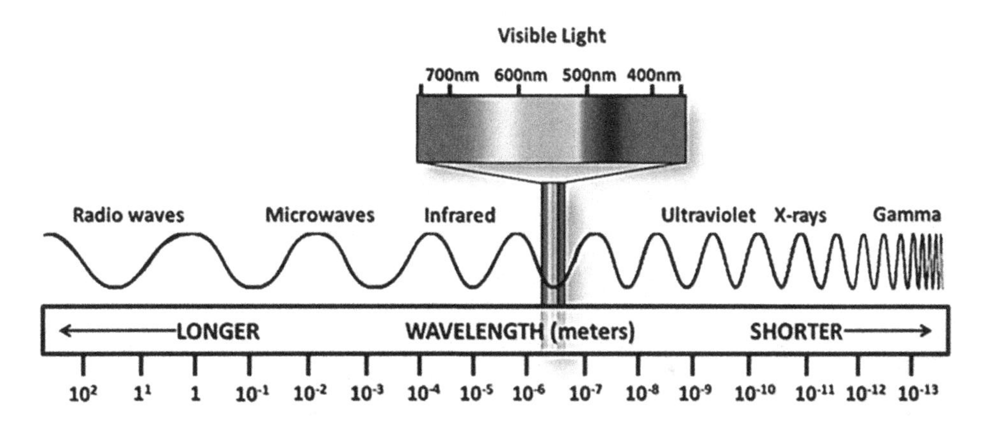

the energy of the electrons. Below a certain frequency of radiation, v_0, no electrons are emitted no matter how intense the radiation (Figure 7).

These facts are explained using the photon model of light. Light (and all EM radiation) is emitted and absorbed in little packets or quanta called photons. The energy of a photon is equal to its frequency multiplied by Planck's constant (Equation 6.1), h = 6.63 x 10^{-34} Js.

Figure 7. Photoelectric concept

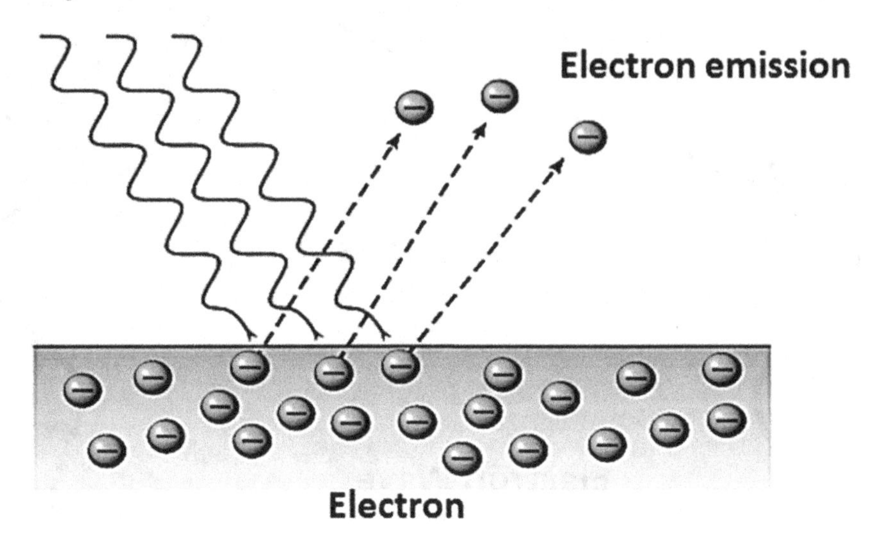

KE_{max} of electrons = hν- Φ where Φ is the energy needed to escape from the metal. This phenomenon introduces the **wave-particle duality** of nature (Figure 8): light behaves as a wave at times (e.g., Young's slits) and as a particle at times (Figure 9). This duality is central to the way quantum mechanics explains nature as it applies to everything.

Figure 8. Wave-particle duality

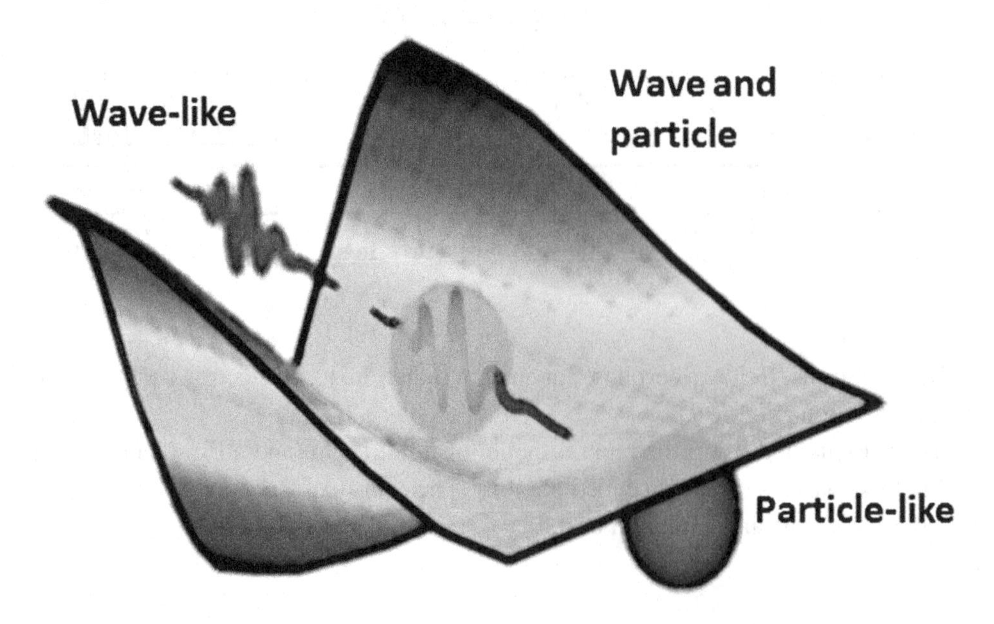

Figure 9. Wave-particle duality propagation

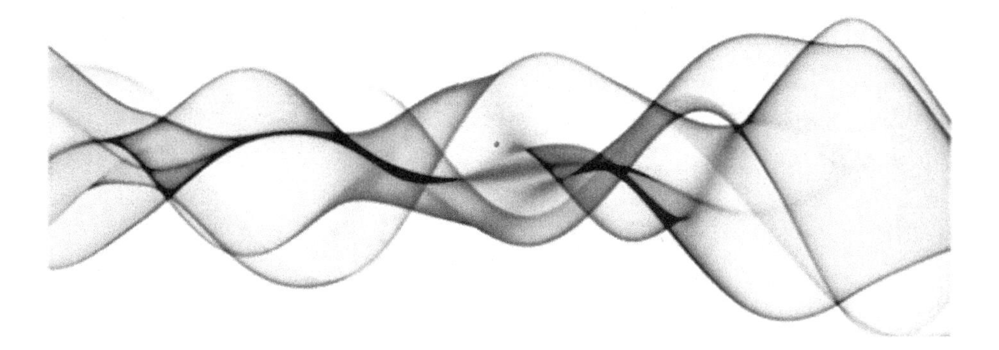

YOUNG'S SLITS

Thomas Young used this experiment to 'prove' that light was a wave at a time when the light was thought to be a particle. The light going through two slits interferes and produces a pattern that is easy to explain using a wave model, but which cannot be explained if light acts like particles (Figure 10).

Wave-particle duality: If the light was just a wave, then the electrons would absorb some energy no matter what the frequency. If the light was just a wave, then the emission of an electron would take longer when a lower intensity light was used, not instantaneously.

Nonetheless, light is not just a wave. It can also behave as though it was made of tiny energy packets or particles. We call these particles, photons (Figure 11). It is one of these photons that will hit one electron on the plate, the electron will absorb the energy and it will fly off the plate. So if the intensity is greater, i.e. there are more photons, then more electrons can be knocked off.

The main question that arises is how tsunami waves and their effects have been imagined from space? To answer this question, we need to understand how remote sensing sensors can imagine this phenomenon which mainly is a function of the interaction of EM radiations with tsunami waves and the debris remains after the disaster. From the point of view of optical remote sensing, two important issues must be well understood: (i) EM-radiation- matter interactions; and (ii) blackbody radiation (Brown 1977; Campbell et al., 2011;Marghany 2018).

Remote sensors determine electromagnetic (EM) radiation that has intermingled with the Earth's surface. Interactions with matter can exchange the bearing, intensity, wavelength content, and polarization of EM radiation. The nature of these changes is reliant on the chemical makeup and physical structure of the material exposed to the EM radiation. Changes in EM radiation resulting from its interactions with the Earth's surface, therefore, provide major clues to the characteristics of the surface materials. Upon striking matter, EM may be transmitted, reflected, scattered, or observed in the proportion that depends upon (i) the combinational and physical properties of the medium; the wavelength of the frequency of the incident radiation; and the angle at which the incident radiation strikes a surface. In other words, the total amount of radiant flux in specific wavelengths (λ) incident to the terrain (Φ_{i_λ}) must be accounted for by evaluating the amount of radiant flux reflected from the surface ($\Phi_{reflected\lambda}$), the amount of radiant flux absorbed by the surface ($\Phi_{absorbed\lambda}$), and the amount of radiant flux transmitted through the surface ($\Phi_{transmitted\lambda}$) (Marghany 2018):

Figure 10. Young's slits

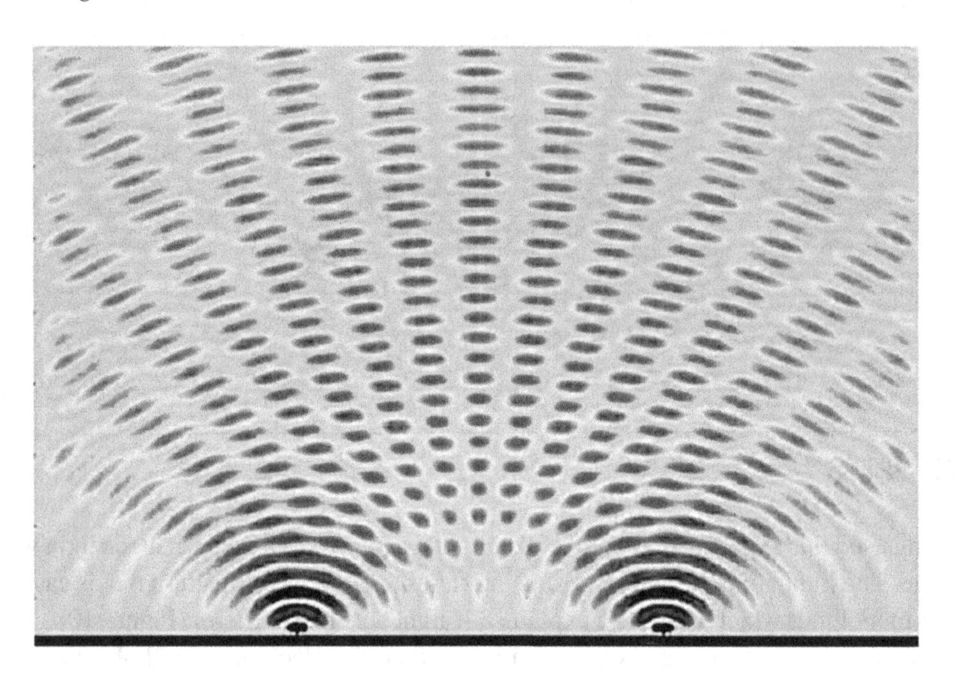

Figure 11. Sketch of photon behaviour

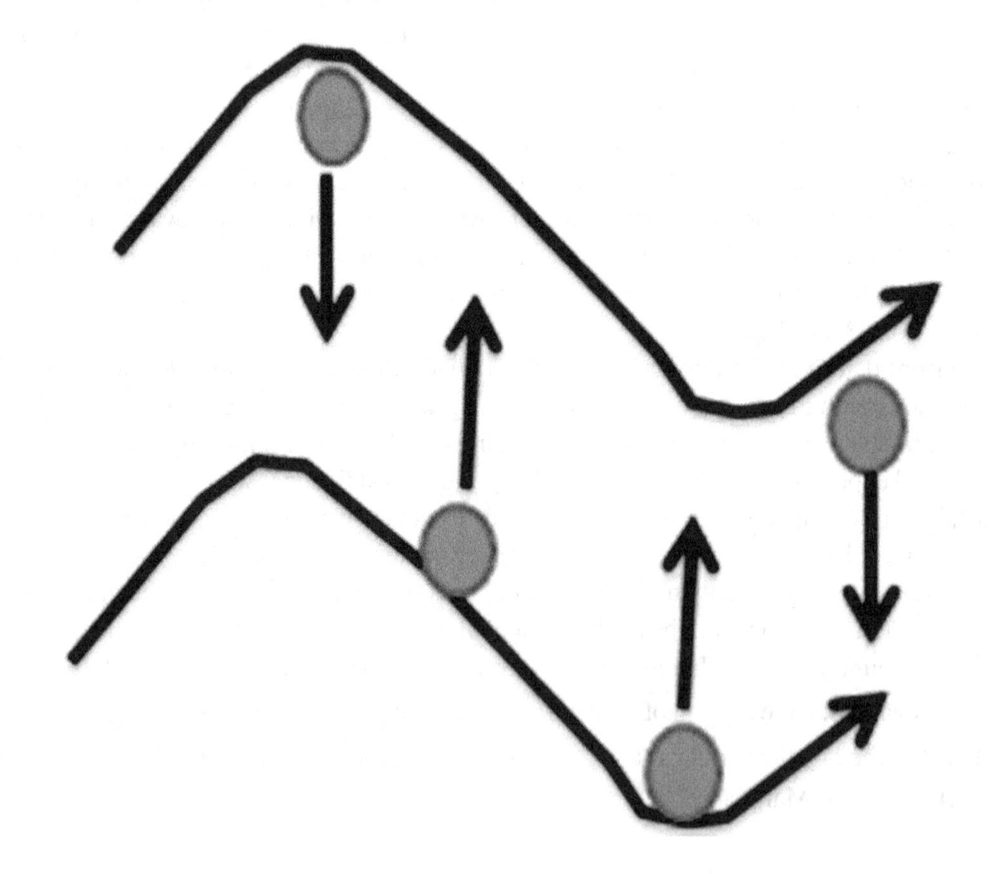

$$\Phi_{i_\lambda} = \Phi_{reflected\,\lambda} + \Phi_{absorbed\,\lambda} + \Phi_{transmitted\,\lambda} \tag{6.10}$$

The radiation passes through a substance without significant attenuation. Transmission through material media of different densities (e.g., air, water) causes radiation to be reflected or deflected from a straight-line path with an accompanying change in its velocity and wavelength; the frequency always remains constant (Chelton et al., 2011).

Let us take the incident beam of EM at an angle θ_i which is deflected toward the normal incident beam in going beam from a low-density medium to a denser one at an angle θ_t. Emerging from the far side of the denser medium, the beam is refracted from the normal at an angle θ_r. The angle relationships in Figure 12 are $\theta_i > \theta_t$ and $\theta_i = \theta_r$ (Figure 12). The change in EMR velocity is explained by the index of reaction, which is the ratio between the velocity of EMR in a vacuum c and its velocity in a material medium V (Lillesand et al., 2007 and Marghany 2018):

Figure 12. Transmission and refraction of EMR

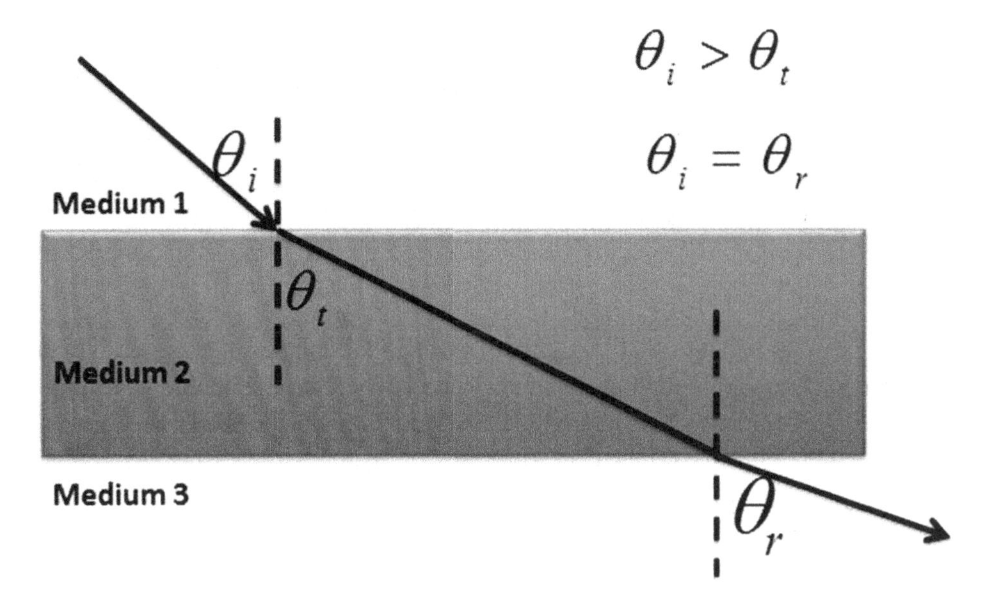

$$n = \frac{c}{V} \tag{6.11}$$

The index of refraction of a vacuum (perfectly transparent medium) is equal to 1, or unity. V has never been greater than c, and n can never be less than 1 for any substance. Indices of refraction vary from 1.0002926 (for the earth's atmosphere) to 1.33 (for water) and 2.42 (for a diamond). The index of refraction leads to Snell's law (Hecht,2001; Lillesand et al., 2007;Marghany 2018).

$$n_1 \sin\theta_i = n_2 \sin\theta_t \tag{6.12}$$

Figure 13 shows the EMR refraction behaviour. Reflection is also known as specular reflection. It describes the process whereby incident radiation bounces off the surface of a substance in single predictable direction. The angle of reflection is always equal and opposite to the angle of incidence $\theta_i=\theta_r$.

Figure 13. EMR reflection

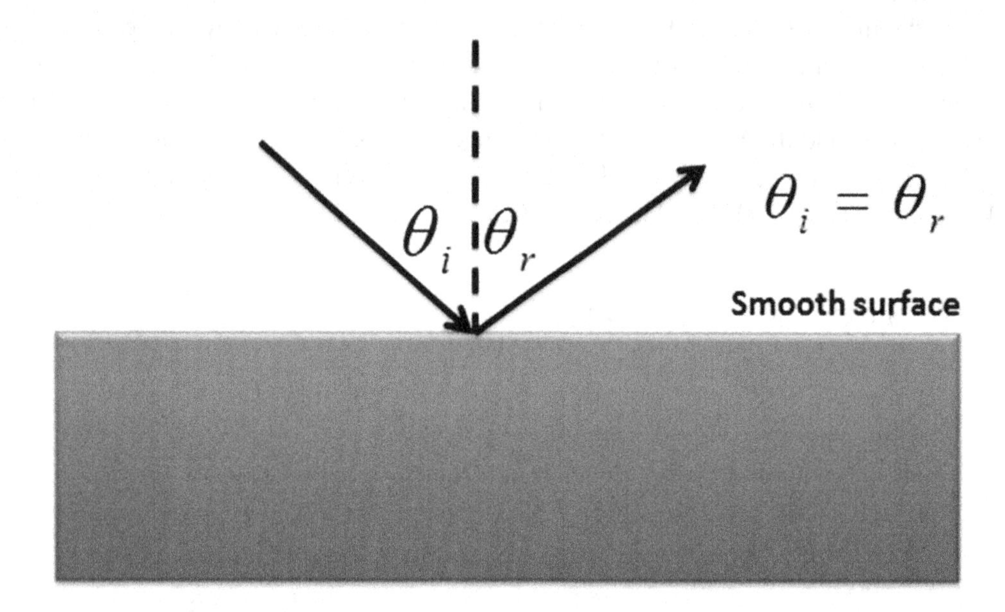

Reflection is caused by surfaces that are smooth to the wavelengths of incident radiation. These smooth, mirror-like surfaces are called specular reflectors. Therefore, specular reflection causes no change to either the EMR velocity or wavelength. Consistent with Hecht (2001), the theoretical amplitude reflectance of a dielectric interface can be derived from electromagnetic theory. In this regard, $\vec{E}$ polarized perpendicular to the plane of incidence:

$$r_{\perp} = \frac{n_1 \cos\theta_i - n_2 \cos\theta_r}{n_1 \cos\theta_i + n_2 \cos\theta_r}; \quad (6.13)$$

For $\vec{E}$ a polarized parallel to the plane of incidence then:

$$r_{\parallel} = \frac{n_2 \cos\theta_i - n_1 \cos\theta_r}{n_2 \cos\theta_i + n_1 \cos\theta_r}; \quad (6.14)$$

where, n_1, θ_i, n_2 and θ_r are the refractive indices and angles of incidences and refraction, respectively. Here r is the ratio of the amplitude of the reflected electric field to incident field. Consequently, the intensity of the reflected EMR is the square of this value (Chelton, 2001 and Marghany, 2018)

Electromagnetic Scattering

Once electromagnetic radiation is created, it is transmitted through the earth's atmosphere almost at the speed of light in a vacuum In other words, in a vacuum, electromagnetic radiation of short wavelengths travels as fast as radiation of long wavelengths. Nevertheless, the atmosphere might disturb not only the velocity of EMR but also its wavelength, intensity, spectral distribution, and/or direction. Nevertheless, scatter differs from the reflection in that the direction associated with scattering is unpredictable, whereas the direction of reflection is predictable. There are essentially three sorts of scattering: (i) Rayleigh; (ii) Mie; and (iii) Non-selective (Figure 14).

Figure 14. EMR scattering

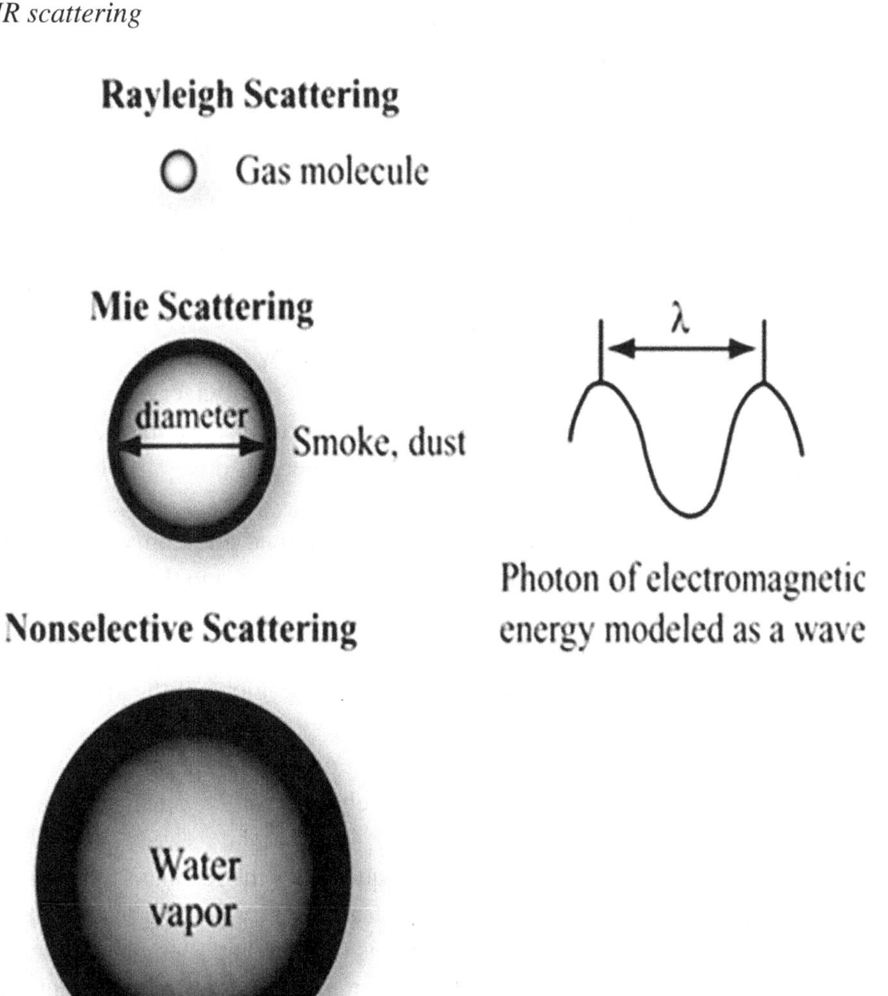

Rayleigh scattering arises when the diameter of the matter (usually air molecules) is many times smaller than the wavelength of the incident electromagnetic radiation. This type of scattering is named after the English physicist who offered the first coherent explanation for it. All scattering is accomplished through absorption and re-emission of radiation by atoms or molecules in the manner described in the discussion on radiation from atomic structures. It is impossible to predict the direction in which a specific atom or molecule will emit a photon, hence scattering.

The energy required to excite an atom is associated with short-wavelength, high-frequency radiation. The amount of scattering is inversely related to the fourth power of the radiation's wavelength. For example, blue light (0.4 mm) is scattered 16 times more than near-infrared light (0.8 mm). The intensity of Rayleigh scattering varies inversely with the fourth power of the wavelength (λ^{-4}) (Chelton, 2001).

Mie scattering takes place when there are essentially spherical particles present in the atmosphere with diameters approximately equal to the wavelength of radiation being considered. For visible light, water vapour, dust, and other particles ranging from a few tenths of a micrometre to several micrometres in diameter are the main scattering agents. The amount of scattering is greater than Rayleigh scattering and the wavelengths scattered are longer. Pollution also contributes to beautiful sunsets and sunrises. The greater the amount of smoke and dust particles in the atmospheric column, the more violet and blue light will be scattered away and only the longer orange and red wavelength light will reach our eyes (Marghany, 2018).

Non-selective scattering is produced when there are particles in the atmosphere several times the diameter of the radiation being transmitted. This type of scattering is non-selective, i.e. all wavelengths of light are scattered, not just blue, green, or red. Thus, water droplets, which make up clouds and fog banks, scatter all wavelengths of visible light equally well, causing the cloud to appear white (a mixture of all colors of light in approximately equal quantities produces white). Scattering can severely reduce the information content of remotely sensed data to the point that the imagery loses contrast and it is difficult to differentiate one object from another. Non-selective scattering is a function of (i) the wavelength of the incident radiant energy; and (ii) the size of the gas molecule, dust particle, and/or water vapour droplet encountered (Lillesand et al., 2007).

Absorption is the process by which radiant energy is absorbed and converted into other forms of energy. An absorption band is a range of wavelengths (or frequencies) in the electromagnetic spectrum within which radiant energy is absorbed by substances such as water (H_2O), carbon dioxide (CO_2), oxygen (O_2), ozone (O_3), and nitrous oxide (N_2O). The cumulative effect of the absorption of the various constituents can cause the atmosphere to close down in certain regions of the spectrum. This is bad for remote sensing because no energy is available to be sensed.

In certain parts of the spectrum, such as the visible region (0.4 - 0.7 mm), the atmosphere does not absorb all of the incident energy but transmits it effectively. Parts of the spectrum that transmit energy effectively are called "atmospheric windows".

Absorption occurs when the energy of the same frequency as the resonant frequency of an atom or molecule is absorbed, producing an excited state. If instead of re-radiating a photon of the same wavelength, the energy is transformed into heat motion and is recruited at a longer wavelength, absorption occurs. When dealing with a medium like air, absorption and scattering are frequently combined into an extinction coefficient (Brown 1977).

Transmission is inversely related to the extinction coefficient of the thickness of the layer. Certain wavelengths of radiation are affected far more by absorption than by scattering. This is particularly true of infrared and wavelengths shorter than visible light (Goddijn-Murphy et al., 2018).

Following Marghany (2018), the *absorption* of the Sun's incident electromagnetic energy in the region from 0.1 to 30 mm is caused by various atmospheric gases(Figure 15). The first four graphs depict the absorption characteristics of N_2O, O_2 and O_3, CO_2, and H_2O, while the final graph depicts the cumulative result of all these constituents being in the atmosphere at one time. The atmosphere essentially "closes down" in certain portions of the spectrum while "atmospheric windows" exist in other regions that transmit incident energy effectively to the ground. It is within these windows that remote sensing systems must function. The combined effects of atmospheric absorption, scattering, and reflectance reduce the amount of solar irradiance reaching the Earth's surface at sea level.

INTERACTION PROCESSES ON REMOTE SENSING

To understand how different interaction processes impact the acquisition of aerial and satellite images, let's analyze the reflected solar radiation that is measured at a satellite sensor. As sunlight initially enters the atmosphere, it encounters gas molecules, suspended dust particles, and aerosols. These materials tend to scatter a portion of the incoming radiation in all directions, with shorter wavelengths experiencing the strongest effect. The preferential scattering of blue light in comparison to green and red light accounts for the blue color of the daytime sky. Clouds appear opaque because of intense scattering of visible light by tiny water droplets. Although most of the remaining light is transmitted to the surface, some atmospheric gases are very effective at absorbing particular wavelengths. The absorption of dangerous ultraviolet radiation by ozone is a well-known example. As a result of these effects, the illumination reaching the

Figure 15. EMR absorption

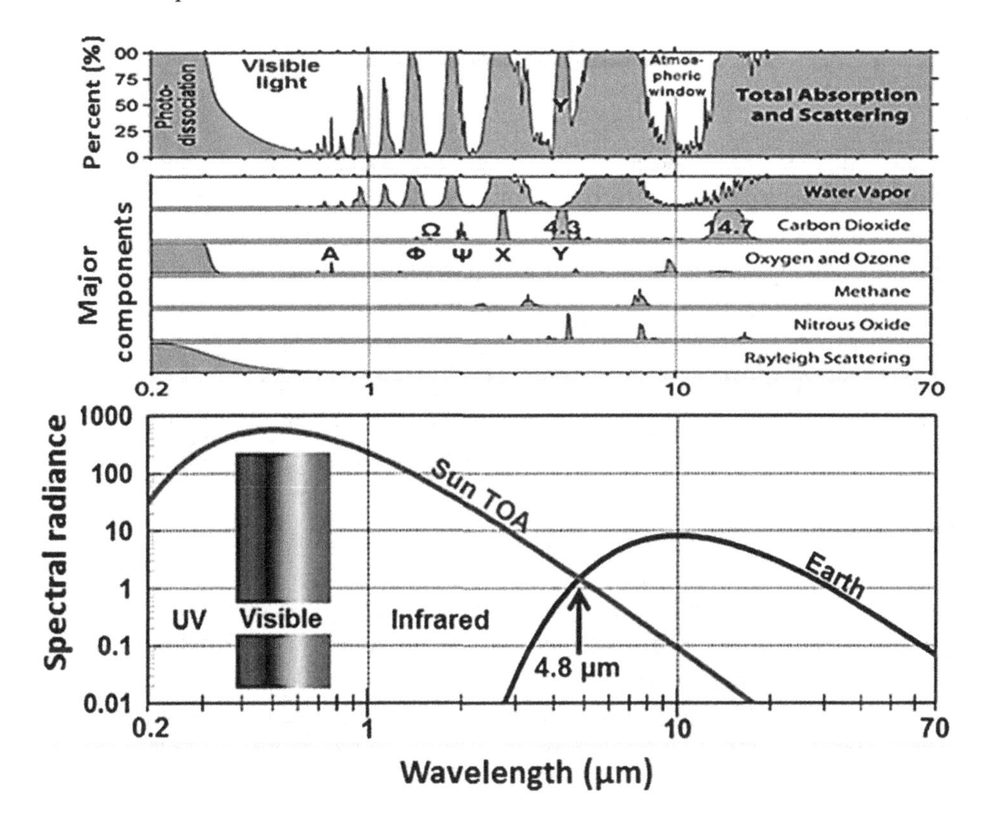

surface is a combination of highly filtered solar radiation transmitted directly to the ground and a more diffused light scattered from all parts of the sky, which helps illuminate shadowed areas (Figure 16).

Figure 16. Typical EMR interactions in the atmosphere and at the Earth's surface

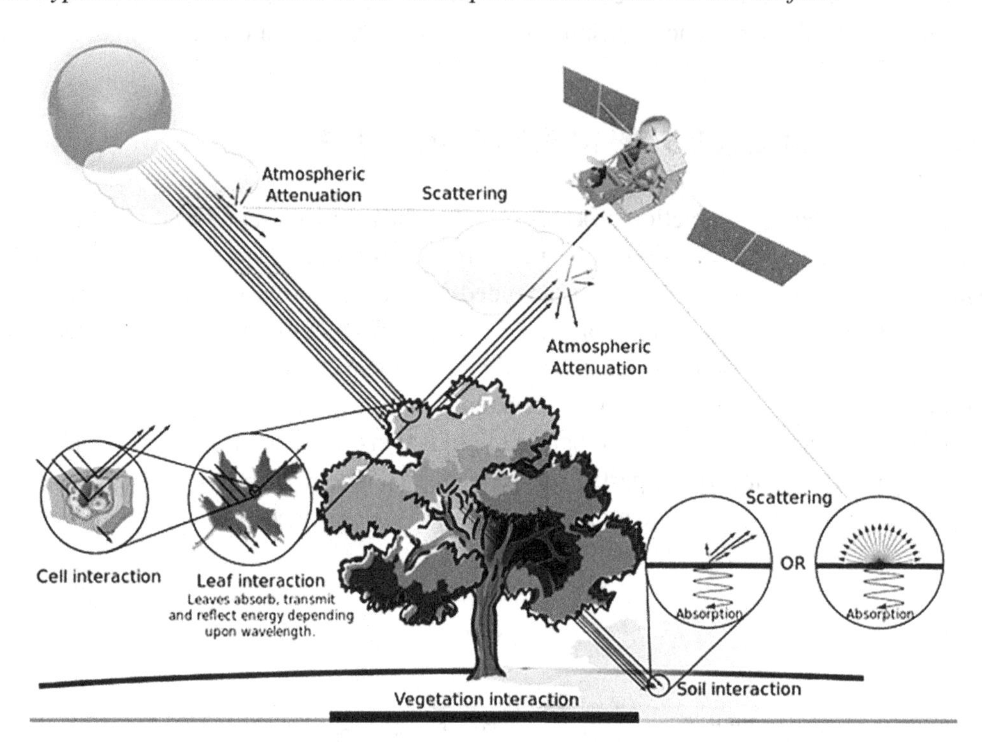

As this altered sunlight based radiation reaches the ground, it might experience soil, shaking surfaces, vegetation, or different materials that absorb a bit of the radiation. The measure of vitality assimilated fluctuates in wavelength for every material naturally, making a kind of spectral signature. The vast majority of the radiation not ingested is diffusely reflected (scattered) moving down into the environment, some of it toward the satellite. Consequently, the upwelling radiation experiences a further round of dissemination and assimilation as it goes through the climate before being distinguished and measured by the sensor. In the event that the sensor is equipped for identifying thermal infrared radiation, it will likewise acquire radiation transmitted by surface materials as a result of solar heating.

BLACKBODY RADIATION

A black body is a sentimentalized physical body which retains all occurrences of electromagnetic beams, irrespective of frequency or angle of incidence. A white body is unified with a coarse surface which mirrors all occurrences beams absolutely and consistently in all paths. Blackbody radiations are transmitted by hot solids, fluid, or thick gases and have a persistent dispersion of emanated wavelength, as shown in Figure 17.

Figure 17. The temperature of a black body decreases, its intensity also decreases and its peak moves to longer wavelengths

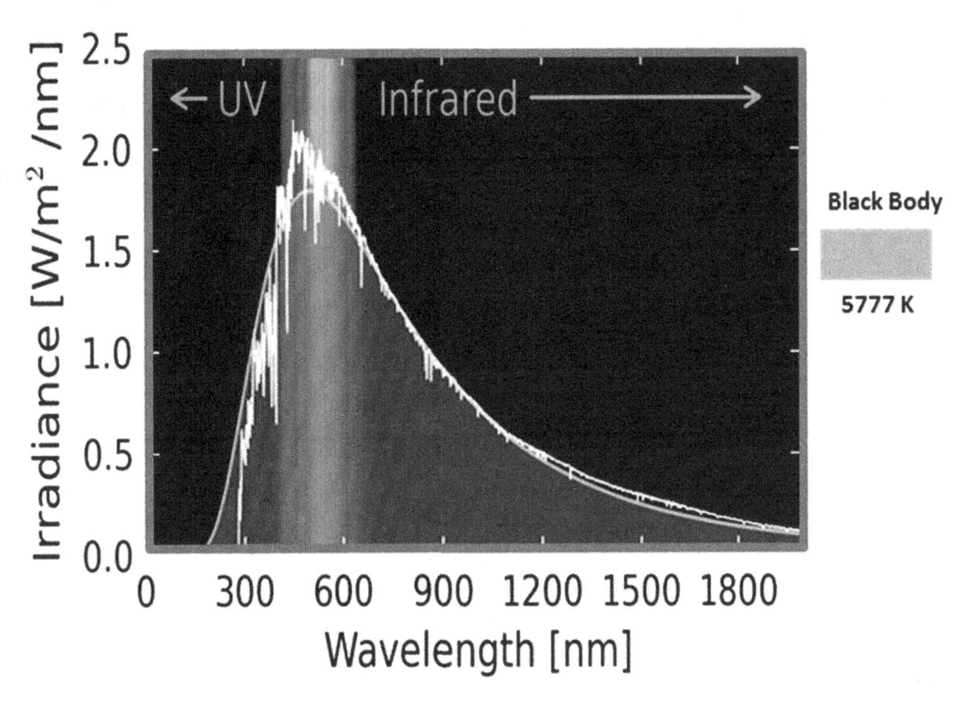

The curve in this figure gives the radiance *L* in the following dimensions:

$$\frac{Power}{unit\ \text{area.wavelength.solid angle}};\tag{6.15}$$

Or units of watts/(m^2 μ ster). The radiance equation is

$$Radiance = L = \frac{2hc^2}{\lambda^5}.\frac{1}{e^{\frac{hc}{\lambda kT}}-1};\tag{6.16}$$

Here, c=3 x 10^8 m/s, h=6.626 x 10^{-34} joules per second (J/s), and k= 1.38 x 10^{-23} joules per Kelvin (J/K). Real materials will differ from the idealized blackbody in their emission of radiation. The emissivity of the surface is a measure of the efficiency with which the surface absorbs (or radiates) energy and lies between 0 (for perfect reflector) and 1 (for a perfect absorber). A body that has ε=1 is called a "black" body. In the infrared, many objects are nearly black bodies-in particular, vegetation. Materials with ε<1 are called grey bodies. Emissivity ε will vary with wavelength (Robinson, 2004 and Goddijn-Murphy et al., 2018).

Another form of Planck's law can be given by

$$Radiant\ exi\tan ce{=}M{=}\frac{2\pi hc^2}{\lambda^5}.\frac{1}{e^{\frac{hc}{\lambda kT}}-1}.\frac{watts}{m^2\mu m}. \tag{6.17}$$

The difference is that the dependence on the angle of the emitted radiation has been removed by integrating of the solid angle. This can be done for black bodies because they are 'Lambertian" surface by definition-the emitted radiation does not rely on the angle, and $M{=}\pi L$ (Chelton et al.,2001). The power radiated is given by the Stefan-Boltzmann law:

$$R = \sigma\varepsilon T^4\ \mathrm{W/m^2}, \tag{6.18}$$

where R is the power radiated per square meter, ε is the emissivity, σ=5.67 x 10^{-8} $\mathrm{W/m^2}K^4$ which is Stefan's constant, and T is the temperature of the radiator in K (Brown,1977 and Marghany 2018). Wien's displacement law gives the wavelength at which the peak in radiation occurs:

$$\lambda_{max} = 2.898x10^{-3}(m/K)T^{-1} \tag{6.19}$$

Wien constant is $2.898x10^{-3}(m/K)$ for known temperature T in K, and λ_{max} is in meters.

SPECTRAL SIGNATURES

The spectral signatures produced by wavelength-dependent absorption provides the key to discriminating different materials into images of reflected solar energy. The property used to quantify these spectral signatures is called *spectral reflectance*: the ratio of reflected energy to incident energy as a function of wavelength. The spectral reflectance of different materials can be measured in the laboratory or in the field, providing reference data that can be used to interpret images. As an example, the illustration below shows contrasting spectral reflectance curves for three very common natural materials: dry soil, green vegetation, and water (Robinson, 2004).

The reflectance of dry soil rises uniformly through the visible and near-infrared wavelength ranges, peaking in the middle infrared range. It shows only minor dips in the middle infrared range due to absorption by clay minerals. Green vegetation has a very different spectrum. Reflectance is relatively lower in the visible range but is higher for green lighter than for red or blue, producing the green colour we see. The reflectance pattern of green vegetation in the visible wavelengths is due to selective absorption by chlorophyll, the primary photosynthetic pigment in green plants (Robinson, 2004 and Marghany 2018). The most noticeable feature of the vegetation spectrum is the dramatic rise in reflectance across the visible-near infrared boundary, and the maximum peak of near-infrared reflectance. Infrared radiation penetrates plant leaves, and is intensely scattered by the leaves' complex internal structure, resulting in high reflectance (Brown 1977). The dips in the middle infrared portion of the plant spectrum are due to absorption by water. Deep clear water bodies effectively absorb all wavelengths longer than the visible range, which results in very low reflectivity for infrared radiation (Lillesand et al., 2007).

Figure 18 Three dimensions for remote sensing

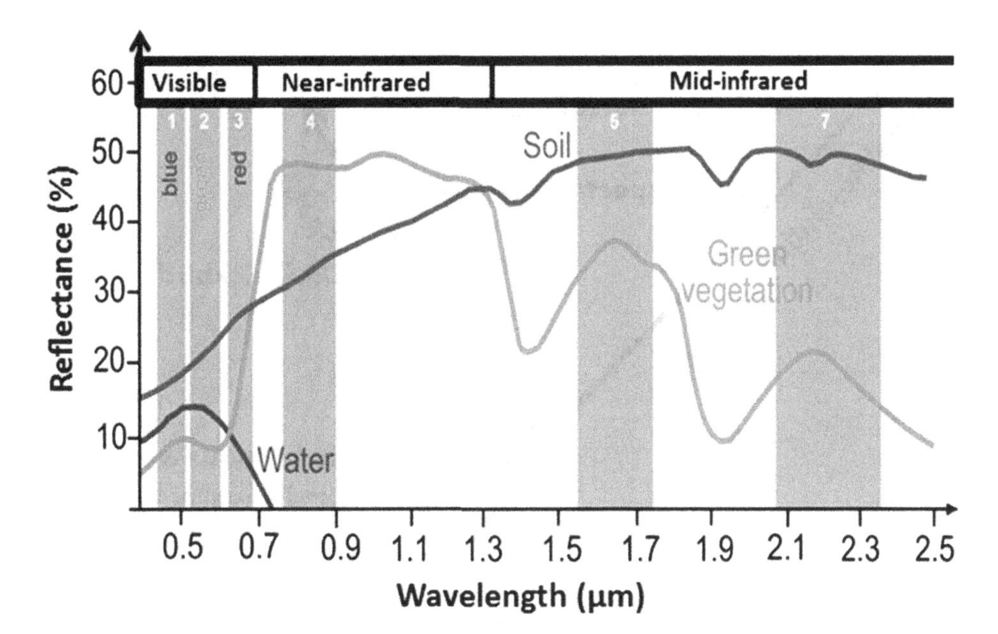

SPATIAL DIMENSIONS

The spatial dimension plays a great role for different satellite remote sensing to capture the clues the Flight MH370 debris, particularly, the three dimensions associated with remote sensing imagery: (i) spectral resolution; (ii) spatial resolution; and temporal resolution (Figure 19). In this regard, the three dimensions identify competing requirements for design and operation.

Spectral Resolution

The *spectral resolution* of a remote sensing system can be described as its capability to differentiate dissimilar parts of the range of determined wavelengths. Essentially, these quantities to the number of wavelength intervals ("bands") that are determined how the narrow spectra for each interval are. An "image" produced by a sensor system can contain one very wide-ranging wavelength band, a few broad bands, or many narrow wavelength bands. The names usually used for these three image categories are *panchromatic*, *multispectral*, and *hyperspectral*, respectively (Goddijn-Murphy et al., 2018).

Multispectral images determine even a finer variety of spectrum or wavelength that is required to replicate colour. That is why the multispectral data have a greater spectral resolution than regular colour data. Spectral resolution is the potential to unravel spectral features and bands into their separate components. The spectral resolution required through the analyst or researcher relies upon the utility involved. For instance, activity analysis for fundamental pattern identification generally requires low/ medium resolution.

The multispectral bands are listed as follows:

Figure 19. Three-dimensional projection of a hyperspectral cube

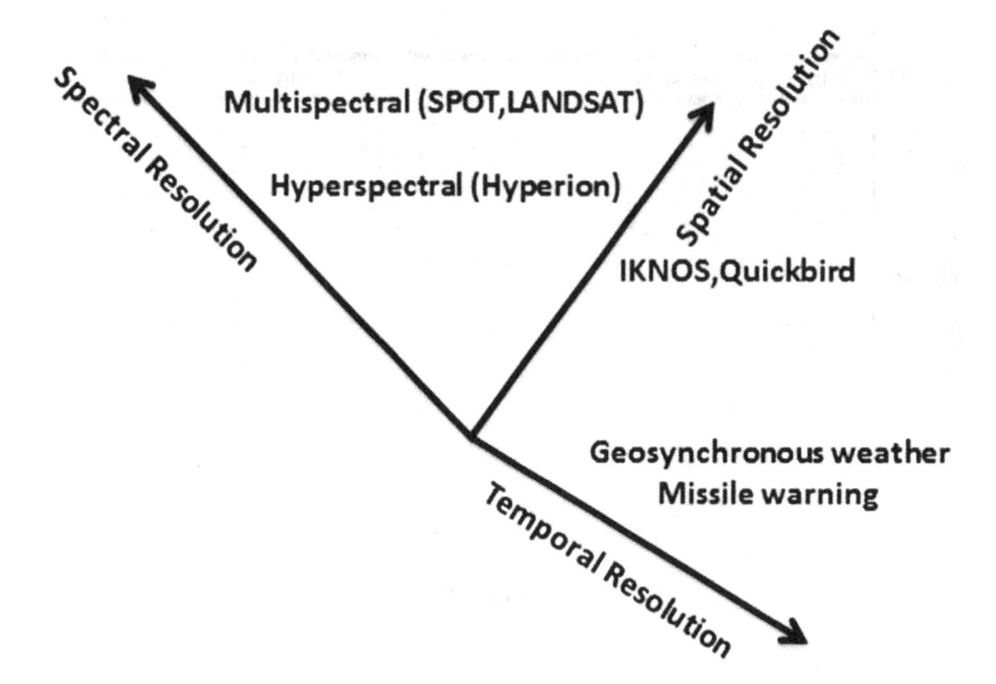

- **Blue**, 450-515 and 520 nm, is used for atmosphere and deep water imaging and can reach depths up to 150 feet (50 m) in clear water.
- **Green**, 515..520-590..600 nm, is used for imaging vegetation and deepwater structures, up to 90 feet (30 m) with clear water.
- **Red**, 600..630-680..690 nm, is used for imaging man-made objects, in water up to 30 feet (9 m) deep, soil, and vegetation.
- **Near-infrared**, 750-900 nm, is used primarily for imaging vegetation.
- **Mid-infrared**, 1550-1750 nm, is used for imaging vegetation, soil moisture content, and some forest fires.
- **Far-infrared**, 2080-2350 nm, is used for imaging soil, moisture, geological features, silicates, clays, and fires.
- **Thermal infrared**, 10400-12500 nm, uses emitted instead of reflected radiation to image geological structures, thermal differences in water currents, and fires, and for night studies.

The HRV sensor aboard the French SPOT (Système Probatoire d'Observation de la Terre) 1, 2, and 3 satellites (20-meter spatial resolution) has this design. The color-infrared film used in some aerial photography provides similar spectral coverage, with the red emulsion recording near infrared, the green emulsion recording red light, and the blue emulsion recording green light. The IKONOS satellite from Space Imaging (4-meter resolution) and the LISS-II sensor on the Indian Research Satellites IRS-1A and 1B (36-meter resolution) add a blue band to provide complete coverage of the visible light range and allow natural-colour band composite images to be created. The Landsat Thematic Mapper (Landsat 4 and 5) and Enhanced Thematic Mapper Plus (Landsat7) sensors add two bands in the middle infrared (MIR).

Like other spectral sensors, Hyperspectral sensor collects and processes data from throughout the electromagnetic spectrum (Figure 20). The conception of the hyperspectral sensor is to attain the spectrum for every pixel in the data of a scene, with the persistence of determining objects, identifying materials, or designing an algorithm (Robinson, 2004). Even though the human eye realizes the color of visible light with commonly three bands (red, green, and blue), spectral imaging divides the spectrum into many extra bands. This method of dividing data into bands can be extended beyond the visible. In hyperspectral imaging, the recorded spectra have fine wavelength resolution and cover a wide range of wavelengths (Marghany, 2018).

Figure 20. Sketch of Instantaneous Field of View (IFOV)

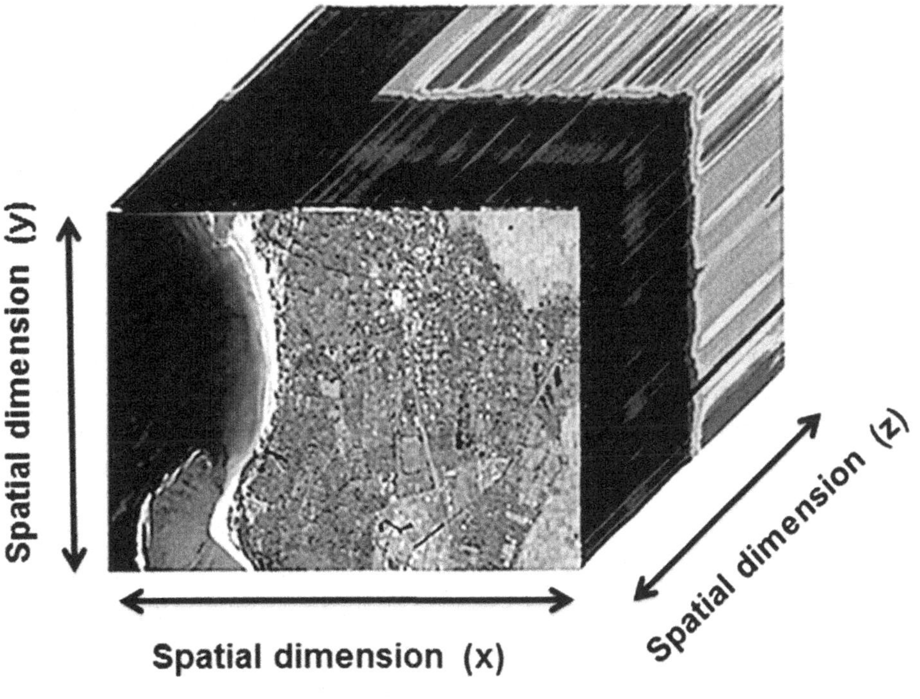

Spatial Resolution

Spatial Resolution pronounces how much information in a remote sensing data is seen with the human eye. The potential to resolve or separate small details are one way of the Flight MH370 is known as spatial resolution. Spatial resolution is a measure of the spatial detail in an image, which is a feature of the design of the sensor and its operating altitude above the surface. Each of the detectors with a remote sensor measures energy acquired from a finite patch of the Earth's surface. The smaller these individual patches are, the greater the detail which can be inferred from the data. For digital images, spatial resolution is most frequently expressed as the ground dimensions of an image pixel.

Further, it relies mainly upon their Instantaneous Field of View (IFOV). The IFOV is the angular cone of visibility of the sensor (A) and determines the area on the Earth's surface which is "seen" from a given altitude at one unique moment in time (B). The dimension of the location considered is determined by multiplying the IFOV by means of the distance from the ground to the sensor (C) (Figure 21). This location on the ground is referred to as the resolution cell and determines a sensor's more spatial resolution. For a homogeneous feature to be detected, its dimension usually has to be equal to or larger than the resolution cell. If the feature is smaller than this, it may not be detectable as the common brightness of all features in that resolution cell will be recorded. Nevertheless, smaller features may also on occasion be detectable if their reflectance dominates within an articular resolution cell, permitting sub-pixel or resolution cell detection (Zweck,2016).

The shape is one visual factor that we can use to recognize and identify objects in an image. The shape is usually discernible only if the object dimensions are several times larger than the cell dimensions. On the other hand, objects smaller than the image cell size may be detectable in an image. If such an object is sufficiently brighter or darker than its surroundings, it will dominate the average brightness of the image cell it falls within, and that cell will contrast in brightness with the adjacent cells.

Finally, there are three types of spatial resolution (i) low resolution; (ii) moderate spatial resolution satellite data; and (iii) high resolution. The low resolution ranges between 30 - > 1000 m while high resolution ranges from 0.41 - 4 m. for example QuickBird has high resolution of multispectral band with 2.4 m for pixel.

Temporal Resolution

Temporal resolution (TR) denotes the precision of a dimension as a function of time. In other words, the temporal resolution specifies the revisiting frequency of a satellite sensor for a particular location. It includes (i) excessive temporal resolution: < 24 hours — 3 days ; (ii) medium temporal resolution: 4 – 16 days; and (iii) low temporal resolution: > 16 days . In this understanding, the earth's surface change rates based on tsunami's effects can be monitored and estimated over different periods

The MODIS instrument, for instance, monitors the complete surface of the Earth every 1-2 days. Nevertheless, MODIS will monitor the poles more regularly than it will monitor a given location on the equator. In this regard, its temporal resolution is greater at the poles. Additionally, the repetition rate and the temporal resolution of the earth looking at satellites is 14-16 days (IKONOS: 14 days, LANDSAT 7: 16 days, SPOT: 26 days). On the contrary, meteorological satellites such as METEOSAT 8 with 15 min have extremely shorter repetition rates.

Finally, the temporal resolution, which defines the period of time required for a satellite platform to revisit a specific geographic location (also known as the revisit period). Therefore, it is an important factor for detecting changes, and the rate of these changes, occurring on the surface of the planet and supports the observations of environmental change, natural hazards, urbanization, deforestation, and weather systems. Thus, the actual temporal resolution of a sensor depends on a variety of factors, including the satellite/sensor capabilities, the swath overlaps, and latitude.

ESTABLISHING A NEW DEFINITION OF REMOTE SENSING FOR MH370 TRACKING

Consistent with the above perspective, it can establish a new definition of remote sensing. It can be said that remote sensing is the quantum information collected from any object on the ground or space due to reflection or backscatter of photons which is a function of the physical properties of the object.

For a specific definition of MH370 remote sensing, it can be said that it is a searching of the quantum information physical properties of MH370, debris, flaperon, and fuselage. This quantum information is presented somewhere in the Indian Ocean.

In other words, the ability to collect imagery of the same area of the Earth's surface at different periods of time is one of the most important elements for applying remote sensing data. In this understanding, spectral characteristics of features may change over time and these changes can be detected by collecting and comparing multi-temporal imagery. Nonetheless, there are not many satellite images that have been delivered during the Flight MH370 vanishing.

CONCLUSION

Remote sensing technology has been implemented to determine the logical debris belonging to MH370. From the point of view of the electromagnetic spectrum to physical characteristics of optical satellite sensors with high and low resolution, the MH370 debris can be identified in satellite images. In this view, the chapter delivers a new definition of remote sensing technology of MH370 as a specific and unique case. This definition is considered particular debris imagines in satellite images as quantum information which is presented somewhere in the Indian Ocean.

REFERENCES

Bakshi, U. A., & Godse, A. P. (2009). *Basic Electronics Engineering*. Technical Publications.

Brown, G. S. (1977). The Average Impulse Response of a rough surface and its applications. *IEEE Trans. Antennas Propag., 25*.

Campbell, J. B., & Wynne, R. H. (2011). *Introduction to Remote Sensing* (5th ed.). The Guilford Press.

Chelton, D. B., Ries, J. C., Haines, B. J., Fu, L. L., & Callahan, P. S. (2001). *Satellite Altimetry, Satellite altimetry and Earth sciences* (L. L. Fu & A. Cazenave, Eds.). Academic Press.

Goddijn-Murphy, L., Peters, S., Van Sebille, E., James, N. A., & Gibb, S. (2018). Concept for a hyper-spectral remote sensing algorithm for floating marine macro plastics. *Marine Pollution Bulletin, 126*, 255–262. doi:10.1016/j.marpolbul.2017.11.011 PMID:29421096

Hecht, E. (2001). *Optics* (4th ed.). Addison Wesley.

Lillesand, T. M. Kiefer, R. W., & Chipman, J. (2007). Remote Sensing and Image Interpretation (6th ed.). New York: John Wiley and Sons.

Marghany, M. (2018). *Advanced Remote Sensing Technology for Tsunami Modelling and Forecasting*. CRC Press. doi:10.1201/9781351175548

Robinson, I. S. (2004). *Measuring the oceans from space: the principles and methods of satellite oceanography*. Springer Praxis Books.

Zweck, J. (2016). Analysis of Methods Used to Reconstruct the Flight Path of Malaysia Airlines Flight 370. *SIAM Review*, *58*(3), 555–574. doi:10.1137/140991996

Chapter 7
Optical Satellite Sensors for Tracking MH370 Debris

ABSTRACT

This chapter reviews the optical satellite data around the tracking of MH370 debris. To this end, limited optical sensors are involved in Gafen-1, Worldview-2, Thaichote, and Pleiades-1A satellite data. Moreover, Google Earth data is also implemented to define debris that likely belongs to MH370. In doing so, automatic target detection based on its spectral signature is implemented to recognize any segment of MH370 debris. Consequently, most of the debris that has shown on satellite images does not belong to MH370. Needless to say, bright spots perhaps belong to the scattering of garbage floating in ocean waters or clouds.

INTRODUCTION

The critical question is, can the remote sensing technology assist to detect the Flight MH370 debris in the Southern Indian Ocean? Further, how does remote sensing technology deliver a logical explanation for the vanishing of MH370?

Particularly, the remote sensing techniques have incredible proficiencies for tracking and detecting the exact location of MH370 vanishing. The further potential of remote sensing is recognizing the capability of acquiring satellite data of huge areas, the satellite remote sensing can provide a synoptic view of the accumulation of the possible MH370 debris. Consequently, satellite technology has been broadly used for tracking the main route of MH370, detection of the search area, mapping, monitoring, and discrimination of MH370 debris from the surrounding environment. Moreover, satellite technology would play an important role in delivering further information about MH370 ending due to its "wide and insightful eyesight" in two observing similar airline crashes.

Contrariwise, satellite technologies have a great footprint and they can capture the large-scale areas of the ocean in a short period. In addition, satellite technology can detect object size less than 0.5 m. The Flight MH370 wingspan is approximately 60 m. In this understanding, the satellite within the resolution of 0.5 m and less than 60m would be to spot it if the debris is floating on the ocean surface. Nonetheless,

DOI: 10.4018/978-1-7998-1920-2.ch007

debris detection and identification are required a high-resolution satellite with less than 0.5 m. Debris appears in several bright dots or pixels in a satellite resolution of 10 m (Gorelick et al., 2017).

This chapter decisively evaluates the existence of satellite images that have spotted the possible MH370 debris across the Southern Indian Ocean. This can bridge the gap found between different theories of MH370 ending and the exact location of flight vanishing. Specifically, the surface results of MH370 possible debris have been analyzed by the use of optical satellite images, evaluating the acquisitions post to the mysterious finish of MH370. The investigation has targeted in the Southern Indian Ocean. This chapter focuses on available data for monitoring MH370 debris, which are Gafen-1, Worldview-2, Thaichote, and Pleiades-1A satellite data. Finally, Google Earth is discussed to track the possible locations of MH370 debris.

THEORETICAL OF TARGET DETECTION IN OPTICAL SENSORS

According to Gorelick et al., (2017), a spectral signature is a keystone for object detection in optical remote sensing images. In this respect, the spectral imaging is categorized into multispectral, hyperspectral, and ultraspectral. More willingly than UV, hyperspectral devices exploit reflections from hundreds of bands in the infrared range of the electromagnetic spectrum. However, cameras (and eyeballs) identify targets by their shape or by contrasts of light and dark, hyperspectral sensors can assemble reflections at various IR wavelengths and automatically determine the material that a target is made of.

MH370 debris detection and, optimistically, classification relies on consistent, exhaustive information about the distinctive spectral features of a debris, which vary from the surrounding environment(Mélin, and Vantrepotte, (2015). The more imaging bands available to examine the image scene, the higher the chances are of detecting and identifying anomalous MH370 debris in a natural environment (Morel, 1988;Babin et al, 2003; Emberton et al., 2016)

MH370 debris floating on the water surface restrain surface, leaving light in a number of behaviours, (i) downwelling light reflects (R) contrarily off debris (RMH370 debris) than off ocean surface (ROcean surface), (ii) transmittance of downwelling light through debris (RMH370 debris), is different from transmittance through the air-water interface (ROt), changing the underwater light (Rws), and henceforth the backscattered upwelling light (Rds), and (iii) subsurface upwelling light diffuses through debris differently than through the water-air interface (Figure 1). In the SWIR, pure water has absorption peaks near 1.45 μm, 1.94 μm and 2.95 μm. A thin film of water on the debris, therefore, can significantly reduce leaving the light of the debris (Babin et al, 2003 and Emberton et al., 2016).

Let us assume that the MH370 debris-covered area is ($A_{MH370\ debris}$), which projected in the field of view (FOV). The MH370 debris area fraction (f) is defined as:

$$f = \frac{A_{MH370\ debris}}{A_{ocean}} \tag{7.1}$$

where A_{ocean} is the total ocean surface area. Then the total radiance reflection $R_{total}(\lambda)$ from the ocean surface and MH370 debris area is received by the sensor as:

$$R_{total}(\lambda) = (1 - \frac{A_{MH370\ debris}}{A_{Ocean}})R_o(\lambda) + (\frac{A_{MH370\ debris}}{A_{Ocean}})R_{MH370\ debris}(\lambda) \quad (7.2)$$

here λ is the wavelength of reflected radiance. Equation 7.2 indicates the reflected radiance from MH370 debris ($R_{MH370\ debris}(\lambda)$) is a function of electromagnetic spectral, which interact with ocean surface (R_o) and MH370 debris.

Figure 1. Schematic of sunlight hitting the ocean surface and MH370 debris

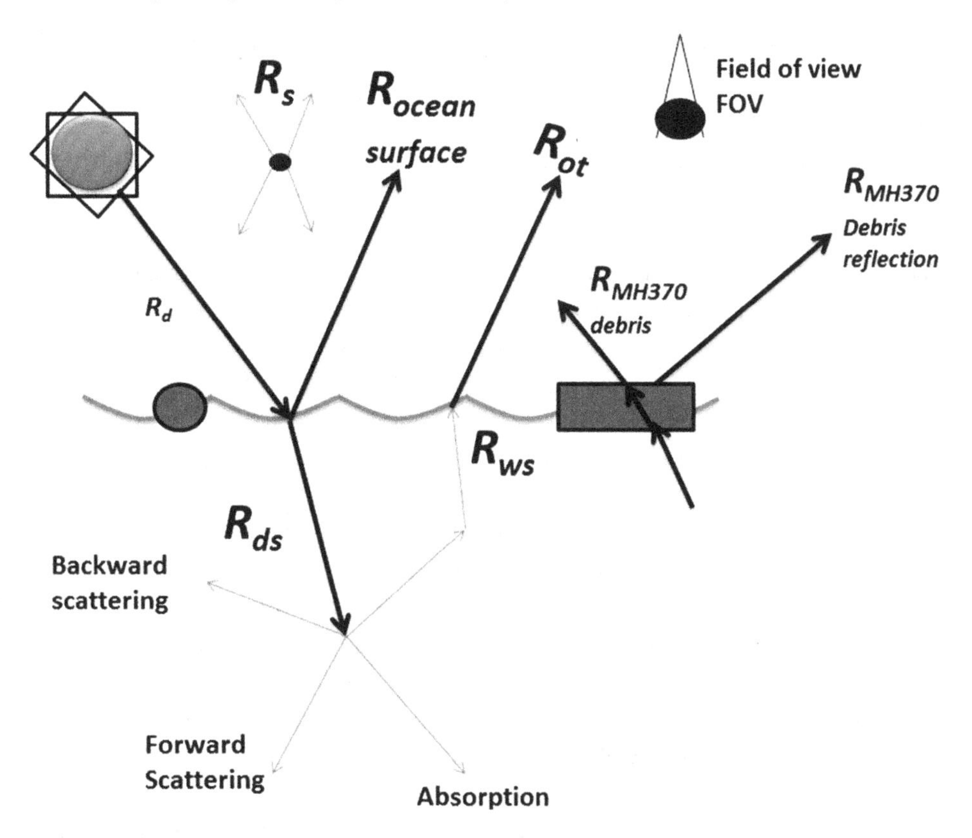

In other words, $R_{MH370\ debris}(\lambda)$ does not only denote MH370 debris reflected sunlight in the air in case of a semi-transparent debris. In this circumstance, underwater light (R_{ws}) transmitted through the debris also contributes to $R_{MH370\ debris}(\lambda)$ as:

$$R_{MH370\ debris}(\lambda) = \left(\frac{R_{debris}}{R_d}\right)R_d + \left(\frac{R_{debris}}{R_{ws}}\right)R_{ws} \quad (7.3)$$

Equation 7.3 demonstrates that $R_{MH370\ debris}(\lambda)$ is computed by the debris shape, surface roughness, and solidness in combination with the electromagnetic beam incident angle. In this understanding, the reflected radiation in the nadir view R_d, where its angular distribution is not uniform.

Conversely, R_d contains the solar beam and diffuse skylight, whose magnitudes rely on sky circumstances. These circumstances include clear, the cloudy and hazy sky in addition to the variation of sun elevation angle. In this retrospect, light diffuses as the sun is not at a zenith angle (Bier et al., 2018). Therefore, subsurface upwelling radiance R_{ws}, which is not reflected in the water – debris interface. In this sense, R_{ws} is transmitted through the debris area where radiance may be vanished owing to absorption, backward and forward scatterings.

ELECTROMAGNETIC SPECTRAL REGIONS FOR MH370 DEBRIS

In other words, when electromagnetic spectral reflects off MH370 debris, it produces a specific chemical signature unique to its material. If the signature is in a database of known materials, then a single pixel can provide enough information to identify a debris substance.

In the nadir direction, debris reflectance increases with an increasing fraction of diffuse skylight. In the NIR, the subsurface reflectance is< 1% OF wavelengths for most natural water types, but high for turbid waters where it is 1–2%.

MH370 debris is three dimensional and could return electromagnetic spectral back in the satellite sensor's view from its flanks, particularly when the debris is plunging and spinning on the ocean waves. Similarly, if 3-D debris shapes vary the illumination circumstance adjacent to the sea surface, they can influence separately other object's radiance reflectance. In these circumstances, the debris reflectance is also dependent on debris concentration (Morel, 1988).

Consequently, the debris surface is a specular reflector as well as a rough surface which is known as a diffuse reflection. Both skylight and solar radiation are scattered in the nadir view. In this understanding, if the debris surface is wet, the ocean water smoothes out the surface, which is reducing diffuse reflectance. Moreover, in the infrared band, water is absorbed. In this circumstance, a small layer of water can diminish the signal out of infrared wavelengths (Lee et al.,1998).

The subsurface water body is occasionally considered as a Lambertian reflector, where the reflected light is entirely diffuse and un-polarised. In this view, the reflectance of debris relies on the incidence angle of electromagnetic spectra. In truth, MH370 debris can have several forms with Indian Ocean surfaces at dissimilar angles, and downwelling radiance

at various angles are imitated in the nadir direction. However, debris floating on the sea surface pitch and roll with the waves, creating additional similarly scattered reflected electromagnetic radiation. Besides, downwelling light is not completely diffused. The angular distribution depends on the composition of sunlight of direct light and diffuse skylight, controlled by the solar elevation angle and sky conditions such as cloud cover (Gorelick et al.,2017).

In the visible region of the spectrum, the camouflage netting and healthy green vegetation is almost identical. In the NIR and SWIR regions, however, the spectra of the man-made materials and natural vegetation differ dramatically. Specifically, Spectral reflectance increases rapidly between 0.5μm and 0.6μm, reaching a maximum of around 30 to 40%. These features tend to support the premise that the fabric material used to assemble the distraction alters the characteristics of the energy reflected from its surface (Goddijn-Murphy et al.,2018).

The most striking similarities occur in the spectra are taken from the same target class their like spectral shape and similar reflectance values are most obvious in the regions from 0.7 μm to 1.23 μm and

1.5 μm to 1.7 μm. The shorter wavelengths are characterized by high reflectance values, where the latter region is dominated by a broad absorption feature (Gorelick et al.,2017 and Goddijn-Murphy et al.,2018).

MH370 Debris materials could be easily discernible in the visible region of the spectrum, but their signatures become less distinct long wards of 0.7 μm. In fact, MH370 fuselage, high intensity in the visible is probably due to the brighter hues

The shape or structure of the spectra at wavelengths near 0.92μm is dominated by electronic transitions in the d-cell electrons. At 1.0 μm, vibrational absorption occurs due to the presence of bound and unbound water molecules in the material. Hyperspectral imaging is generally described as a measurement of energy from both natural and man-made surfaces. The measured intensity as a function of wavelength (λ) creates a spectral record of a given material in hundreds of contiguous bands within a specific portion of the electromagnetic spectrum. For reflected solar energy, this spectral record is based upon the variations in reflectance between 0.4 μm to 2.5 μm. Knowledge of these physical processes allows us to understand the variations in spectral observed in MH370 debris (Goddijn-Murphy et al.,2018).

The hyperspectral devices identify MH370 debris makeup, not colour. It could spot the debris materials because of the difference in reflection between man-made and natural materials. A hyperspectral image will pop out at a different wavelength, which can discriminate between a false white colour of the white cape and the one that belongs to MH370 fuselage.

Specifically, near-infrared and shortwave infrared bands are the best at distinguishing between natural terrain and man-made objects. Other uses for these bands include finding areas of the disturbed earth—that would help scanners find tunnels and hidden improvised explosive devices by the moisture content of the soil. In disaster response, hyperspectral sensors could determine what areas have been flooded, trace the spread of spilled hazardous materials, and spot campfires of survivors in need. Long-wave IR sensors could be used to characterize effluents, allowing for the verification of missile launches, mapping toxic clouds, and spying factory plumes—that could prove useful in environmental monitoring, or making sure partners in international treaties are keeping their word (Lee et al.,1998 and] Emberton et al., 2016).

SATELLITE REMOTE SENSING

Satellite-based multispectral and hyperspectral ocean colour remote sensors play a tremendous role to discriminate between MH370 debris, man-made objects, and surrounding sea environment. In this respect, satellites that convey ocean colour devices regularly involve other devices that are advantageous for man-made object detection. For instance, the ocean and land colour instrument (OLCI) sensor with 21 bands in the visible-near infrared (VISNIR) (0.4–1.02 μm) on Sentinel-3, which functions in co-operation with Sentinel-3s SLSTR device containing 9 bands in the visible-short-wavelength infrared (VIS-SWIR) 0.55–12 μm.

In the NIR or SWIR, the water-leaving radiance is equal to zero. Nevertheless, the most commonplace atmospheric correction technique is the black pixel method. Conversely, the obtained measurements from these bands only involving aerosol atmospheric and ocean surface belongings. In this view, the correction would consequently expect to mask the indication in the NIR and SWIR from MH370 debris in the ocean. Nonetheless, an alternative atmospheric and sun glint rectification algorithm.

For instance, POLYMER algorithm derives ocean colour parameters in the entire sun glint spectrum and does not necessitate trivial water spectral reflectance in near-infrared bands. Specifically, POLY-

MER is established on a model, extended from 0.7 μm to 0.9 μm by exploiting the matched spectrum for turbid waters, which can remove the optical signal from sea surface debris.

However, the MH370 debris reflection can be corrupted by oceanic whitecaps, which are reflected in the solar spectral range. In this circumstance, it is preferred to use satellite data under wind speed is < 3 or 4 ms^{-1} and negligible whitecaps for MH370 debris detection.

Gaofen-1 Satellite

The Chinese government has launched two satellites:(i) Gaofen-1; and (ii) Gaofen-2 in 2013 and 2014, respectively. Both satellites are launched into space for a high-definition earth observation system (HDEOS). In the next ten years, conversely, another three or four satellites in HDEOS are presumed to be departed. At present, the images attained by GF-1 are accessible to the public in pursuit of endorsement.

On-board of the GF-1 satellites, there are four high spatial resolution (16 m) wide-field-of-view (WFV) cameras offering a repeating period of 4 days owing to their wide combined coverage (4 × 200 km). Currently, Feng et al., (2018) stated that the GF-1 satellite data have been implemented in various uses, involving exploring for proof in criminal circumstances and observing disasters, among various others, as stated by numerous majority media.

The GF-1 is in sun-synchronous orbits with a descending node, and its overpass times are approximately at 10:30 am local time of China. In this regard, a sun-synchronous orbit is achieved by having the osculating orbital plane precess (rotate) approximately one degree eastward each day concerning the celestial sphere to keep pace with the Earth's movement around the Sun (Figure 2).

Four WFV cameras are on board the GF-1 satellites with a combined swath of ~ 800 km. The revisiting period for GF-1 is ~ 4 days at the equator, which is enabling the detection of short-term changes in land surface features (Chen et al., 2018). Therefore, Chen et al., (2018) reported that GF-1 conveys two resolutions: (i) 2 m panchromatic; and (ii) 8 m multispectral high-definition cameras, and four 16 m resolution wide-angle cameras.

PMC (PAN and Multispectral Camera)

Furthermore, GF-1 has four spectral bands covering visible to NIR spectral ranges that are configured in the GF-1 WFV instruments (Table 1). In other words, GF-1 has the capability for observing in the visible range at a resolution of 2m in PAN, and of 8m in multispectral mode. The radiometric resolution of these bands is enumerated over 10-bit digital numbers (DN) (Chen et al., 2018).

Wide Field Imager (WFI)

Wide-field imager (WFI) is a medium-resolution push-broom camera set (4 cameras) with TDI (Time Delay Integration) capability observing in the VNIR range at a spatial resolution of 16m in multispectral mode (Table 7.2) (Padwick et al., 2010).

GF-1 Data for Possible MH370 Debris

China claimed that the GF-1 satellite was capable to track the MH370 debris at approximately 04:00 GMT on 18th March 2014. The suspected debris was located about 120 km south of the west of the first

Figure 2. Sun-synchronous orbits

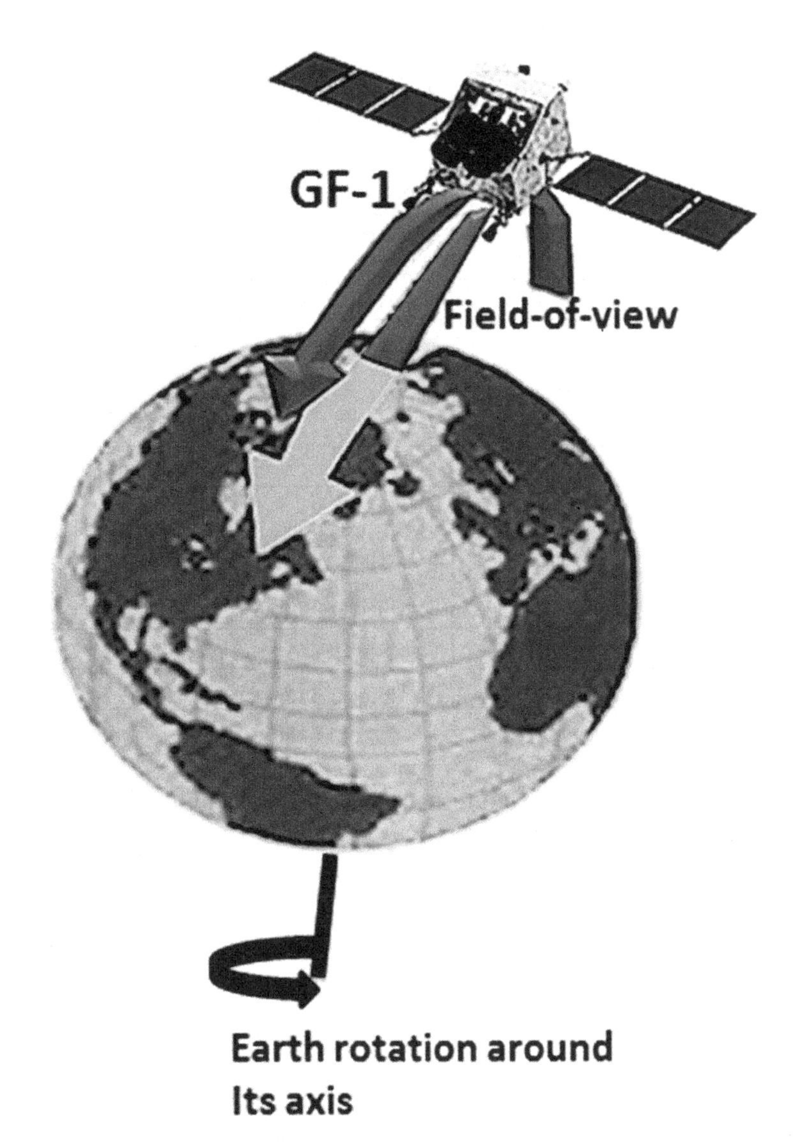

search area. The object was located along the latitude of 44° 57′ 30″ S between the longitude of 90° 13′ 40″ E and 90° 13′ 45′E (Figure 3). On the contrary, on 18 March 2014, this object was not clear in the search area.

Figure 3 indicates PAN band of GF-1 which covers the wavelength of 0.45-90 μm (Table 1). In this understanding, PAN band has a 2 m resolution, which allows detecting 22.5 m object length. In the upper-right part of the image, there is the existence of dense water vapor, which is imaged between the latitude of 44° 57′ 25″ S and 44° 57′ 30″ S. Over the uncontaminated ocean zone, which is far away from the mainland of Perth, Australia, the water and aerosol have a small effect on the radiance at the Top of Atmosphere (TOA) in the 0.443–0.670 μm wavelength bands. In this understanding, the TOA signal is largely caused by the Rayleigh scattering of gas molecules. The GF-1 sensor can be calibrated

Table 1. Physical Characteristics of GF-1 PMC

Physical Characteristics	Bands
Spectral bands	PAN: 0.45-90 μm B1/blue: 0.45-0.52 μm B2/green: 0.52-0.59 μm B3/red: 0.63-0.69 μm B4/NIR: 0.77-0.89 μm
Ground sample distance at the nadir	PAN: 2.0 m MS: 8 m
Swath width at the nadir	60 km

Table 2. Physical Characteristics of GF-1 WFI

Physical Characteristics	Bands
Spectral bands	PAN: 0.45-0.90 μm B1/blue: 0.45-0.52 μm B2/green: 0.52-0.59 μm B3/red: 0.63-0.69 μm B4/NIR: 0.77-0.89 μm
Ground sample distance at the nadir	16 m
Swath width at the nadir	800 km

as a function of the high precision of Rayleigh scattering radiative computations (Padwick et al., 2010 and Chen et al., 2018).

The length of an object is approximately 24 m and its width less than 12 m. However, this object cannot represent the flaperon. On the contrary, the size of the flaperon is less than the object size, which is imaged in GF-1 satellite data. Indeed, the flaperon dimension is 1.6 m x 2.4 m (Bier et al.,2018).

Digital Global for Tracking MH370 Debris

DigitalGlobe preserves and functions the agilest and complicated constellation of high-resolution commercial earth imaging satellites. Together, WorldView-1, GeoEye-1, WorldView-2, WorldView-3, and WorldView-4 are capable of assembling meticulously over one billion square kilometers of superiority imagery per year and propounding short period revisits around the globe. Consequently, DigitalGlobe confirmed that it delivered the Australian Maritime Safety Authority (AMSA) with the satellite data that reveal debris that perhaps are related to MH370. These satellite images are delivered by WorldView-2 within 12 days post MH370 missing.

WorldView-2 Satellite

The WorldView-2 sensor offers a high-resolution panchromatic band and 8 multispectral bands; 4 widespread hues (red, green, blue, and near-infrared 1) and four (four) new bands (coastal, yellow, red part, and near-infrared 2), complete-color images for improved spectral evaluation (Table 3), mapping and

Figure 3. Suspected MH370 debris by the GF-1 satellite

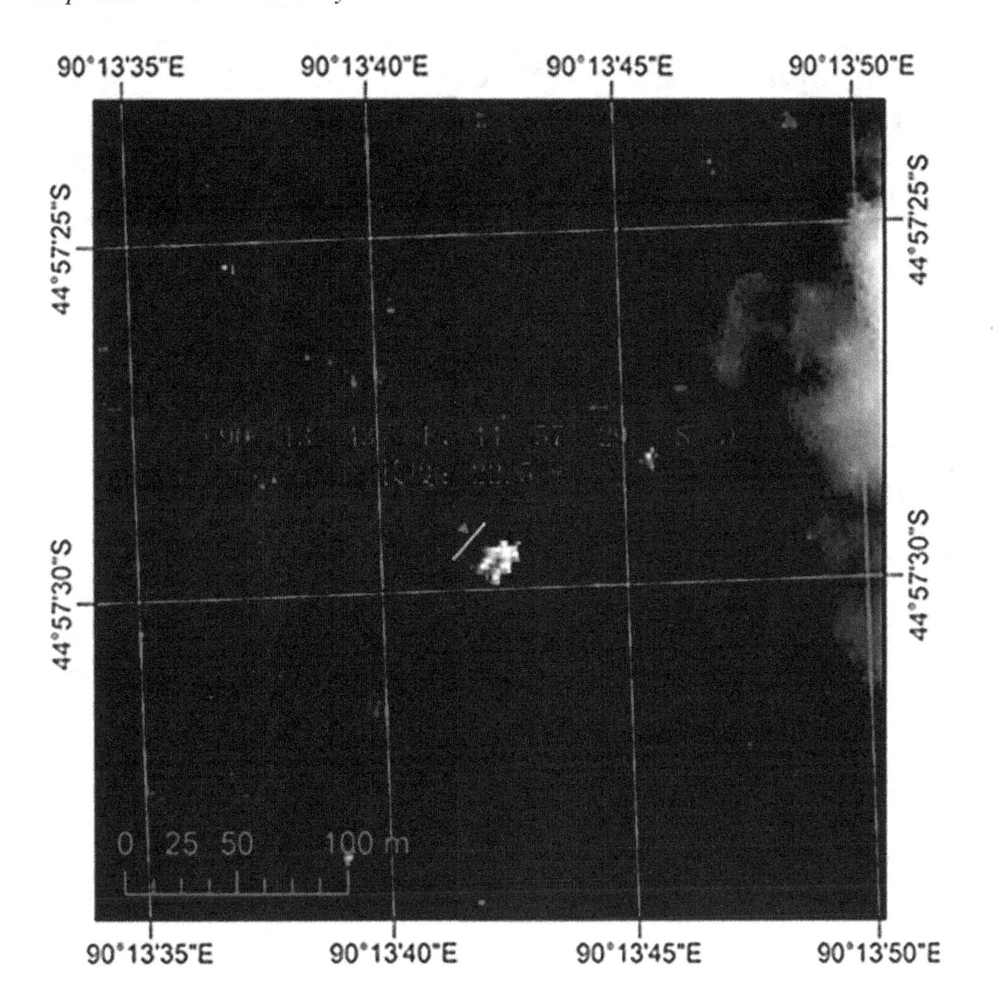

tracking applications, land-use planning, catastrophe alleviation, exploration, defense and intelligence, and visualization and simulation environments.

With its stepped forward agility, WorldView-2 is capable of acting as a paintbrush, sweeping backward and forward to gather very huge regions of multispectral imagery in a single bypass. WorldView-2 alone is capable of gathering almost 1 million km^2 each day, doubling the gathering potential of our constellation to almost 2 million km^2 per day. The aggregate of WorldView-2's expanded agility and excessive altitude permits it to generally revisit any place on the earth in 1.1 days, revisit time drops underneath one day, and never exceeds two days, offering the most identical-day passes of any commercial excessive-decision satellite (Padwick et al., 2010; Wolf, 2012; Novack et al.,2011).

The high-resolution spectral bands of Worldview-2 can discuss as follows (Figure 5):

Coastal Band (0.4-0.45 μm):

This band is used to identify the vegetation and analysis and offering information regarding bathymetric studies, which is based on its chlorophyll and water penetration characteristics. Similarly, this band

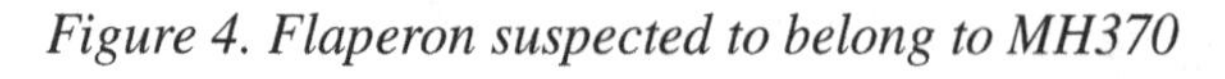

Figure 4. Flaperon suspected to belong to MH370

Table 3. Physical Characteristics of WorldView-2

Physical Characteristics	PAN	Multispectral
Spectral bands	0.45-0.8 μm	0.4-0.45 μm (coastal)
		0.45-0.51 μm (blue) 0.51-0.58 μm (green) 0.585-0.625 μm (yellow) 0.63-0.69 μm (red)
		0.705–0.745 μm (red edge)
		0.77–0.895 μm (near IR-1)
		0.86-0.9 μm (near IR-2)
Swath-width 16.4 km at nadir at the nadir	16.4 km at nadir	
Spatial Resolution	0.46 m GSD at Nadir 0.52 m GSD at 20 degrees off-Nadir	1.84 m GSD at Nadir 2.08 m GSD at 20 degrees off-nadir
Revisit Time	1.1 days at 1m GSD or less 3.7 days at 20 degrees off-nadir or less (0.52 meter GSD)	
Orbital Altitude	770 km	

is subject to atmospheric scattering and can be used to investigate atmospheric correction techniques (Padwick et al., 2010).

Yellow Band (0.585-0.625 μm):

Figure 5. Worldview-2 multispectral bands

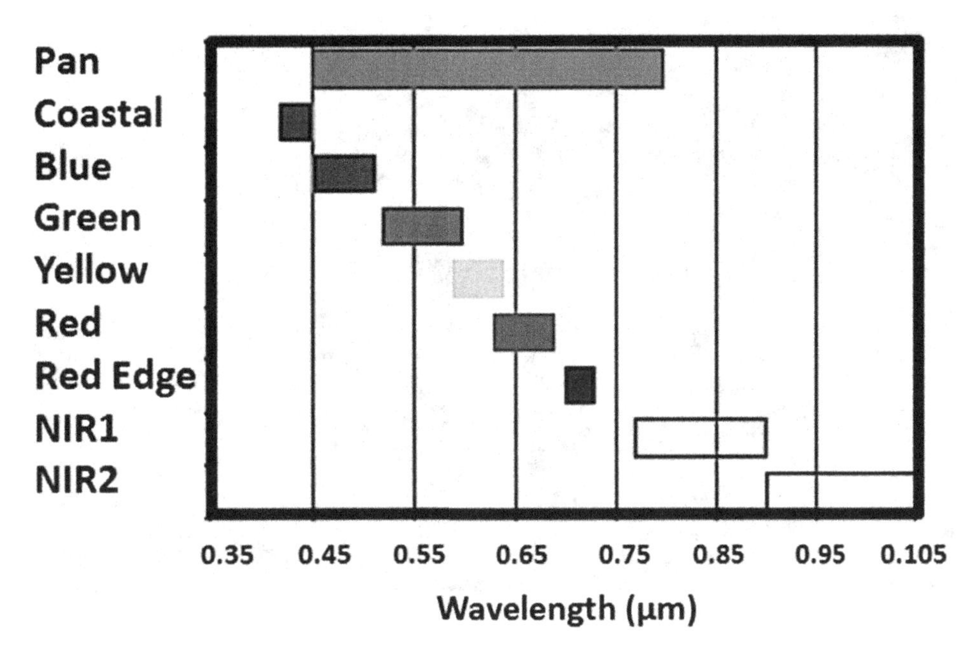

This band is used to recognize the "yellowness" condition of targets, important for vegetation applications. Correspondingly, this band supports the development of "true-colour" hue correction for human vision representation (Novack et al., 2011).

Red Edge Band (0.705-0.745 µm):

Red edge band-aids in the examination of vegetation characteristics. It is precisely related to plant health, which is revealed through chlorophyll production (Novack et al., 2011).

Near-Infrared Band (0.86- 0.104 µm):

This band overlaps the NIR 1 band, however, it is less precious by the atmospheric impact. It supports, therefore, vegetation investigation and biomass analyses (Padwick et al., 2010 and Novack et al., 2011).

Suspected MH370 Debris by Worldview-2

Tomnod, a company owned by the satellite provider DigitalGlobe, started a crowdsourcing campaign in which over two million volunteers have studied WorldView-2 images of the area. The search area was sliced up in many small images which every user was able to see and tag with four types: wreckage, oil slick, life raft, and other. Like other micro-tasking platforms, Tomnod uses triangulation to calculate areas of greatest consensus by the crowd [94]. In this context, Google Earth shows numerous suspected locations of MH370 debris (Figure 6).

A PAN band of Worldview-2 was acquired on 16th March 2014, shows three bright objects, which are suspected to be MH370 debris. The three objects are clearly distinguished from the surrounding sea

Figure 6. Worldview-2 data from suspected MH370 debris in Google earth

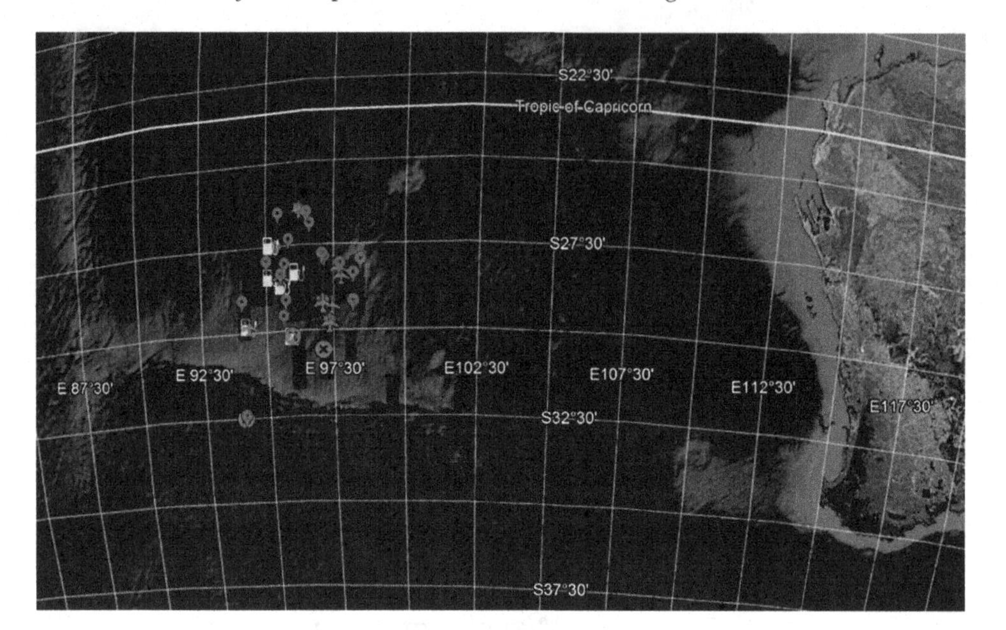

environment. The three objects are located between the latitude of 45° 58′ 34″ S and 45° 58′ 30″ S and longitude of 90° 57′ 37″ E and 90° 57′ 40″ E. Nevertheless, Gähler and Marghany (2013) stated that the information provided by the Charter was also wrong. In this view, it is very difficult to search small items even if the imagery has a spatial resolution of 0.5m in an unspecific area with the current Earth observation technology

Worldview-2 multispectral bands have pointed bright spot on March 16, 2014, along 43° 53′ 10″ S and 92° 19′ 54.94″E. The length of this object is about 5 m. Worldview-2 involves 0.40-0.45 μm of a coastal and blue band of 0.45-0.51, which can detect any object floating on the sea surface. Besides, the spatial resolution of the Worldview-2 is 1.84 m GSD at nadir, and 2.08 at 20° off-nadir, respectively, which allows imaging an object of 5m.

THAICHOTE SATELLITE

Thaichote is the first Earth observation satellite in Thailand. Three products are delivered from Thaichote satellite: (i) panchromatic; (ii) pan-sharpened; and (iii) Multispectral products. Thaichote multispectral products provide a 15m resolution (at nadir) and 8 bits information depth. All four bands are delivered as one file. The output scene is a square scene of 90 km. x 90 km. On the contrary, Thaichote imagery products are also available in a 4-band pan-sharpened product option. These products combine the visual information of 4 multispectral bands (blue, green, red, Near IR), with the spatial information on the panchromatic band. Pan-Sharpened products are available as a product option for level 2A only. The output scene is a square scene of 22 km. x 22 km.

Finally, Thaichote panchromatic products provide a 2m resolution (at nadir) and 8 bits of information depth. The output scene is a square scene of 22 km. x 22 km.

Figure 7. Suspected MH370 debris in Worldview-2 PAN band data

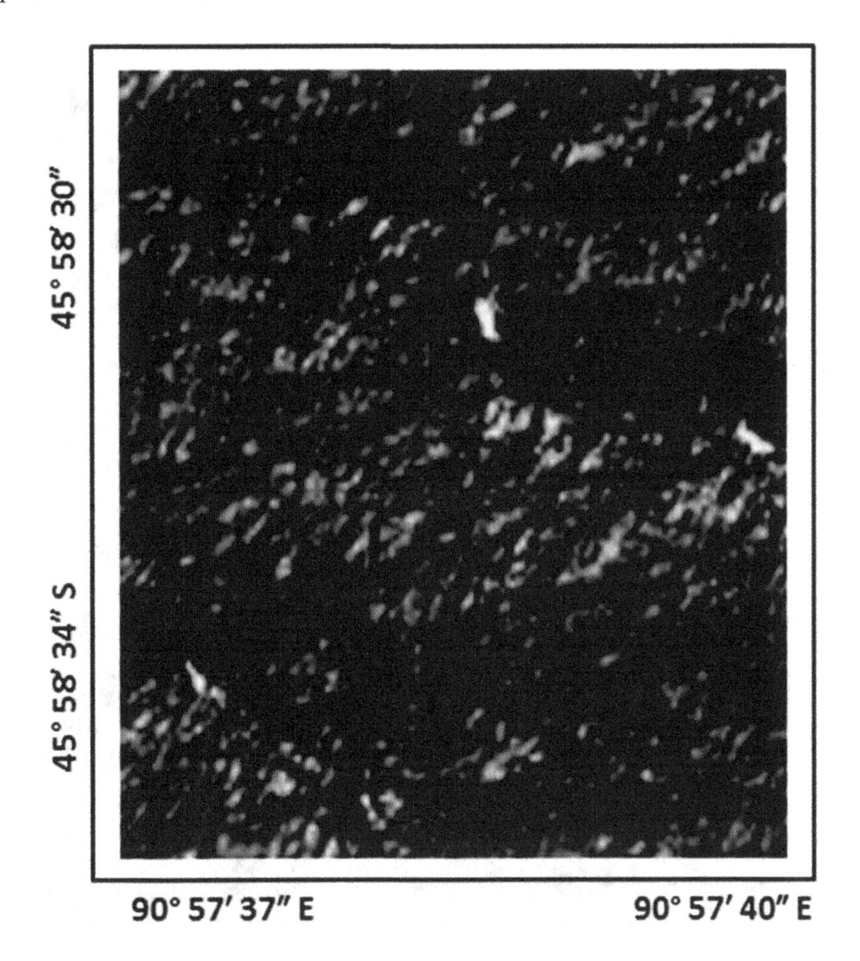

Possible Debris in Thaichote Satellite

Thaichote satellite spots suspected debris - about 300 objects floating in the southern Indian Ocean, where missing Flight MH370 may have crashed. The floating objects were identified by Thailand's Geo-Informatics and Space Technology Development Agency. It said that the objects ranged from 2m to 15 m in size, scattered over an area about 2,700 km southwest of Perth (Figure 9). The Thaichote image was acquired on March 24, 2014. However, the image indicates heavy cloud covers and stormy conditions. This makes great difficulties to identify any object belong to MH370. On the other hand, these bright spots, perhaps belong to the scattering of garbage floating in ocean waters. Despite Thaichote satellite spots approximately 300 objects (Figure 10), but no one of them belong to MH370 debris

It seems the Thaichote satellite data shown in Figures 7.9 and 7.10 one of the multispectral bands, i.e. NIR band (0.77 - 0.90 μm). In fact, this band able to track any object floating in the sea surface. This band in multispectral which has 15 m resolution. The spectral reflectance of this band is approximately 80% (Figure 11), which is suitable for metal and plastic trackings.

Figure 8. Possible MH370 in Worldview-2 multispectral data

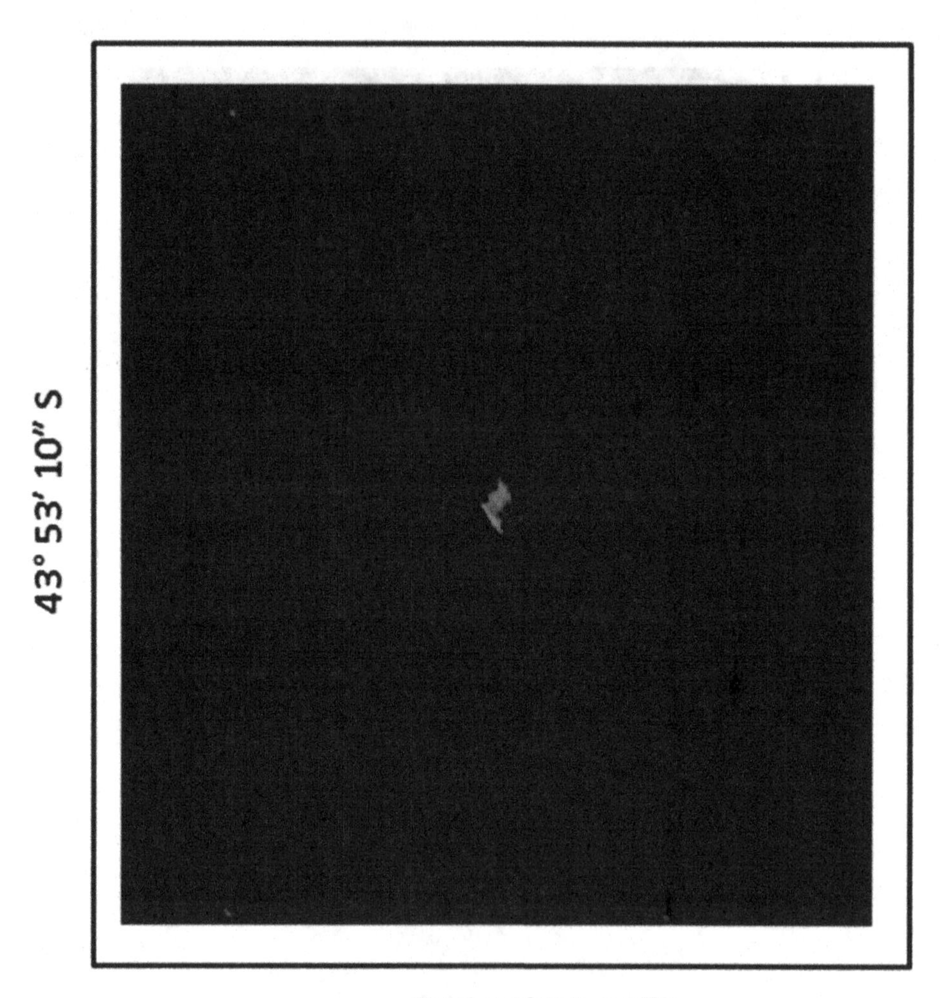

Table 7. Physical Characteristics of Thaichote Satellite

Parameters	Panchromatic	Multispectral
Wavelength	0.45-0.90 μm	B0 (blue): 0.45 - 0.52 μm B1(green): 0.53 - 0.60 μm B2 (red): 0.62 - 0.69 μm B3 (NIR): 0.77 - 0.90 μm
Resolution	2m	15 m
Swath-width	22 km. (Nadir)	90 km. (Nadir)

Figure 9. Possible MH370 debris in Thaichote satellite

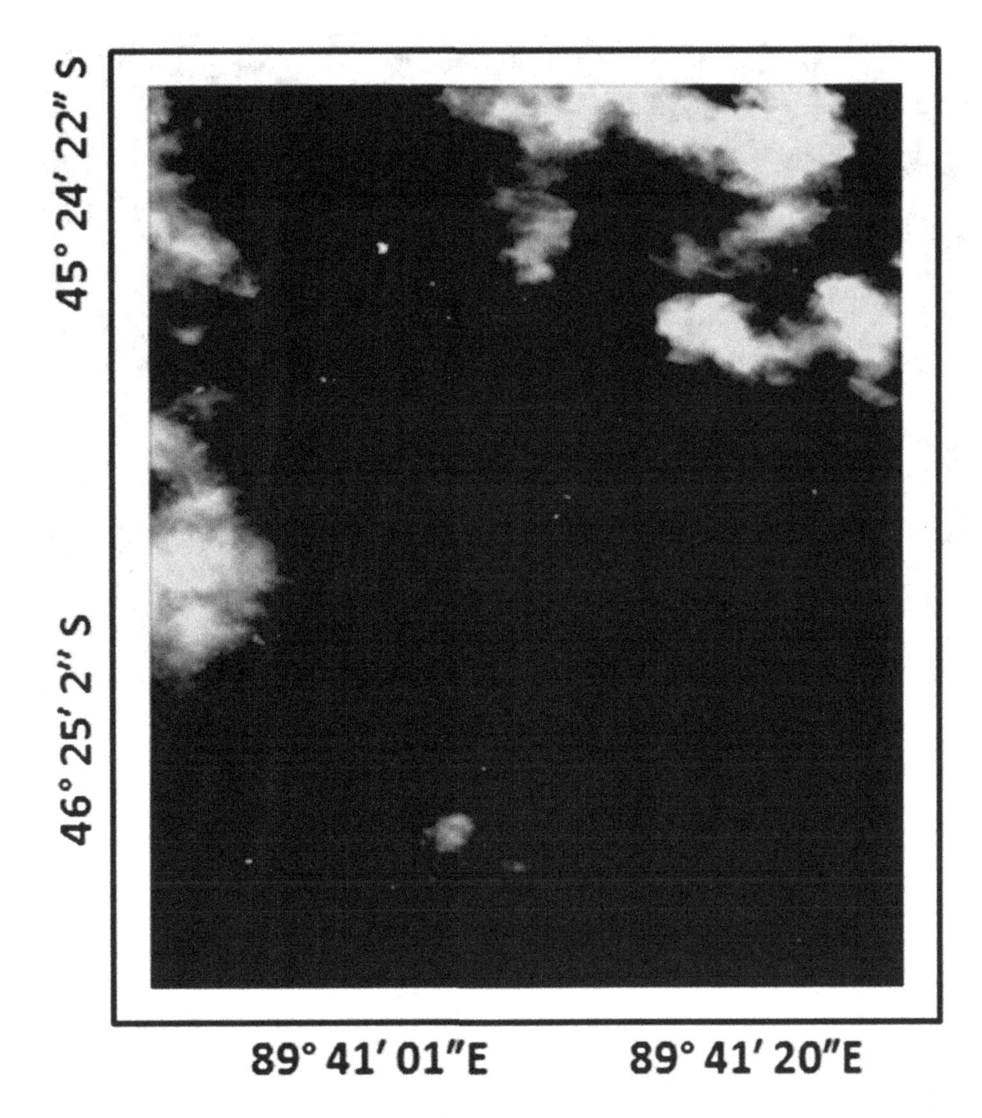

Pleiades Imagery

The Pléiades constellation is composed of two very-high-resolution optical Earth-imaging satellites. Pléiades-HR 1A and Pléiades-HR 1B can provide the coverage of the Earth's surface with a repeat cycle of 26 days. Designed as a dual civil/military system, Pléiades meet the space imagery requirements of European defense as well as civil and commercial needs.

The Pleiades-1A satellite is capable of providing orthorectified colour data at 0.5-meter resolution (roughly comparable to GeoEye-1) and revisiting any point on Earth as it covers a total of 1 million square kilometers (approximately 386,102 square miles) daily. Perhaps most importantly, Pleiades-1A is capable of acquiring high-resolution stereo imagery in just one pass and can accommodate large areas (up to 1,000 km x 1,000 km).

Figure 10. Bright spots scattered in Thaichote satellite

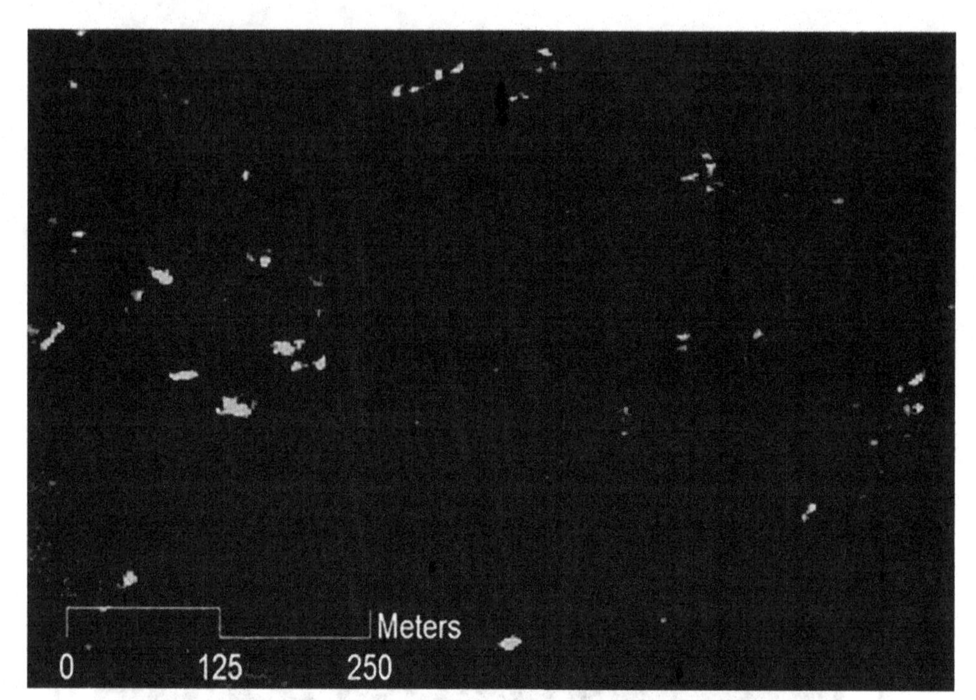

Figure 11. Spectral reflectance of Thaichote satellite

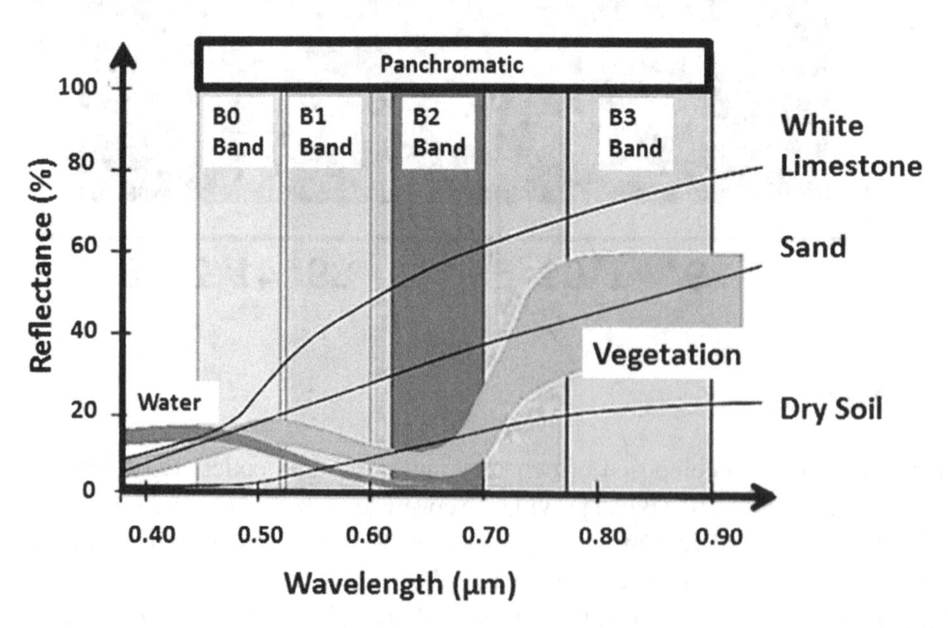

The Pleiades-1A satellite features four spectral bands (blue, green, red, and IR) (Table 5), as well as image location accuracy of 3 meters (CE90) without ground control points. Image location accuracy can be improved even further — up to an exceptional 1 meter — by the use of GCPs. Because the satellite has been designed with urgent tasking in mind, images can be requested from Pleiades-1A less than six

Table 5. Physical Characteristics of Pleiades Satellite

Parameters	Panchromatic	Multispectral
Wavelength	0.48-0.82 μm	B0(Blue): 0.450-0.530 μm B1(Green):0.510-0.590 μm B2(Red):0.620-0.700 μm B3(NIR): 0.775-0.915 μm
Resolution	0.7 m	2.8 m
Imaging Swath	20 km at nadir	

hours before they are acquired. This functionality will prove invaluable in situations where the expedited collection of new image data is crucial, such as crisis monitoring.

Provision of stereo imagery (up to 350 km x 20 km or 150 km x 40 km) and mosaic imagery of size up to 120 km x 120 km. More than 250 images/day is expected from each spacecraft of the constellation.

Possible MH370 Bebris in Pleiades 1A Satellite

Griffin et al. (2017) have analysed four images of 0.5m-resolution Airbus Pleiades 1A visible-channel images, each 25x20km, 100km apart and captured at ~4Z (midday Perth time) on 23 March 2014. These images (referred to here as PHR1-4) show 70 objects of interest, ranging from a 2m to a 12m in size, which are sorted into 5 categories: "probably natural", "possible natural", "uncertain", "possible man-made" and "probably man-made". Griffin et al. (2017) found the greatest number (9) of probable man-made objects in image PHR4 near 34.5°S, 91.3°E, but all 4 of the images contained some of the 28 "possibly man-made" objects. These images are all within 150km of the 7 the arc (to the west), between latitudes 34.5°S and 35.3°S (Figure 12). This is to the west of the southern half of the new search area recommended by the First Principles Review and proposed by ATSB in December 2016.

It may never be possible to know how many, if any, of these objects, originated from the accident aircraft, 9M-MRO. And if they were from the aircraft, we must also remember that the bulk of the other objects may not have been very close to the locations of the images. In the absence of any explanation of what these items are, we will postulate that at least some of them are pieces of 9M-MRO, and try to determine where they were on 8 March 2014, as a way of potentially refining our estimate of the location of the crash site.

In this context, the detected objects are claimed as probably man-made. The objects present geometric characters that do not match with wave patterns (Figure 13), or other probable natural phenomena, and are different from their surroundings (Griffin et al.,2017) .

It is difficult to conclude any confirmed knowledge according to MH370 debris across the confirmed crash area. Optical data were overpasses the crash site, which is dominated by the heavy cloud covers. This cloud cover obscures the aerial view in some neighborhoods. On the contrary, visual interpretation can be used to deliver precise information. Therefore, these digital data are required standard image processing algorithms to confirm either this debris is belonging to MH370 or not.

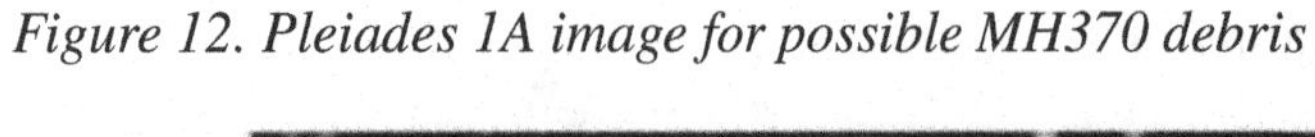

Figure 12. Pleiades 1A image for possible MH370 debris

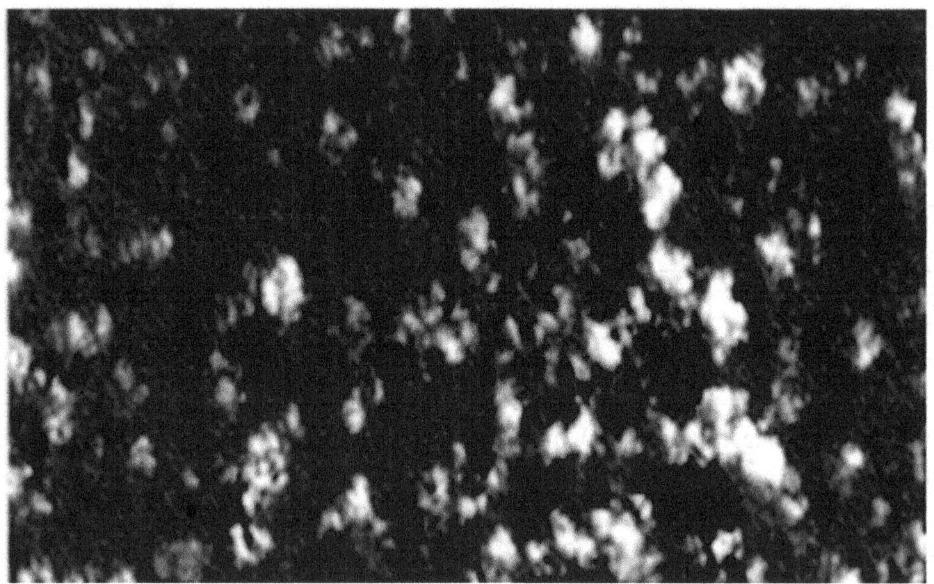

Figure 13. Possible man-made objects in Pleiades 1A data

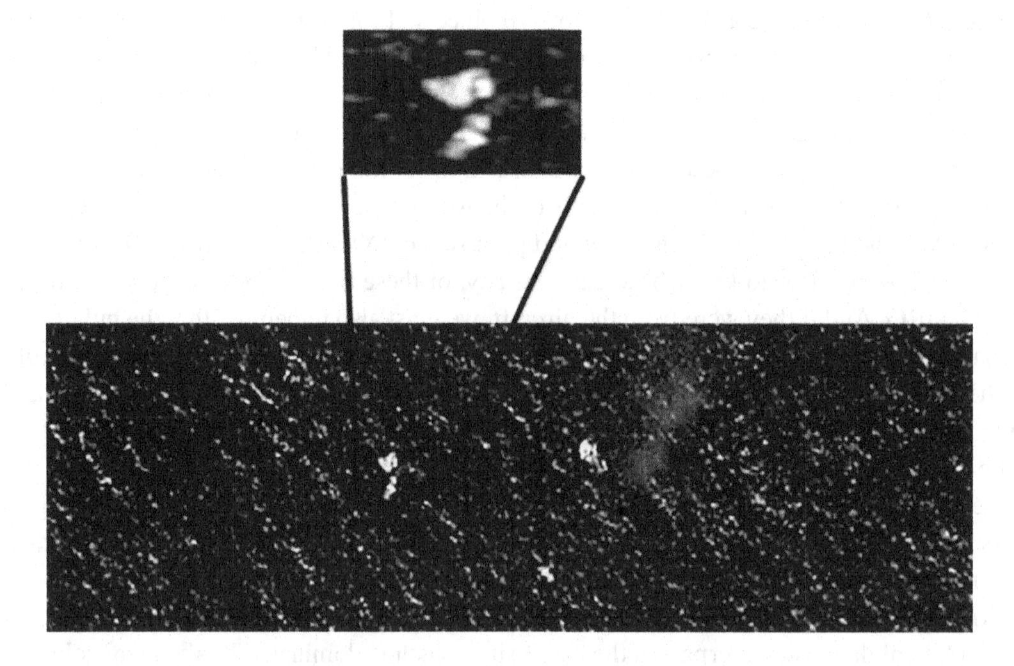

GOOGLE EARTH FOR POSSIBLE DEBRIS DATA

Google earth is a computer application that renders a 3-d illustration of earth primarily based on satellite imagery. This system maps the earth by using the usage of superimposing satellite pictures, aerial photographs, and GIS records onto a 3-D globe, allowing clients to look at towns and landscapes from numerous angles. Customers can explore the globe by moving into addresses and coordinates, or by using a keyboard or mouse. This device can also be downloaded to a cell phone or tablet, using a touchscreen or stylus to navigate. Customers might also use this gadget to add their very own data to the usage of keyhole markup language and upload them through numerous assets, which includes forums or blogs. Google earth is in a function to expose numerous varieties of images overlaid on the surface of the earth and is likewise a web map provider customer (Clarke et al., 2010; Brown, 2006; Patterson 2007).

Remote sensing provides a cheaper and real-time method mapping and monitoring of vast areas. A product such as Google Earth enables users to view remote areas anywhere on earth. The user can zoom into the complex to get a clearer view and get the physical address of the area. Further, interactive broadcasts-news, weather forecast, more recently Malaysia Airlines MH370 disappearance news to passenger relatives.

Virtual globes, Google Earth, in particular, permit scientists across the world to communicate the data and research discoveries in a spontaneous three-dimensional (3D) global viewpoint. Dissimilar from conventional GIS, practical globes are low cost and effluent to exploit in data visualization, gathering, and an investigation (Yu and Gong 2012).

In this analysis, visual inspection was utilized to identify areas with possible MH370 debris. Visual inspection is intuitive and the maps provided sufficient information to possible debris investigations (Figure 14).

Figure 14. The location of different satellite images scattered over the Indian Ocean in Google Earth

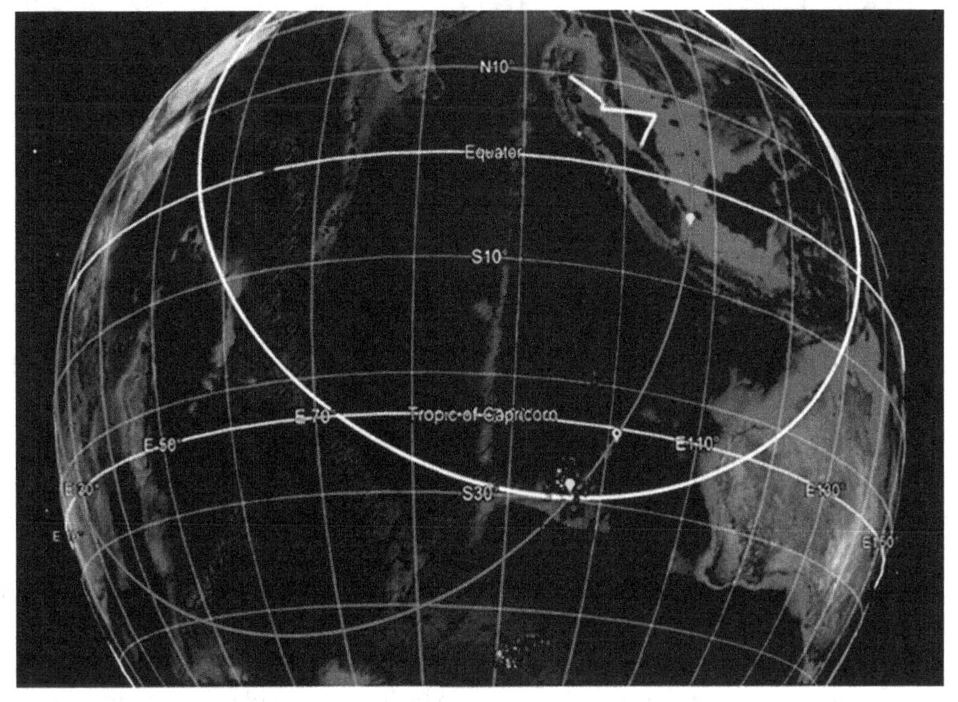

● **Location of different satellite data**

Figure 15. Black box pings location on Google Earth

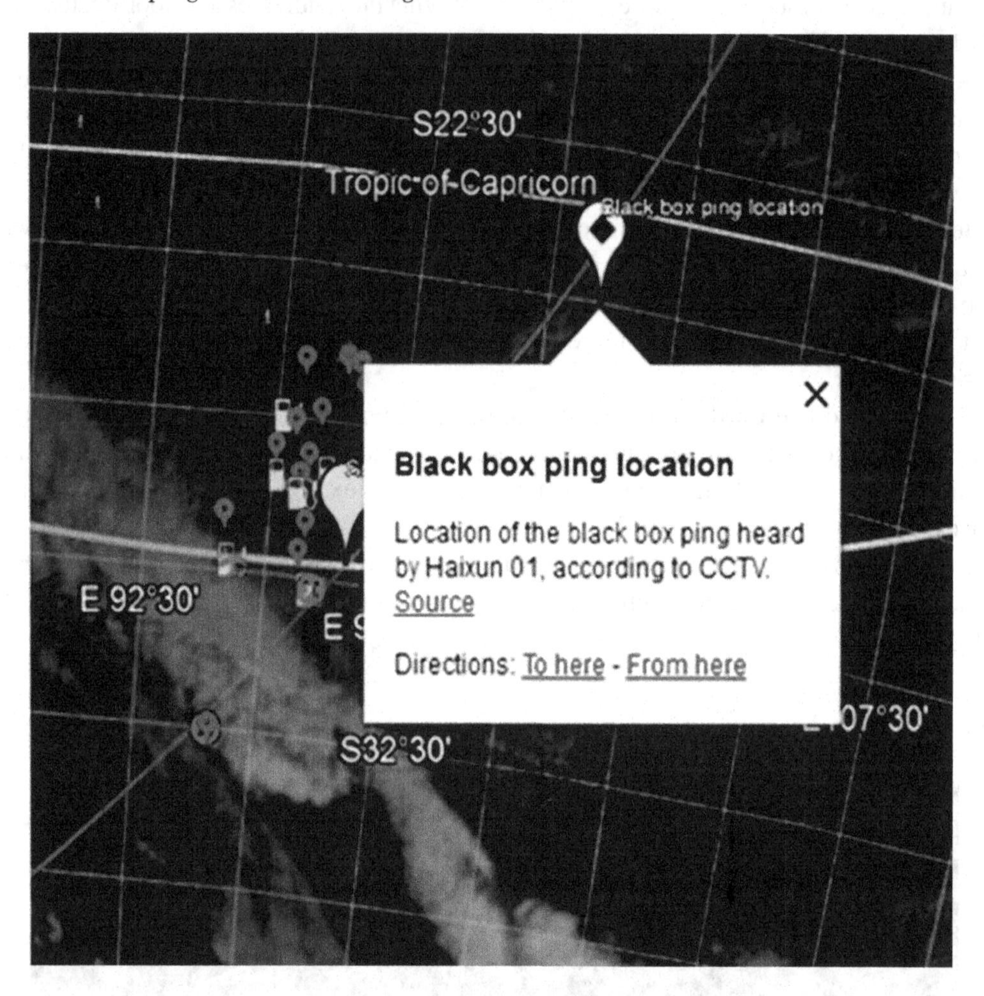

Moreover, Google Earth provides also the geographical location of the black box of MH370 which is claimed based on its ping location (Figure 15). However, the black –box cannot be found since the start of MH370 on March 8, 2014, until now. Furthermore, Google Earth also spots the patches of the oil slick, which are scattered across the search area (Figure 16).

Nevertheless, the Google Earth archive data require the surplus phases of image screen capture, the formation of a digital mosaic of the images, and georeferencing of the images with handheld GPS units. Conversely, these phases could potentially be circumvented if georeferenced possible debris images were already available (possibly as a donation from imagery providers) or possibly by writing or obtaining software to screen capture the whole image. Though should such georeferenced images or the screen capture software is not available, these phases are one decision to permit for georeferencing and integration of available satellite images. Moreover, this technique of creating a base map of MH370 data can be exercised utilizing any category of digital imagery, which are available, for instance, aerial photos.

Furthermore, an additional advantage of accomplishment georeferencing is that the data point accuracy can be improved to a positional accuracy of 5–6 m. Google Earth has a positional accuracy of 39.7 m (root-mean-squared error) (Chang et al., 2009).

Figure 16. Oil slick patches spot on Google Earth

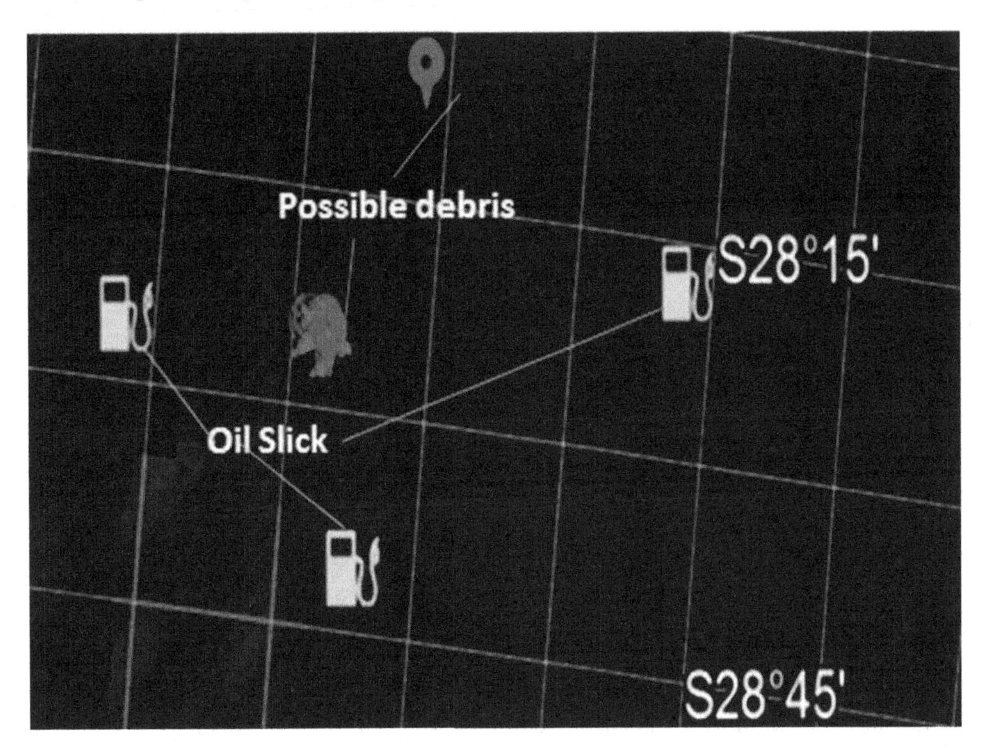

Therefore, Google Earth is often considered as a rapid source of information. It includes all the imagery for each country of the world. In other words, Google Earth is a database viewer that combines satellite data to reconstruct a three-dimensional model (3-D) of the ocean and the Earth's surface (Gorelick et al., 2017).

CONCLUSION

Few optical sensors have claimed the image of MH370 debris. These optical sensors involved Gafen-1, Worldview-2, Thaichote, and Pleiades-1A satellite data. Even Google Earth data was used to determine possible debris belonging to MH370. The investigation of the debris was done by using automatic target detection based on its spectral signature. Therefore, most of the debris that has existed on satellite images does not belong to MH370. In this view, bright spots, perhaps belong to the scattering of garbage floating in ocean waters or clouds.

The great difficulty concerning MH370 is that the scenario of its crash is still ambiguous. Besides, it is difficult to determine its debris in the Southern Indian Ocean due to the huge coverage of man-made debris. The next chapter will investigate the debris that is spotted by different optical satellite sensors using the advanced genetic algorithm.

REFERENCES

Babin, M., Morel, A., Fournier-Sicre, V., Fell, F., & Stramski, D. (2003). Light scattering properties of marine particles in coastal and open ocean waters as related to the particle mass concentration. *Limnology and Oceanography*, *48*(2), 843–859. doi:10.4319/lo.2003.48.2.0843

Bier, L. M., Park, S., & Palenchar, M. J. (2018). Framing the flight MH370 mystery: A content analysis of Malaysian, Chinese, and US media. *The International Communication Gazette*, *80*(2), 158–184. doi:10.1177/1748048517707440

Brown, M. C. (2006). *Hacking google maps and google earth*. Wiley Pub.

Chang, A. Y., Parrales, M. E., Jimenez, J., Sobieszczyk, M. E., Hammer, S. M., Copenhaver, D. J., & Kulkarni, R. P. (2009). Combining Google Earth and GIS mapping technologies in a dengue surveillance system for developing countries. *International Journal of Health Geographics*, *8*(1), 49. doi:10.1186/1476-072X-8-49 PMID:19627614

Chen, X., Xing, J., Liu, L., Li, Z., Mei, X., Fu, Q., Xie, Y., Ge, B., Li, K., & Xu, H. (2017). In-flight calibration of GF-1/WFV visible channels using Rayleigh scattering. *Remote Sensing*, *9*(6), 513. doi:10.3390/rs9060513

Clarke, P., Ailshire, J., Melendez, R., Bader, M., & Morenoff, J. (2010). Using Google Earth to conduct a neighborhood audit: Reliability of a virtual audit instrument. *Health & Place*, *16*(6), 1224–1229. doi:10.1016/j.healthplace.2010.08.007 PMID:20797897

Emberton, S., Chittka, L., Cavallaro, A., & Wang, M. (2016). Sensor capability and atmospheric correction in ocean colour remote sensing. *Remote Sensing*, *8*(1), 1. doi:10.3390/rs8010001

Feng, L., Li, J., Gong, W., Zhao, X., Chen, X., & Pang, X. (2016). Radiometric cross-calibration of Gaofen-1 WFV cameras using Landsat-8 OLI images: A solution for large view angle associated problems. *Remote Sensing of Environment*, *174*, 56–68. doi:10.1016/j.rse.2015.11.031

Gähler, M., & Marghany, M. (2016). Remote sensing for natural or man-made disasters and environmental changes. In *Environmental applications of remote sensing* (pp. 309–338). InTech. doi:10.5772/62183

Goddijn-Murphy, L., Peters, S., Van Sebille, E., James, N. A., & Gibb, S. (2018). Concept for a hyperspectral remote sensing algorithm for floating marine macro plastics. *Marine Pollution Bulletin*, *126*, 255–262. doi:10.1016/j.marpolbul.2017.11.011 PMID:29421096

Gorelick, N., Hancher, M., Dixon, M., Ilyushchenko, S., Thau, D., & Moore, R. (2017). Google Earth Engine: Planetary-scale geospatial analysis for everyone. *Remote Sensing of Environment*, *202*, 18–27. doi:10.1016/j.rse.2017.06.031

Griffin, D., Oke, P. R., & Jones, E. (2017). The search for MH370 and ocean surface drift-Part II. *CSIRO, EP172633*.

Lee, Z., Carder, K. L., Mobley, C. D., Steward, R. G., & Patch, J. S. (1998). Hyperspectral remote sensing for shallow waters. I. A semianalytical model. *Applied Optics*, *37*(27), 6329–6338. doi:10.1364/AO.37.006329 PMID:18286131

Mélin, F., & Vantrepotte, V. (2015). How optically diverse is the coastal ocean? *Remote Sensing of Environment, 160*, 235–251. doi:10.1016/j.rse.2015.01.023

Morel, A. (1988). Optical modeling of the upper ocean in relation to its biogenous matter content (case I waters). *Journal of Geophysical Research. Oceans, 93*(C9), 10749–10768. doi:10.1029/JC093iC09p10749

Novack, T., Esch, T., Kux, H., & Stilla, U. (2011). Machine learning comparison between WorldView-2 and QuickBird-2-simulated imagery regarding object-based urban land cover classification. *Remote Sensing, 3*(10), 2263-2282.

Padwick, C., Deskevich, M., Pacifici, F., & Smallwood, S. (2010, April). WorldView-2 pan-sharpening. In *Proceedings of the ASPRS 2010 Annual Conference, San Diego, CA, USA (Vol. 2630)*. Academic Press.

Patterson, T. C. (2007). Google Earth as a (not just) geography education tool. *The Journal of Geography, 106*(4), 145–152. doi:10.1080/00221340701678032

Wolf, A. F. (2012, May). Using WorldView-2 Vis-NIR multispectral imagery to support land mapping and feature extraction using normalized difference index ratios. In Algorithms and Technologies for Multispectral, Hyperspectral, and Ultraspectral Imagery XVIII (Vol. 8390, p. 83900N). International Society for Optics and Photonics. doi:10.1117/12.917717

Yu, L., & Gong, P. (2012). Google Earth as a virtual globe tool for Earth science applications at the global scale: Progress and perspectives. *International Journal of Remote Sensing, 33*(12), 3966–3986. doi:10.1080/01431161.2011.636081

Chapter 8
Aeroplane Detection Techniques in Optical Satellite Sensors

ABSTRACT

This chapter demonstrates an automatic detection approach for aeroplanes in optical satellite data. This chapter hypothesizes that aeroplane fuselage can be retrieved in satellite images. Aeroplane detection is a challenging task in remote sensing images due to its variable sizes, colours, complex backgrounds, and orientations. To this end, principle component analysis (PCA) and a deep belief network (DBN) are used to detect the MH370 flight. Needless to say that all detected targets are not segments of MH370.

INTRODUCTION

Maritime operational surveys for searching the disappearing Flight MH370 are notoriously challenging in the harsh ocean, but the satellite sensors could assist in closing the gap in detecting and distinguishing the MH370 debris from the surrounding environment. There is no doubt that the MH370 debris has remarkable shapes than other materials floating on the Indian Ocean s' surface. In continuing with Chapter 7, the different algorithms can detect an aeroplane from the optical satellite sensor. Can aircraft detect in satellite images? This chapter aims to answer this critical question.

This chapter hypothesizes that aeroplane fuselage can retrieve in satellite images. Aeroplane detection is a challenging task in remote sensing images due to its variable sizes, colours, complex backgrounds, and orientations. Conventional aircraft recognition methods focus on extracting the overall shape features of aircraft for recognition, which is too idealistic for targets in remote sensing images. In this chapter, numerous aeroplane detection approaches and feature extraction algorithms are investigated

It is needless to mention that remote sensing is highly useful for monitoring, identifying and mapping the accurate geographical location of an aeroplane on the earth's surface. It is a great potential for appraising, monitoring and mapping various parameters relating physical properties of the aeroplane fuselage.

DOI: 10.4018/978-1-7998-1920-2.ch008

GENERAL PROCEDURES FOR DETECTION OF AEROPLANE IN SATELLITE DATA

In high-resolution satellite images, aeroplane recognition is proved to be a challenging task because of its multifaceted structure, variable dimensions, colours, and orientations (Deepa and Kala 2017). In this view, geometrical shape, image background, and image gradient across the aeroplane are such parameters, which impact the detection of the aeroplane through an image processing tool.

The main approach for aeroplane detection is a function of the aeroplane shape features. This method is considered extremely idealistic for object detection in remote sensing data. On the other hand, this approach is controlled by different aeroplane shape types. In this context, this approach must be based on the specific templates, which involve physical characteristics for each sort of aeroplane. In this regard, it can be easy to match the detected object to the different template kinds (Vaijayanthi and Vanitha, 2015). In this procedure, the similarity between objects does not rely on the overall shape detection. Consequently, this approach can detect aeroplane persistently deprived of impeccable abstraction of the frame or shape of targets in the function of a circumstance. For instance, it relies on the circumstance of parts missing and shadow disorder.

Generally, the recognition approach involves: (i) possibly refocusing targets are primarily notorious on time-series satellite images; (ii) the object is then retrieved by computing both spectral and spatial features; and (iii) lastly, matching procedure uses to distinguish an aeroplane for precise detection (Figure 1). Moreover, there are also other identification procedures, which is based on computing the direction post binarization (Wuet al., 2015), and then classify the aeroplane kind. Nonetheless, these procedures also require the binary satellite images of each aeroplane kind as a prerequisite for direction evaluation, which reduces the feasibility. Moreover, aeroplane recognition often experiences numerous disorders, for example, dissimilar contrasts, clutter, and homogeneity strength (Dudani e al., 1997).

DIMENSIONALITY REDUCTION

Another approach is based on dimensionality reduction. In this regard, it is the procedure of the reducing quantity of random variables contained in the target. Both feature selection and extraction are keystone procedures. According to this approach, feature selection attempts to locate a separation of the original changeable. For example, the detection of aeroplane tire is constrained by lower DN or grey value than another part of an aeroplane (Figure 2). Feature extraction, consequently, transforms the higher dimensions, images into lower dimensions. To this end, the Principal Component Analysis (PCA) can achieve dimensionality reduction of the satellite data. PCA transforms a set of possibly correlated variables into a set of values of linear interrelated variables. In this circumstance, the PCA quantity must be less than or equal to the number of original variables. However, PCA cannot detect automatically the MH370 debris in the PLEIADES 1 A satellite data (Figure 3a). The true colour image is shown in Figure 3a, while the result of PCA is shown in Figure 3b. However, PCA highlight objects that are different from their surroundings, but they do not belong to Flight MH370.

Figure 1. Identification procedures of an aeroplane in satellite images

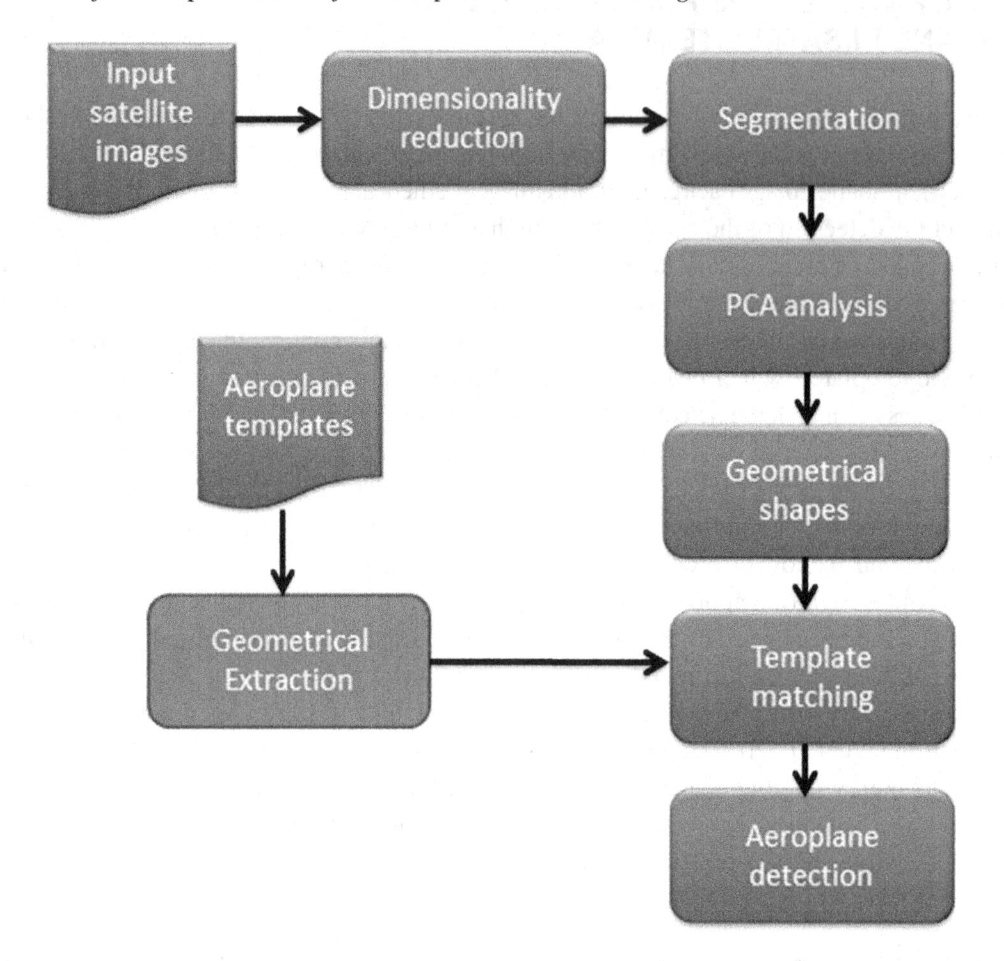

OTSU THRESHOLDING ALGORITHM

The automatic detection of the aeroplane can be achieved by image segmentation. Image segmentation divides the satellite remote sensing images (8.4a) into several segments (Figure 4b). In other words, segmentation is implemented to simplify and alternate the illustration of satellite data to acquire accurate image analysis. In this regard, unsupervised Otsu's thresholding technique is taken into consideration to partition an image into non-intersecting clustering. Otsu's approach, therefore, is automatically detected in several clusters, which, is based on image thresholding or reduction of the grey level image to a binary image. Formerly, histogram probability thresholding locates the specific object from the heritage.

In Otsu's approach, we systematically explore for the threshold, which, minimizes the variance within the cluster, expressed as a weighted sum of variances of the two clusters, which, are defined as:

$$\sigma_w^2(t) = Q_1(t)\sigma_1^2(t) + Q_1(t)\sigma_2^2(t) \quad (8.1)$$

here σ_1 and σ_2 are variances of the two clusters, Q_1 and Q_2 the cluster probabilities, which are estimated via:

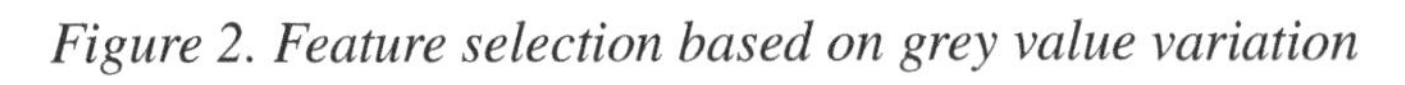

Figure 2. Feature selection based on grey value variation

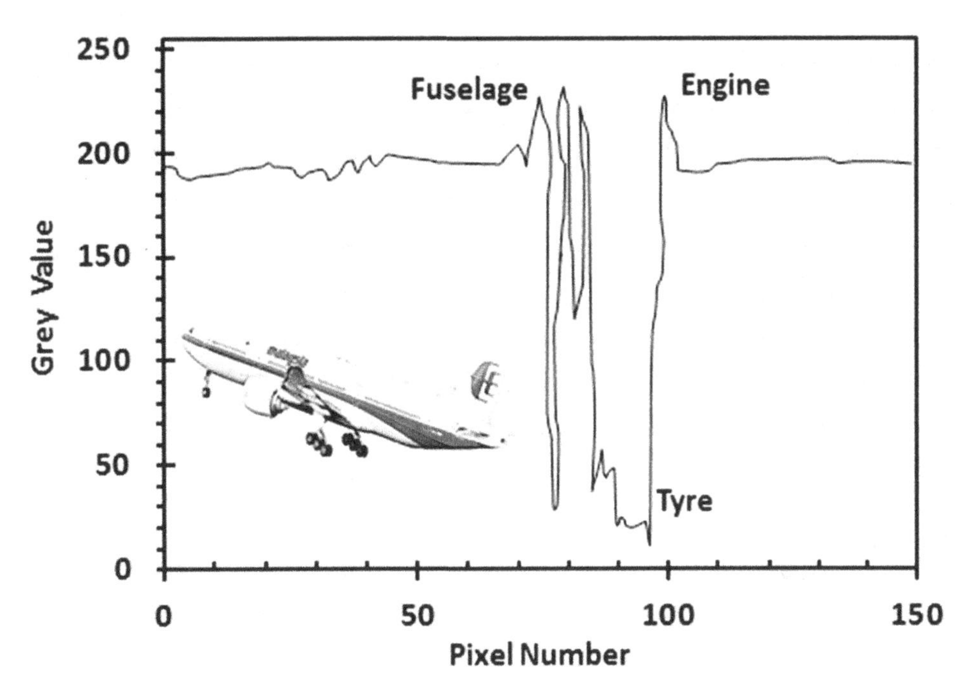

Figure 3a. Feature selection based true colour image

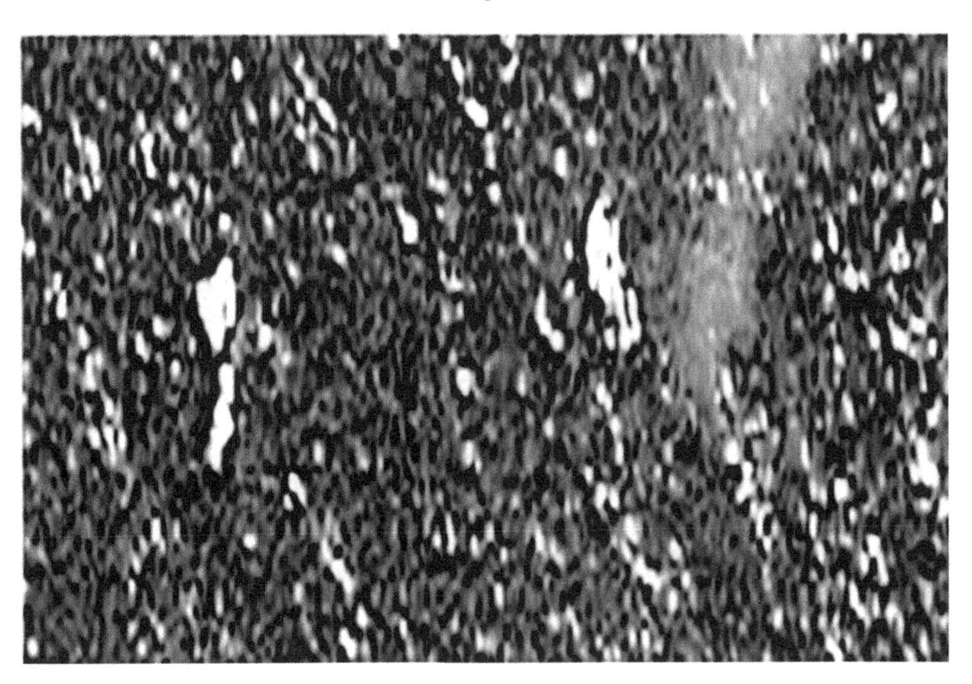

Figure 3b. PCA

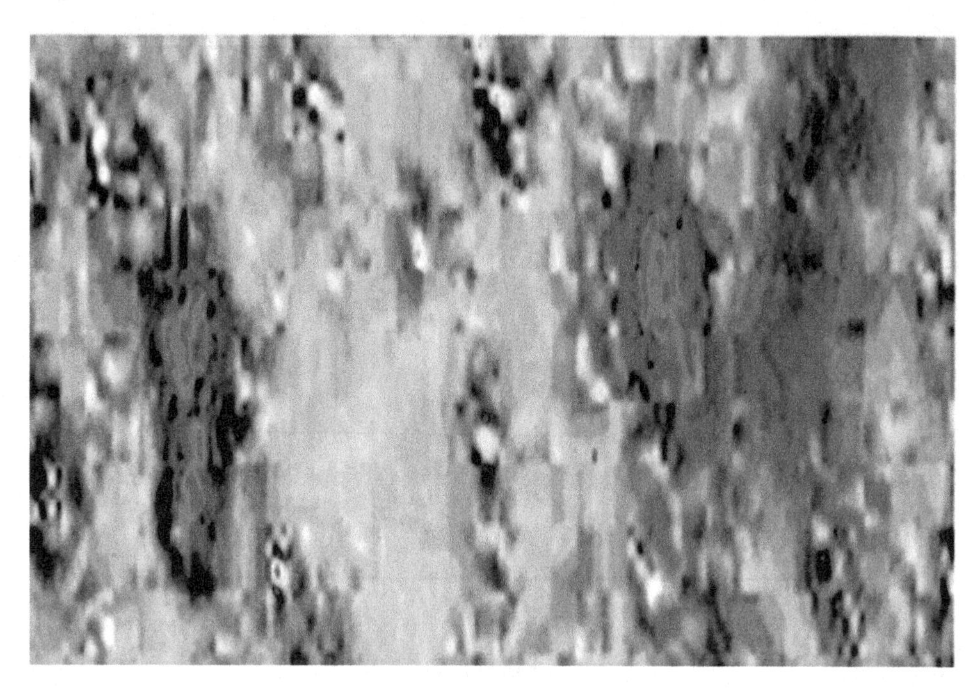

Figure 4a. Example of image segmentation true colour image

Figure 4b. Segmentation

$$Q_1 = \sum_{i=0}^{t-1} p(i) \tag{8.2}$$

$$Q_2 = \sum_{i=t}^{L-1} p(i) \tag{8.3}$$

where p is the probability, and t is a threshold, which is estimated from L bins of the histogram? In other words, Otsu's approach can formulate in term of cluster means μ as:

$$\mu_1(t) = \frac{\sum_{i=0}^{t-1} ip(i)}{\sum_{i=1}^{t} p(i)} \tag{8.4}$$

$$\mu_1(t) = \frac{\sum_{i=0}^{L} ip(i)}{\sum_{i=t+1}^{L-1} p(i)} \tag{8.5}$$

being μ_1 and μ_2 are the means of two clusters, respectively. Equations 8.4 and 8.5 demonstrate that the cluster probabilities and their means can be calculated iteratively (Otsu, 1979). The pseudo-code of the Otsu algorithm is shown in Figure 5.

Calculate histogram and probabilities of every grey level,

Figure 5. The pseudo-code of the Otsu algorithm

Calculate histogram and probabilities of every grey level,
Create initial $Q_i(0)$ *and* $\mu_i(0)$,
Determine through all conceivable thresholds $t=1, \ldots\ldots$ *maximum intensity*
Apprise Q_i *and* μ_i
Calculate $\sigma_w^2(t)$
Preferred threshold matches to the maximum $\sigma_w^2(t)$

Create initial $\omega i_(0)$ and $\mu i_(0)$,
Determine through all conceivable thresholds t=1,…… maximum intensity
Apprise ωi and μi
Calculate $\sigma_w^2(t)$
Preferred threshold matches to the maximum $\sigma_w^2(t)$

Since the entire is continuous and self-determining of *t*, the result of changing the threshold is merely to transfer the influences of the two terms back and forth. Hence, minimizing the within-cluster variance is similar to maximizing the between-cluster variance.

Figure 6. Histogram of the Flight MH370 grey level

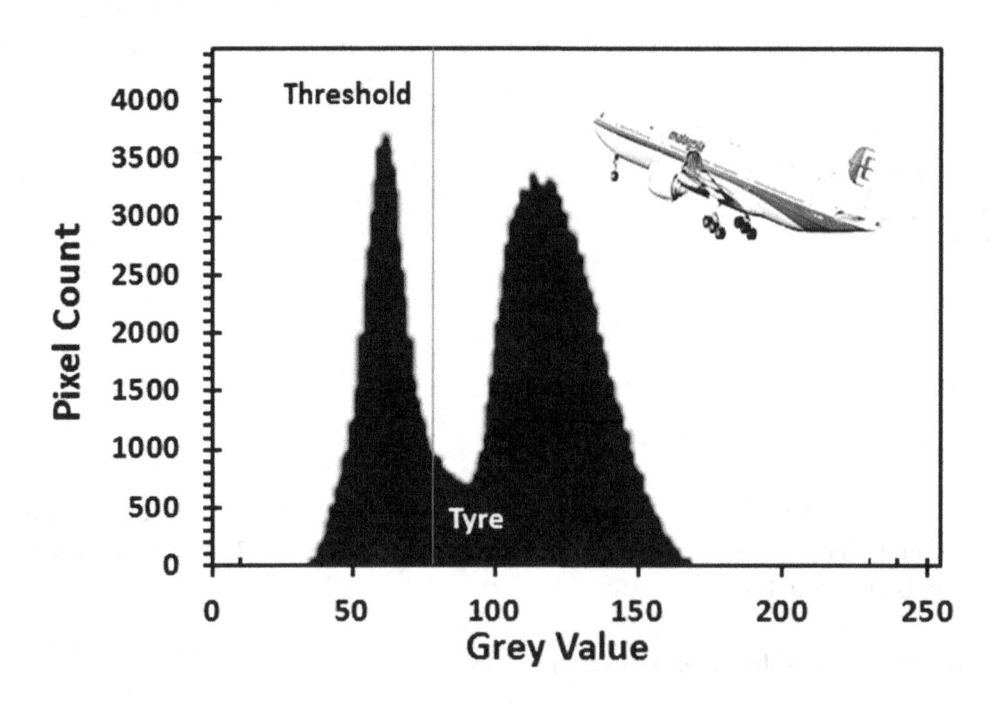

In other words, thresholding is a simple form of segmentation wherein each pixel in an image is compared with the threshold value. If the pixel lies above the threshold, it will likely be marked as foreground and if it is beneath a threshold as heritage. The threshold will most customarily be intensity or colour value. In this approach, the selection of initial threshold value relies upon the histogram of an image and the grey scale of an image. For instance, the aeroplane tyre has a lower peak than aeroplane fuselage (Figure 6). After the segmentation, a related element evaluation is achieved to group similar belongings objects (Figure 7).

The connected element evaluation is used here to extract the nearby object shape descriptors for figuring out the preferred target. The template is implemented as an identical model. Subsequently, correlation measurement is exercised for measuring the similarity between item vicinity functions and simulation verified that the functionality of object monitoring in remote sensing images with an assist of used techniques.

Figure 7. Tyre segmentations through the threshold

MULTI-SPECTRAL SATELLITE IMAGES FOR AEROPLANE DETECTION

Multi-spectral signatures can be used to detect an aeroplane, which is flying far away from the ground station. In this circumstance, the aeroplane must be invisible. In this view, its spectral signature has a similar intensity to background clutter (Maire and Sidonie 2015). In this regard, Mahalanobis transforms can be implemented to reduce a high false alarm rate (Chang et al., 2002 and Karlholm and Ingmar 2002). Mahalanobis accounts for both spectral and spatial dispersions, which lead to excellent statistical detection (Wang et al., 2012). In this context, the two objects in PLEIADES 1 A satellite data (Figure 8a) can be isolated from the background clutter using Mahalanobis transformation (Figure 8b). However, Mahalanobis transformation cannot decide either these two objects belong to the MH370 debris or not.

Other method based on the geometric structure of an aeroplane, which is implemented using Support Vector Machines (SVM) classification. In this approach a supervised learning algorithm analysis the PLEIADES 1 A satellite data and distinguishes different patterns for classification (Figure 9). In

Figure 8a. Mahalanobis transformation of true satellite data

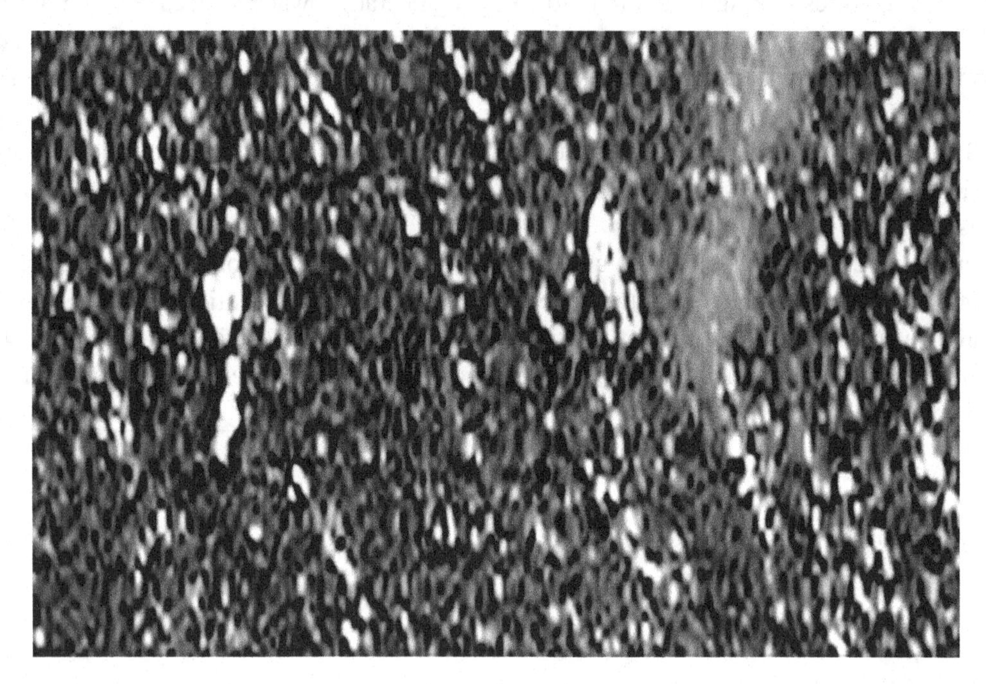

Figure 8b. Object classifications

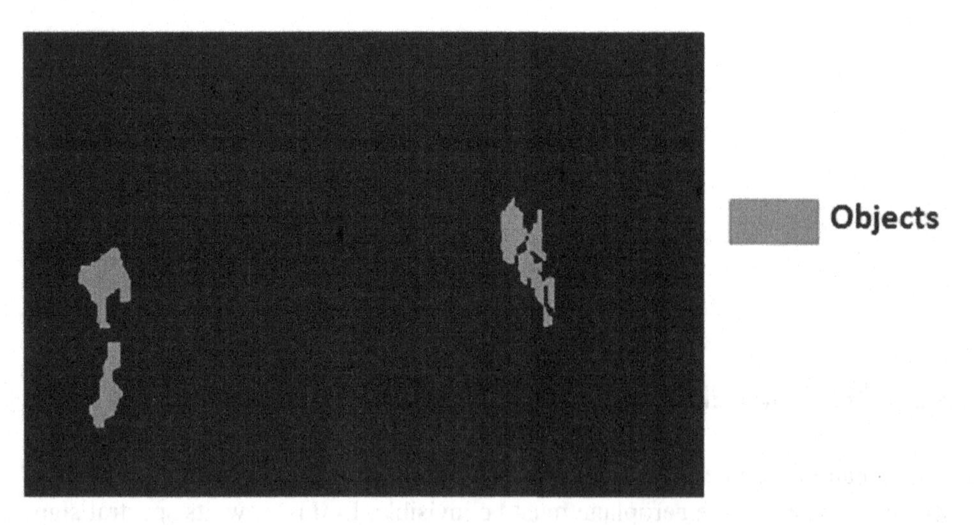

this circumstance, The SVM recognizes different two classes. Nevertheless, the SVM cannot also conclude either both objects are debris of MH370 or not. On the contrary, the SVM is used for stationary aeroplane detection.

ROBUST ALGORITHM FOR AIRPLANE DETECTION IN SATELLITE DATA

Figure 9. SVM for MH370 debris detection

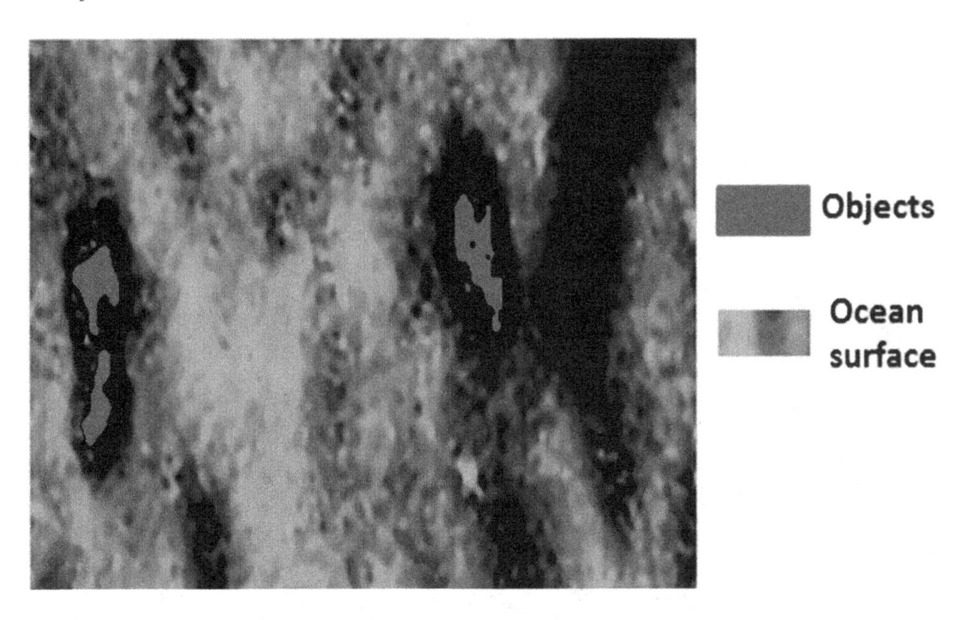

Saliency computation and symmetry detection are keystone algorithm for automatic target detection. These techniques have excellent advantages of the stability of object location and orientation detection. Moreover, they are computed target kind, pose, and size only once, which are saving time without missing any objects. In this context, the distance of each pixel to the surrounding pixels in the current image is estimated. Then the minus values are added together. In this regard, let assume that the pixel value is I and I_i is the rest of the pixel value in the same frame, then Saliency is formulated as:

$$\text{SALS}(I_k) = \text{å}|I_k - I_i| \quad (8.6)$$

being I_k is the current image grey level, which is in the range of the [0,255]. Equation 8.6 can be expended as:

$$\text{SALS}(I_k) = |I_k - I_1| + |I_k - I_2| + ... + |I_k - I_N| \quad (8.7)$$

here the total number of pixels in the current image is N. Using the histogram distribution of pixel variations, one can estimate the frequency variations $F_{n,m}$ between several frames I_m and I_n, respectively. The scientific explanation of Saliency Map (SALS) can mathematically be written as:

$$\text{SALS}(I_k) = \text{å}\, F_n \times |I_m - I_n| \quad (8.8)$$

where F_n is expressed in the form of a histogram, and the computational time of histogram is $O(N)$ time complexity. Therefore, This saliency map algorithm has $O(N)$ time complexity. Since the computational time of histogram is O (N) time complexity in which N is the number of the pixel's number of a frame. Moreover, the minus value and multiply the value of equation 8.8 require 256 times of operation. Consequently, the time complexity of this algorithm is O ($N+256$) which equals to O (N). In this approach, the input image (Figure 10a) is processed to acquire a Saliency Map (Figure 10b).

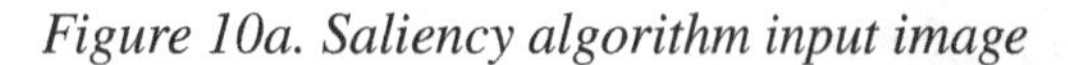
Figure 10a. Saliency algorithm input image

Figure 10b. Saliency map

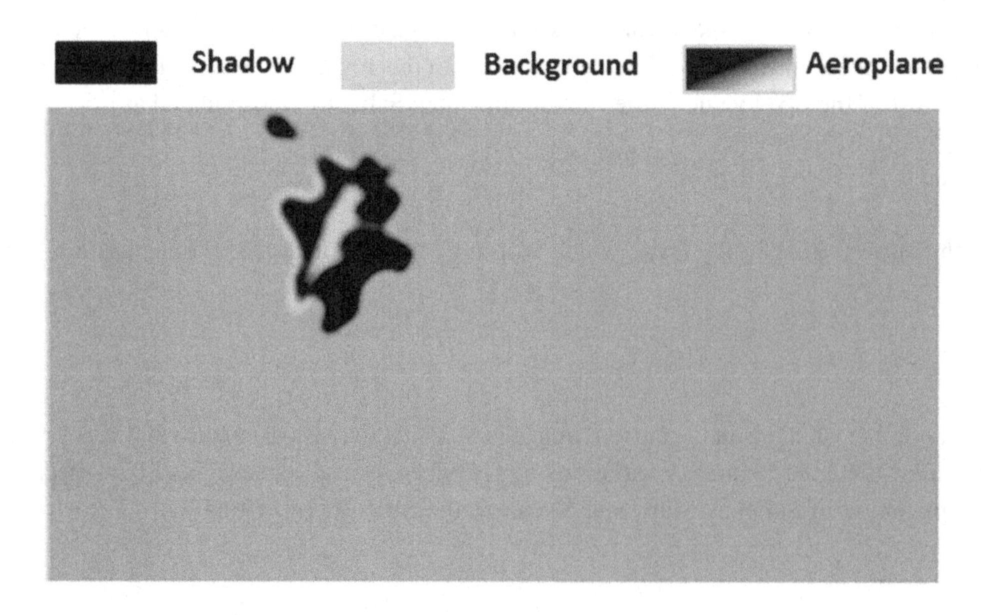

A saliency map masks the surrounding environment of the aeroplane. In this sense, the saliency map delivers detailed information for extracting the aeroplane. In other words, it determines the flight edge and also its shadow (Figure 10b). This works for a stationary aeroplane at the airport, but cannot detect such floated MH370 debris on the surface of the ocean. On the contrary, the improper threshold leads to a redundant saliency map (Figure 11).

Figure 11. Discarded objects in the saliency map

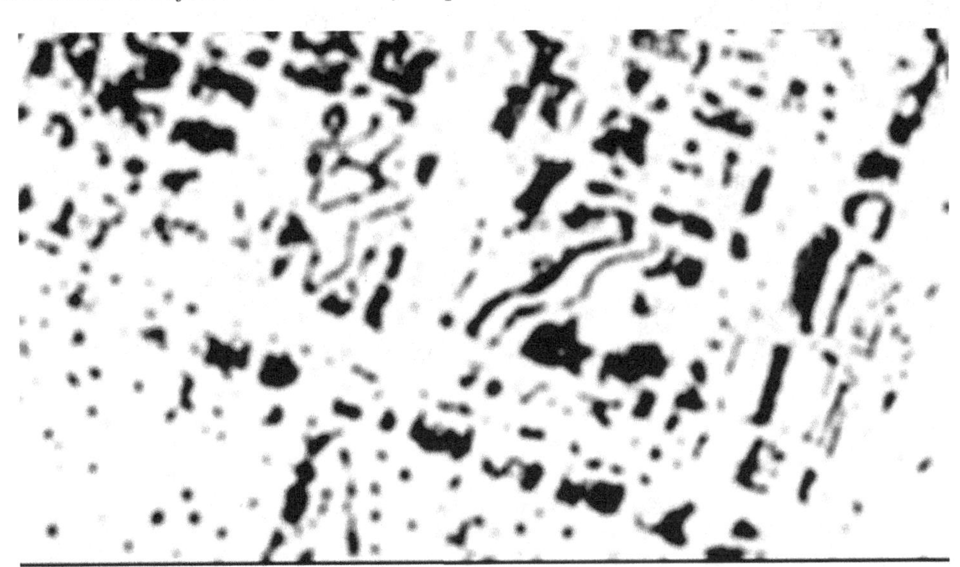

BOUNDARY FEATURE FOR AEROPLANE DETECTION

Boundary features can only use with high-resolution satellite data such as Quickbird with the infrared band (Figure 12a). Many class numbers are produced using a boundary feature algorithm (Figure 12b). This leads to the ambiguity of aeroplane detection. On the other hand, several steps are required to determine the proper boundary feature for the specific aeroplane. These procedures involve labeling, contour tracking, calculation of geometric feature parameters, and then extraction of the aeroplane. In other words, the aeroplane pixels are set for 1, while the surrounding pixels are set for 0. This is known as a binarized technique (Jebara et al., 2004).

However, this technique is required for an automatic threshold to eliminate unwanted background clutter and objects. Moreover, it is considered more time consuming for feature extraction. It is required to combine the supervised classification with geometric feature parameters. In these circumstances, the threshold value for every geometric feature must be computed.

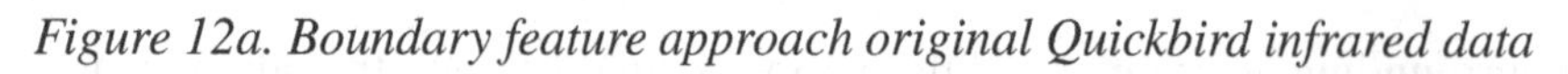

Figure 12a. Boundary feature approach original Quickbird infrared data

Figure 12b. Many class members

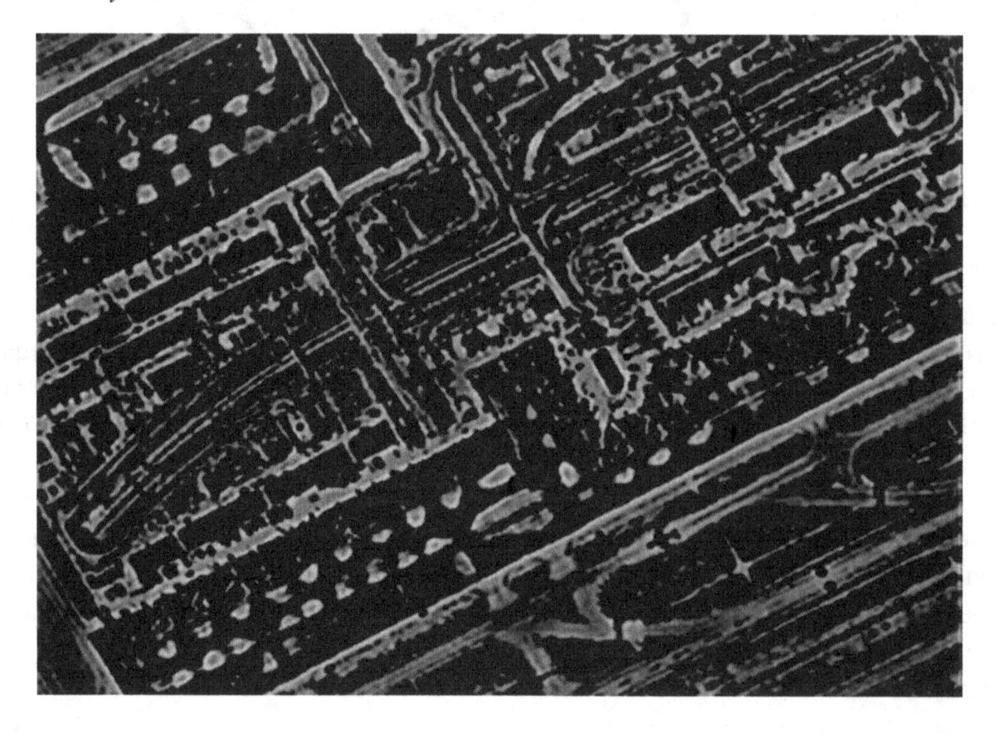

Figure 13. Boundary feature-based binarized technique

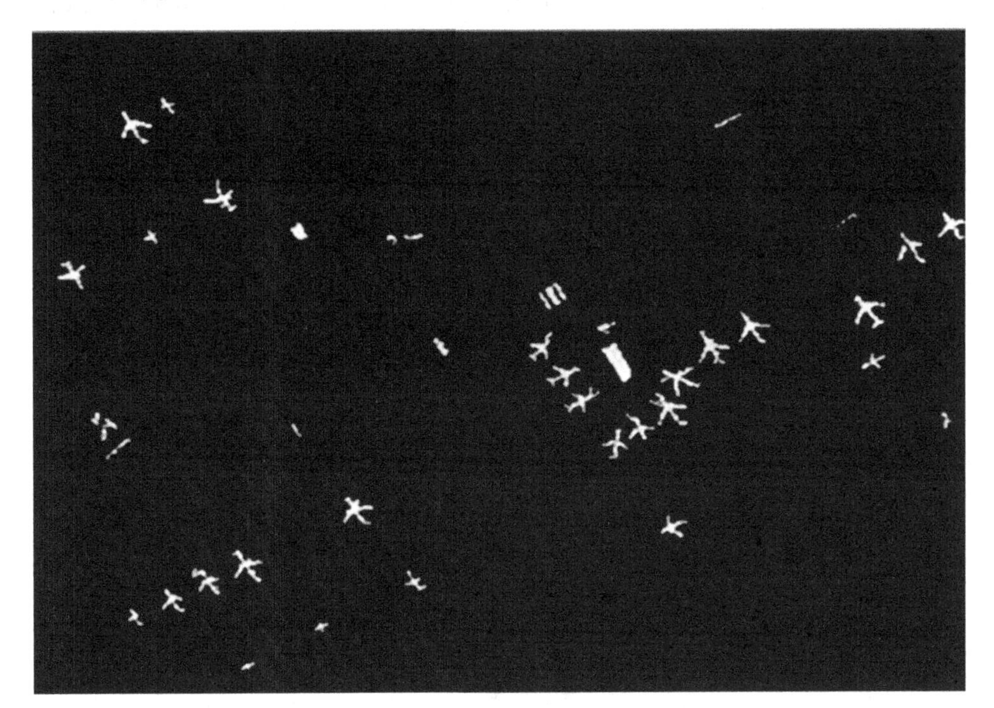

DEEP BELIEF NETS ALGORITHM FOR AEROPLANE DETECTION

A deep belief network (DBN) is a machine learning algorithm, which is an alternative class of deep neural networks. In other words, it is a generative graphical model. Conversely, the machine learning involves a generative method and discriminative method, which are determined clusters in the different degree of statistical classification.

Let us assume that the input satellite data is X and wanted the target variable is *Y*, then a generative algorithm is a statistical estimation of the joint probability distribution (*P*) on *X x Y*, *P (X,Y)*. Therefore, a discriminative algorithm is a computing of the conditional probability of the aeroplane detection *Y*, which is defined as *P(Y|X=x),* where *x* is a given observation. On the contrary, discriminative refers to clusters that are computed without involving an estimated probability. In principle, increasingly probabilistic, delivering more domain information and probability theory to be applied. In practice, different approaches are used, depending on the particular problem, and hybrids can combine the strengths of multiple approaches .

Moreover, deep neural network involves multiple layers of, visible layers of $v_i \in \{0,1\}^C$ and series of hidden units $h_1^1 \in \{0,1\}^{C_1}, \ldots\ldots\ldots, h_i^\ell \in \{0,1\}^{C_l}$, i.e. latent variables. The probability assigned to a vector v_i is formulated as

$$P(v,h^1,.....,h^\ell) = \left(\prod_{k=0}^{\ell-2} P(h^k \mid h^{k+1} \right) P(h^{\ell-1},h^\ell) \tag{8.6}$$

being $h = \left\{ h_i^{(1)}, h_i^{(2)}, h_i^{(3)} \right\}$ is the set of hidden units and k is the level of hidden units of Restricted Boltzmann Machines (RBM) (Figure 14).

Figure 14. Overview of DBN

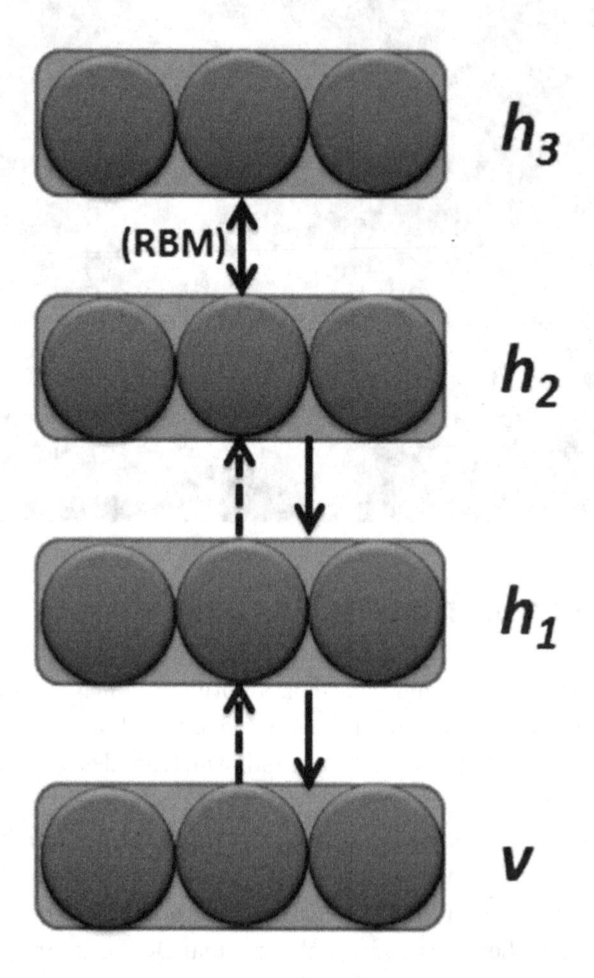

Conversely, DBN links only between layers not between units of each layer. For aeroplane detection, the set of train layer is examined without supervision. In this circumstance, a DBN can learn to probabilistically recreate its participation. In this view, the input layers as feature detectors. In practice, a DBN can be further trained in supervision to detect aeroplane in satellite images. In other words, DBNs can be considered as unsupervised networks such as restricted Boltzmann or autoencoders.

In this understanding, every sub-network hidden layer operates as the perceptible for the subsequent. This composition directs to a debauched, layer-by-layer unsupervised training procedure of the aeroplane. In fact, contrastive divergence directs each sub-network, which allows starting from the lowest visible layer of the aeroplane trained pixels. This leads to considered DBNS as one of the first effective deep learning algorithms.

In an explanation of Vaijayanthi and Vanitha (2015), to train, the DBN, multi-images, including gradient, 50% and 25% thresholding images are used as the input of DBN (Figure 15). Subsequently pretraining, DBN can fine-tune to be a robust detection of an aeroplane. In this circumstance, the DBN dimension has high accuracy. In this regard, DBNs can detect the tiny blurred aeroplanes appropriately in many satellite images. The pseudo-code of the DBN algorithm is demonstrated in Figure 16.

Figure 15. Procedures of DBN based on thresholding and weight between the first hidden layer and visible layer

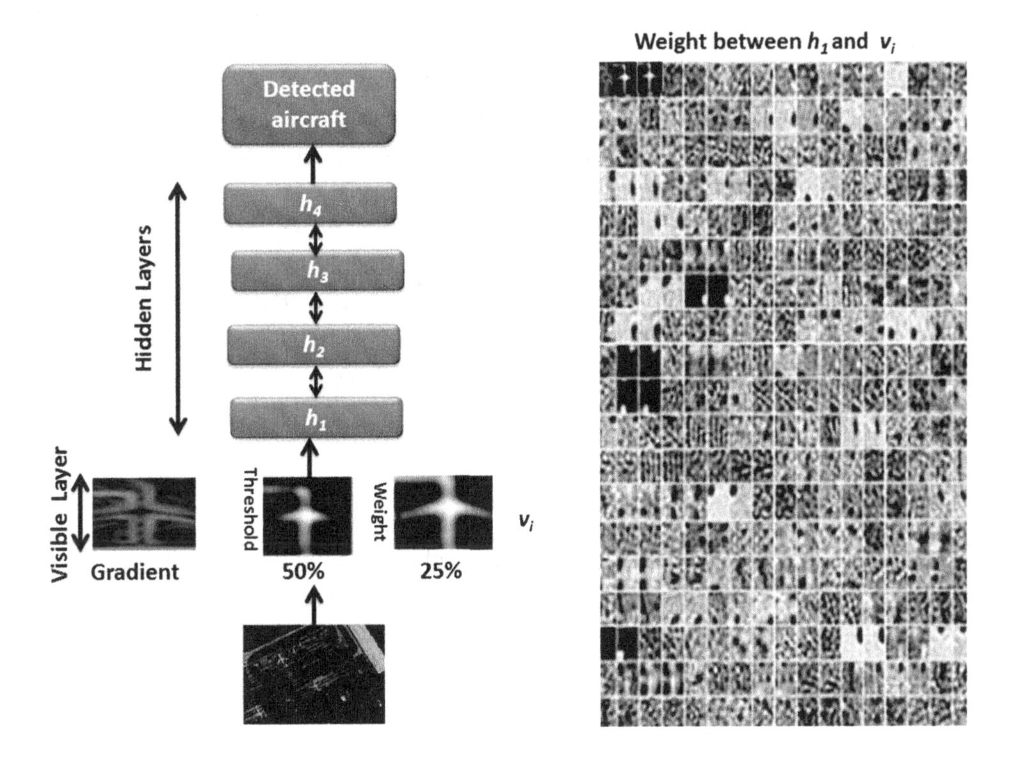

First, initialize an RBM with the desired number of visible and hidden units.

rbm = RBM(num_visible = 6, num_hidden = 2)

Next, train the machine:

training_data = np.array([[1,1,1,0,0,0],[1,0,1,0,0,0],[1,1,1,0,0,0],[0,0,1,1,1,0], [0,0,1,1,0,0],[0,0,1,1,1,0]]) #

A 6x6 matrix where each row is a training example and each column is a visible unit.

r.train(training_data, max_epochs = 5000) #

Don't run the training for more than 5000 epochs.

CONCLUSION

Following the above perspective, such conventional image processing algorithms are extremely useful to detect full aeroplane fuselage. In the circumstance of the stationary phase of the aircraft, i.e. aeroplane at the airport. The above-mentioned tools are considered as semi-automatic tools which are still relying on the threshold technique. The imperfect threshold can reduce the accuracy of target detection. In other words, it can produce extreme false alarm information.

Aeroplane detection exploiting multispectral signature provides restored results compared to other techniques. Multispectral data have less than 10 bands, which can be upgraded by using hyperspectral images. Indeed, hyperspectral images deal with dividing images into spectral various bands. However, there are no any hyperspectral satellite sensors overpass the search area of missing MH370 and archived. There is a limited number of satellites have claimed the existence of MH370 debris, for instance, Pleiades 1 A satellite and Worldview 2. Moreover, spectral analysis of remotely sensed data delivers extra precise information even for small targets. In fact, spectral signature information improves vision and discrimination accuracy of surface material or object being imaged.

Flight MH370 crashed into the Southern Indian Ocean, which broke into several pieces of debris. The debris must be varied in different shape and orientation and could be mixed with other man-made objects, i.e. water grabs. In this understanding, such PCA technique or DBN is not able to identify the debris feature boundaries. This requires an optimization, automatic detection procedure to precise declaration either the detected objects belong to MH370 or not.

REFERENCES

Chang, C. I., & Chiang, S. S. (2002). Anomaly detection and classification for hyperspectral imagery. *IEEE Transactions on Geoscience and Remote Sensing*, *40*(6), 1314–1325. doi:10.1109/TGRS.2002.800280

Deepa, V., & Kala, L. A. (2017). Review on aircraft detection techniques and feature extraction algorithms using digital image processing. *International Journal of Innovative Research in Computer and Communication Engineering*, *5*(2), 2347–2353.

Dudani, S. A., Breeding, K. J., & McGhee, R. B. (1977). Aircraft identification by moment invariants. *IEEE Transactions on Computers*, *100*(1), 39–46. doi:10.1109/TC.1977.5009272

Jebara, T., Kondor, R., & Howard, A. (2004). Probability product kernels. *Journal of Machine Learning Research*, *5*(Jul), 819–844.

Karlholm, J., & Renhorn, I. (2002). Wavelength band selection method for multispectral target detection. *Applied Optics*, *41*(32), 6786–6795. doi:10.1364/AO.41.006786 PMID:12440532

Maire, F., & Lefebvre, S. (2015). Detecting Aircraft in Low-Resolution Multispectral Images: Specification of Relevant IR Wavelength Bands. *IEEE Journal of Selected Topics in Applied Earth Observations and Remote Sensing*, *8*(9), 4509–4523. doi:10.1109/JSTARS.2015.2457514

Otsu, N. (1979). A threshold selection method from gray-level histograms. *IEEE Transactions on Systems, Man, and Cybernetics*, *9*(1), 62–66. doi:10.1109/TSMC.1979.4310076

Tsukasa, H. (2010). Airplane extraction from high resolution satellite image using boundary feature. *Proc. of ISPRS Technical Com. VIII Symposium*.

Vaijayanthi, S., & Vanitha, N. (2015). Aircraft Identification in high-resolution remote sensing images using shape analysis. *International Journal of Innovative Research in Computer and Communication Engineering*, *3*(11), 11203–11209.

Wang, Y., Wang, Y., Xiang, J., Han, Y., & Rao, R. (2012, April). Wavelength Band Selection Method for Target Detection Considering Surface Reflectivity. In *Proceedings of the 2012 Second International Conference on Electric Information and Control Engineering-Volume 01* (pp. 1645-1648). Academic Press.

Wu, Q., Sun, H., Sun, X., Zhang, D., Fu, K., & Wang, H. (2014). Aircraft recognition in high-resolution optical satellite remote sensing images. *IEEE Geoscience and Remote Sensing Letters*, *12*(1), 112–116.

Chapter 9
Optical Satellite Multiobjective Genetic Algorithms for Investigation of MH370 Debris

ABSTRACT

At present, there is no precise method that can inform where the lost flight MH370 is. This chapter proposes a new approach to search for the missing flight MH370. To this end, multiobjective genetic algorithms are implemented. In this regard, a genetic algorithm is taken into consideration to optimize the MH370 debris that is notably based on the geometrical shapes and spectral signatures. Currently, there may be three limitations to optical remote sensing technique: (1) strength constraints of the spacecraft permit about two hours of scanning consistently within the day, (2) cloud cover prevents unique observations, and (3) moderate information from close to the ocean surface is sensed through the scanner. Needless to say that the objects that are spotted by different satellite data do not scientifically belong to the MH370 debris and could be just man-made without accurate identifications.

INTRODUCTION

Despite advanced space technology and maritime technology MH370 fuselage and debris have been declined to track. It is a complicated task to identify Flight MH370 in the Southern Indian Ocean due to its huge water body size. In fact, it stretches for more than 6,200 miles (10,000 km) between the southern tips of Africa and Australia and, without its marginal seas, has an area of about 28,360,000 square miles (73,440,000 square km). Using optical remote sensing technology, which spotted plenty of objects could belong to the MH370 debris (Trinanes et al., 2016). In continuing with Chapter 8, the numerous satellite data with suspected MH370 debris are requisite precise image processing algorithms to classify and discriminate the MH370 debris from the surrounding environment.

Chapter 8 cannot satisfy either the spotted objects belong to MH370 or not. It is an intention of this chapter to solve this analytical problem. There is no doubt that the MH370 debris has incredible shapes

DOI: 10.4018/978-1-7998-1920-2.ch009

than other materials floating on the Indian Ocean s‘ surface. Can a multiobjective genetic algorithm decide the existence of MH370 debris in the Southern Indian Ocean?

This chapter hypothesizes that MH370 debris can also be retrieved by the use of the optimization multiobjective genetic algorithm from varieties of optical satellite sensors. In this regard, a genetic algorithm is taken into consideration to optimize the MH370 debris which is notably based on the geometrical shapes and spectral signature. Currently, there may be three limitations to optical remote sensing technique: (i) strength constraints of the spacecraft permit about two hours of scanning consistently within the day; (ii) cloud covers prevent unique observations, and (iii) moderate information from close to the ocean surface is sensed through the scanner. On this understanding, identification of MH370 debris, particularly based on the geometrical shape and spectral signatures are not always feasible with optical satellite sensors.

REVIEW OF MH370 DEBRIS FLOATED IN SOUTHERN INDIAN OCEAN

There is plenty of MH370 debris collected from the Southern Indian Ocean. The former is a plane flaperon which was found on Réunion Island, on 29 July 2015. The latter discoveries of debris were in Mozambique, South Africa, Rodrigues Island (Mauritius), and Tanzania, which have confirmed belong to the missing MH370. Conversely, two debris found in Mozambique in December 2015 and February 2016, respectively, are almost confirmed from the missing aircraft. Correspondingly, other confirmed fragments of debris initiated in South Africa and Rodrigues Island. Moreover, the wing flap found in Tanzania has been confirmed as instigating from MH370 (Figure 1).

Figure 1. Summary of MH370 debris found

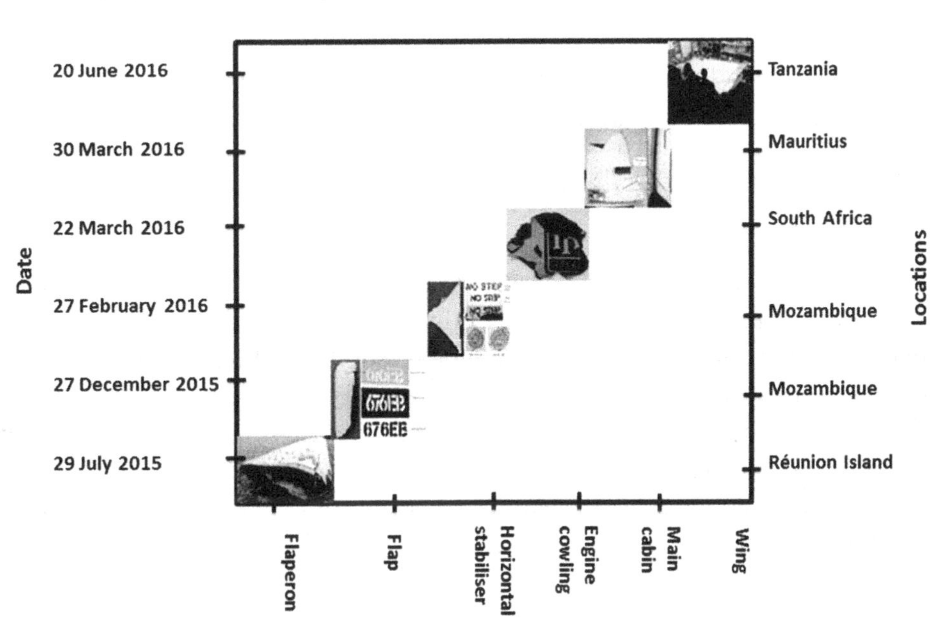

The flaperon is a 2.7 metre-long moveable part on the trailing edge of the wing, used to increase drag for takeoff and landing, and to bank the aircraft. The meter long-metal piece found in Mozambique in December 2015. Then the horizontal stabilizer, which is the right-hand tail section of a Boeing 777 found on 28 February 2016 in Mozambique. Moreover, the Engine cowling segment of 72 cm long found in South Africa on 21 March 2016.

The only interior part of the plane yet discovered, this piece, known as part four, was judged by experts to be a panel segment from the main cabin, associated with the Door R1 closet. It is found on Rodrigues Island, Mauritius, on 30 March 2016. On 23 June 2016, the wing flap found in Pemba Island, Zanzibar. The latter two unconfirmed debris is found by a local fisherman in Southeast Madagascar, on September 2016 (Minchin et al., 2017).

BRIEFING OF SATELLITE IMAGES TRACKED DEBRIS

Four optical satellite sensors with different spectral signature and resolutions have claimed to survey countless MH370 debris in the Southern Indian Ocean. These sensors involve WorldView-2 (WV2), Gaofen (GF-1), Pléiades 50 cm global high-resolution satellite, and Thaichote satellite (Table 1).

Table 1. Summary of Satellite Sensors Specifications Tracked MH370 Debris

Satellite	Resolution
WorldView-2 (WV2)	0.46 m panchromatic, 1.85m multi-spectral at nadir; 0.9 m pan/3.58m multispectral at 45°
Gaofen (GF-1)	2m panchromatic, 8m multispectral, 16 m wide-angle multispectral
Pléiades 1A	0.5m panchromatic, 2m multispectral; 5 acquisition modes
Thaichote	2 m panchromatic, 15 m multispectral

It is generally accepted that those satellite sensors (Table 9.1) provide incontrovertible evidence of a massive and recent "debris event"- (Chapter 7) at a time and place where the missing aircraft plunged into the Southern Indian Ocean. Although these satellite sensors have a high resolution of fewer than 0.5 m and a moderate resolution of 16 m, the sensors cannot identify any objects belongs to the fuselage of MH370. This perhaps due to a conventional image processing tool such as Principle Component Analysis (PCA) (Minchin et al., 2017 andTrinanes et al., 2016).

As individual satellites do not cover the entire surface area of the globe, they need to be tasked to scour certain areas if they are not covering them already.

EXPECTED MH370 DEBRIS

If the Flight MH370 plunges into the ocean, it must be broken into multiple pieces. The expected debris can be used as references to confirm the one spotted in satellite sensors. If the Flight MH370 plunges into the ocean, it must be broken into multiple pieces. The expected debris can be used as references to confirm the one spotted in satellite sensors. In this context, the emergency exit door, luggage, a life vest and dozens of bloated bodies are essential debris must float on the ocean surface. Moreover, Items from the plane, including an oxygen tank and an intact suitcase must also reveal.

The most important evidence of the MH370 crashed into the ocean is an oil spill. However, this evidence does not declare in the search area, which is determined by Inmarsat F3 satellite (Chapter 5). In this context, the oil spill must stick around the debris spotted in satellite images. In other words, spotted debris and an oil spill on the edge of the search area for MH370 must be confirmed. However, ships in the area have been unable to find anything on the water.

On 16 March 2104, Worldview (WV2) spotted debris in 5° 39′ 08.5″ N and 98° 50′ 38.0′E the Malacca Straits. This debris is associated with oil spills (Figure 2) with a length of less than 20 m. This information leads to ambiguities in determining the scenario and exact location of MH370 vanishing. Did satellite sensors spot uncertainty objects? This chapter is devoted to answering this critical question. To this end, an advanced image processing algorithm based on the multiobjective genetic algorithm is used.

Figure 2.

SPECTRAL SIGNATURE

Features on the Earth reflect, absorb, transmit, and emit electromagnetic energy from the sun. Special digital sensors have been developed to measure all types of electromagnetic energy as it interacts with objects in all of the ways listed above. The ability of sensors to measure these interactions allows us to use remote sensing to measure MH370 debris features and the surrounding environment.

A measurement of energy commonly used in remote sensing of the marine debris has reflected energy (e.g., visible light, near-infrared, etc.) coming from its feature structures. The amount of energy reflected from these surfaces is usually expressed as a percentage of the amount of energy striking the objects.

Therefore, every band represented by the wavelength. In this context, the digital numbers (DNs) are switched first to radiance $R(\lambda)$ and planetary reflectance $R_p(\lambda)$. The spectral reflectance can mathematically be written as:

$$R(\lambda) = R_{\min}(\lambda) + [R_{\max}(\lambda) - R_{\min}(\lambda)]\left(\frac{R_p}{R_{p,\max}}\right) \quad (9.1)$$

Here min and max are referred to as radiance minimum and maximum, respectively. Consequently, the $R_p(\lambda)$ is calculated using

$$R_p(\lambda) = \frac{\pi R\ (\lambda) r^2}{\cos\theta S_0(\lambda)} \quad (9.2)$$

where r is the Earth-Sun radius vector, θ is the solar illumination angle, and $S_0(\lambda)$ is the exoatmospheric solar irradiance in the similar wavelength range. Equation 9.2 demonstrates that $R_p(\lambda)$ can exceed 1.0 due to the effects of non-Lambertian reflectance and local slope.

Reflectance is 100% if all of the light striking objects and the object bounces off and is detected by the sensor. If none of the light returns from the surface, the reflectance is said to be 0%. In most cases, the reflectance value of each object for each area of the electromagnetic spectrum is somewhere between these two extremes. Across any range of wavelengths, the percentage of reflectance values for debris and water features can be plotted and compared. Such plots are called "spectral response curves" or "spectral signatures." Differences among spectral signatures are used to help classify remotely sensed images into classes of different features since the spectral signatures of like features have similar shapes (Marghany, 2013 and Minchin et al., 2017).

Multiple scattering and absorption of incident beams are a function of the radiative transfer model, which is expressed as:

$$R_p(\psi) = \frac{\int_0^{2\pi}\int_0^{1} \psi R_u(\psi,\phi)\, d\psi\, d\phi}{\cos\theta I_d} \quad (9.3)$$

where $R_u(\psi,\phi)$ is the radiance in the upward direction at angle cross ψ and azimuth ϕ. However, I_d is the direct irradiance at the top of the object as a function of incident angle $\cos\theta$. Equation 9.3 explains that the spectral "directional-hemispherical" reflectance is well known as the ratio of the reflected flux, (ψ $R_u(\psi,\phi)$ $d\psi d\phi$) the radiance (intensity) integrated over all angles, divided by the incoming direct irradiance ($\cos\theta I_d$) (Dozier, 1989 and Weng, 2011).

IMAGE PRE-PROCESSING

Precise preprocessing of satellite images (Table 9.1) containing the atmospheric corrections, which is the first significant pre-processing step. The image preprocessing also involves radiometric calibration, dark subtraction, and Internal Average Reflectance (IAR) reflectance calibration, layer stacking, georeferencing, image enhancement, and Region of Interest (ROI) findings. The output of this process provides suitable images, which must be integrated with the multiobjective genetic algorithm (Figure 3). All these stages are briefly highlighted in the following sub-sections.

Figure 3. Image pre-processing phases

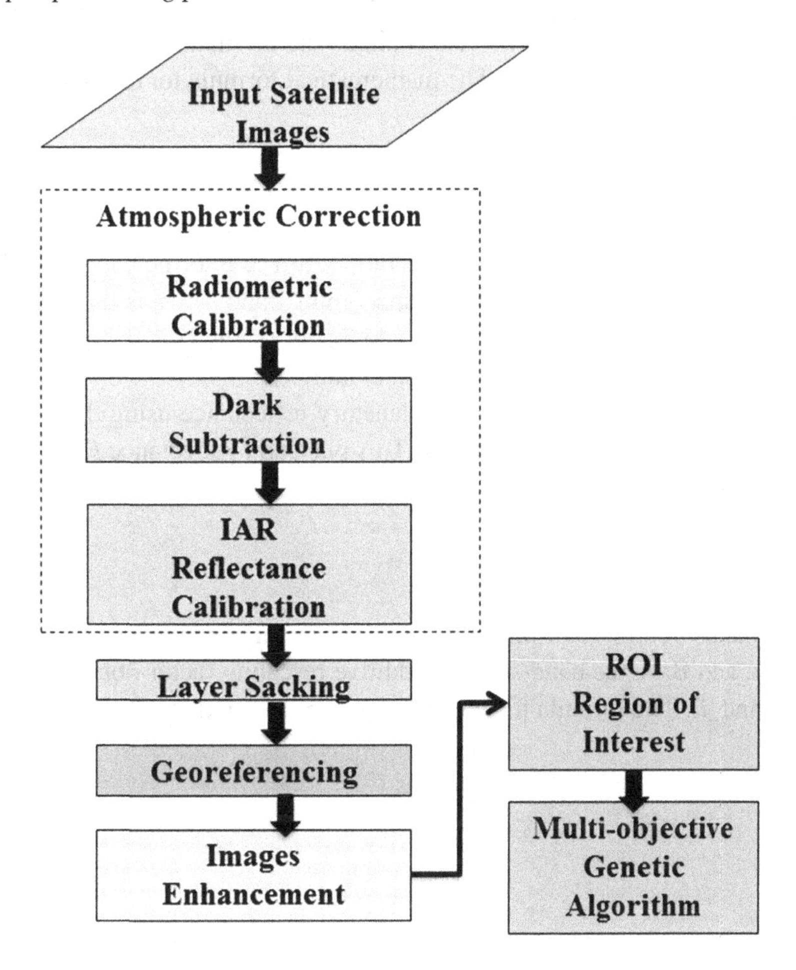

Atmospheric Correction

This pre-processing phase aims to remove the influence of the atmosphere on the surface reflectance. These effects involve the amount of water vapour, aerosols and darkness. These properties used to constrain highly accurate models of atmospheric radiation transfer to get an estimate of the true surface reflectance. There are two atmospheric correction model tools for retrieving spectral reflectance from hyperspectral and multispectral radiance images such as Fast Line of sight Atmospheric Analysis of Spectral Hypercubes (FLAASH) and Quick Atmospheric Correction (QUAC). FLAASH is a first principle atmospheric correction tool that corrects wavelengths in the visible through NIR and shortwave infrared (SWIR) regions up to 3 μm. It works with most hyperspectral and multispectral sensors. Weng (2011) reported that QUAC is an atmospheric correction method for multispectral and hyperspectral imagery that works with the visible spectrum and NIR through SWIR (VNIR-SWIR) wavelength range.

Radiometric Calibration

The satellite images (Table 9.1) are quantized, calibrated and scaled Digital Numbers (DN). These data products are delivered in 16-bit unsigned integer format and rescaled to the Top of Atmosphere (ToA) reflectance and/or radiance using radiometric rescaling coefficients provided in the product metadata file (MTL file). Furthermore, the MTL file also contains the thermal constants needed to convert the satellite data to the brightness temperature. The mathematical formula for conversion of satellite bands to ToA spectral radiance is given by:

$$R(\lambda) = B_s \times \mathrm{C} + \mathrm{B}_l \tag{9.4}$$

here $R(\lambda)$ is the ToA spectral radiance (W × m^{-2} × s rad^{-1} × μm^{-1}); B_s is the band-specific multiplicative rescaling factor obtained from the MTL file (radiance_mult_band_*i*); B_l is the band-specific additive rescaling factor present in the MTL file (radiance_add_band_*i*), and C is the quantized and calibrated standard product pixel values (DN). Here, *i* is the band number.

The image bands, are then converted to ToA planetary reflectance using the reflectance rescaling coefficients provided in the product MTL file. The ToA planetary reflectance $R_p(\lambda)$ without solar angle correction can mathematically be expressed by:

$$R_p(\lambda) = B_p \times \mathrm{C} + \mathrm{B}_p \tag{9.5}$$

where $B_p\mathrm{M}_\rho$ is the band-specific multiplicative rescaling factor obtained from the MTL file (reflectance_mult_band_*i*); and B_p is the band-specific additive rescaling factor obtained from the MTL file (reflectance_add_band_*i*). The formula for ToA reflectance $R(\lambda)$ with solar angle correction yields:

$$R\,(\lambda) = \frac{R\,(\lambda')}{\cos \mathrm{Z}} \tag{9.6}$$

where $R\ (\lambda')$ reflected spectral and Z is local solar zenith angle, which provided in the MTL file called Sun_elevation. For more accurate reflectance calculations, per-pixel solar angles can be used instead of the scene centre solar angle. The radiance as based on DN is computed using:

$$R(\lambda) = \frac{R_{max}(\lambda) - R_{min}(\lambda)}{255}\ \text{x}DN + R_{min}(\lambda) \tag{9.7}$$

where $R(\lambda)$ is the radiance expressed in $Wm^{-2}\ sr^{-1}$, $R_{max}(\lambda)$, $R_{min}(\lambda)$, are the minimum and the maximum spectral radiance, respectively, which are corresponding to the grey level range from 0 to 255 (Weng, 2011).

Dark Subtraction

Dark subtraction also called Dark Object Subtraction (DOS) is the second phase of atmospheric correction for satellite images. It is converting ToA reflectance of satellite imagery to surface reflectance.

Internal Average Reflectance Calibration

Internal average reflectance (IAR) is reflectance calibration used for images normalization to a scene average spectrum. It is effective for reducing hyperspectral data to relative reflectance in an area where no ground measurements exist and little known about the scene. An average spectrum is usually calculated from the entire scene and then used as the reflectance spectrum before being divided into the spectrum of each pixel of the image.

Layer Stacking

Layer stacking is a method to build a new multi-bands file from georeferenced images of various pixel sizes, extents and projections. Layer stacking of satellite images is based on integrating several bands to build a new multi-bands file from the georeferenced image. The output images depend on the inclusion and exclusion of the selected bands from layer stacking.

Georeferencing

Auto registration used for automatically georeference the raster dataset to a referenced one. It is the automated links between unreferenced and referenced raster dataset based on spectral signatures. It is used for satellite imagery in different season and time, image scale, image orientation, band and geographical location. The georeference is possible to obtain the link between raster to raster and raster to vector. In this view, the automated links between an unreferenced raster dataset of satellite images and referenced raster dataset one are achieved.

Images Enhancement

It is an important method for enhancing satellite images. Filters are used to sharpen and equalize the image, where images modification and enhancement is achieved. The output results of filtered and equalized images are better than the pre-processed one.

Region of Interest (ROI)

Finally, ROI is the loaded or selected samples or sub-layer of a raster data. In the view, ROI is used to identify the expected debris in the different satellite images. This phase is the last step of images pre-processing, which is used to transfer all satellite data to the multi-objective genetic algorithm to acquire MH370 debris.

GENETIC ALGORITHM

In the explanation of Sivanandam and Deepa (2008), the genetic algorithm (GA) is a powerful tool in the field of artificial intelligence in computer science. The GA is considered to be an optimal search and evolutionary algorithm that mimics the processes of natural selection. In other words, the GA spawns solutions to optimization problems using techniques inspired by natural evolution, such as inheritance, mutation, selection, and crossover. The schematic diagram of GA procedures is revealed in Figure 4.

Figure 4. Schematic diagram of GA procedures

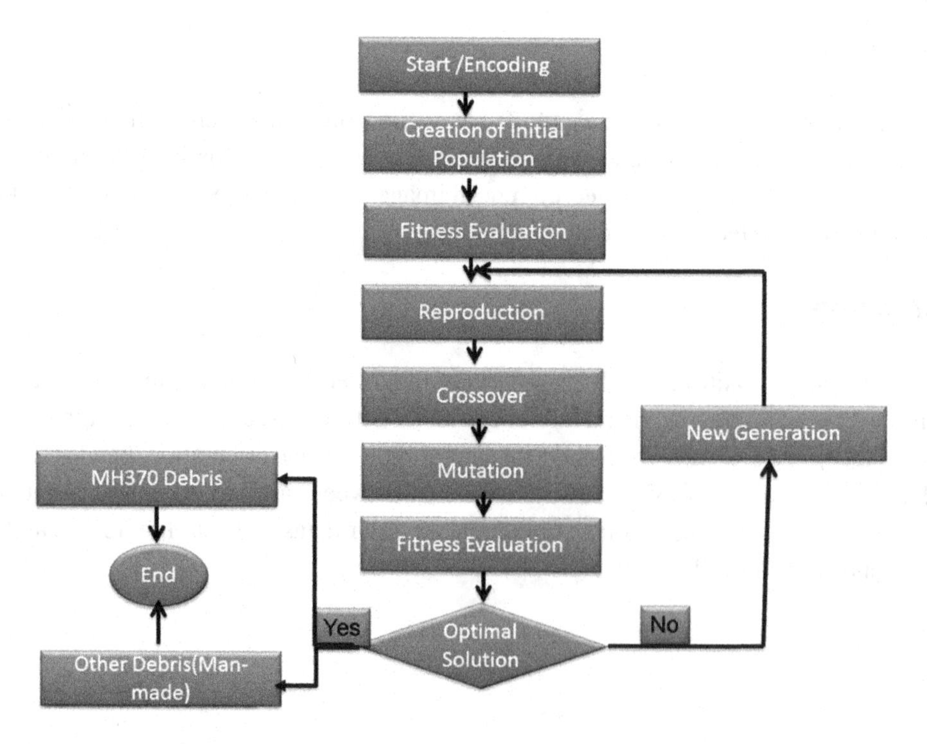

As described by Kahlouche et al. (2002), the genetic algorithm (GA) differs from a classification algorithm. In a classification algorithm, a single point is generated in each iteration. Moreover, a classification algorithm chooses the next point in the classification using a deterministic computation. In contrast, the genetic algorithm (GA) generates a population of cells in each iteration of which one particular cell in the population most closely approaches the optimal solution. Moreover, GA implements probabilistic transition rules, not deterministic rules as in classification algorithms (Mrghany 2013).

A large population of random chromosomes of different satellite spectral signatures is created at the beginning of one iteration of a genetic algorithm. Each one, when decoded, will represent a different solution to the problem at hand. There are N chromosomes in the initial population. Then, the following steps are repeated until a sufficiently accurate solution is obtained: (i) Test each chromosome of the satellite image spectral signatures (Table 9.1) to determine how effective it is as a solution to the problem at hand, and assign a fitness score accordingly. (ii) Select two members of the current population. The probability of being selected is proportional to the chromosomes' fitness. (iii) Following the crossover rate, cross over the bits from each chosen chromosome at a randomly chosen point. (iv) Loop over the bits of the chosen chromosomes and modify them by the mutation rate. (v) Repeat steps ii, iii and iv until a new population of N members has been created.

Data Organization

Let the entire spectral signature set of the area of interest in satellite multispectral images and boundary features in panchromatic data be $[S_1,S_2,S_3,\ldots,S_K]$, and $[B_1,B_2,B_3,\ldots,B_M]$, respectively. In this view, K *and* M are the total numbers of spectral signature and boundary features in the multispectral and panchromatic images (Table 9.1), respectively. Therefore, K *and* M are composed of genes that represent the spectral signature, the boundary feature of suspected objects and their surrounding environment. In this view, the genetic algorithm begins with the population's initialization step.

Following Marghany (2013), a constrained multi-objective problem for MH370 debris discrimination in multiple satellite data encompasses more than one objective and constraint, namely, ships, plastic debris, man-made debris, wave breaking, aircraft carriers, oil spills, and oil platforms. These constrain are considered as a false alarm in multiple satellite images.

The general form of the problem is adapted from Sivanandam and Deepa (2008) and can be described as follows:

$$\text{Minimise } f(S,B)=[f_1(S,B),f_2(S,B),\ldots,f_{K,M}(S,B)]^T \tag{9.8}$$

subject to the following constraints:

$$g_i(S,B)\leq 0,\ i=1,2,3,\ldots,I \tag{9.10}$$

$$h_j(S,B)\leq 0,\ j=1,2,3,\ldots,J \tag{9.11}$$

$$S_{\min}\leq S\leq S_{\max} \tag{9.12}$$

$$B_{min} \leq B \leq B_{max} \tag{9.13}$$

where, $f_i(S,B)$ is the *i-th* pixel of spectral signature S and boundary features B in the multispectral and panchromatic data, $g_i(S,B)$ and $h_j(S,B)$ represent the *i-th* and *j-th* constraints on the spectral signature and boundary feature in the row direction and the column direction, respectively. $(S,B)_{min}$ and $(S,B)_{max}$ are the minimum and maximum values of spectral signatures and boundary features.

The transition rules for the automatic MH370 debris detection capability of the cells are designed based on the input of various spectral signature and boundary feature values (S,B) to define the conditions required for a given pixel to be identified as MH370 debris pixel or not among the neighbouring pixels of a kernel with a window size of 5x5 pixels and lines. These rules can be summarised as follows:

1. If the test pixel represents the sea surface, OR current boundary features THEN $S \geq 0.6$, and the test pixel is not MH370 debris pixels.
2. IF the test pixel represents a bright patch and $S \leq 0.6$, THEN the pixel is identified as MH370 debris.

Population Initialization

Let P_i^j be a gene that corresponds to the spectral signatures of suspected MH370 debris pixels and its surrounding pixels. Consequently, the distribution of randomly selected P_i^j represents the spectral signatures and boundary features variations of both the suspected objects and their surrounding environmental pixels. Furthermore, i varies from 1 to K and j varies from 1 to N, where N is the population size.

The Fitness Function

Following Kahlouche et al. (2002) and Marghany (2013), a fitness function is selected to determine the similarity of each object, spectral signatures corresponding to bright patches both multispectral and panchromatic data. The spectral signature of the selected patches in the multispectral data are denoted by S_i, where i=1,2,3, …, K, and the initial population P_i^J, where j =1,2,3, …, N and i =1,2,3…, K. Formally, the fitness value $f(P^j)$ of each member of the population is computed as follows:

$$f(P^j) = [\sum_{i=1}^{K} \left| P_i^j - \beta_i \right|]^{-1}, j=1,\ldots,N \tag{9.14}$$

where N is the total number of individuals in the population and K is the number of individuals from the population considered in the fitness determination. Generally, Equation (9.14) is used to determine the level of similarity amongst bright patches that correspond belong to MH370 in multispectral and panchromatic satellite data.

The Selection Step

The key element of the selection step of the genetic algorithm is the selection of the fittest individuals $f(P^j)$ from the population P_i^j. The threshold value τ is determined by the maximum fitness value of the population Max $f(P^j)$ and the minimum fitness value of the population, Min $f(P^j)$. In subsequent generations, this step defines the populations *P*. The fittest individuals those that are most likely to present bright patches with high spectral signatures in multispectral and panchromatic data are those with values greater than the threshold τ which is defined as:

$$\tau = 0.5\ [Max\, f(P^j) + Min\, f(P^j)] \tag{9.15}$$

Equation 9.15 is used for the selection step to determine the maximum and minimum acceptable values for the fitness of the population. This is considered to be the step in which the spectral signature and boundary feature populations are generated in the GA procedure.

The Reproduction Step

Consistent with Sivanandam and Deepa (2002), the bulk of the calculation involved in the genetic algorithm lies in the reproduction step, which involves the implementation of the crossover and mutation processes in the spectral signature and boundary feature populations P_i^j determined from the multispectral and panchromatic satellite data, respectively. The crossover operation constructs a population P_i^j that will converge to solutions with high fitness. Thus, the closer the crossover probability is to 1, the more rapid is the convergence (Mrghany 2013). In the crossover step, genes are interchanged between the chromosomes. A local fitness value is assigned to each gene as follows:

$$f(P_i^j) = \left|\beta_i - P_i^j\right| \tag{9.16}$$

The crossover between two individuals serves to preserve all individual populations of the first parent that have a local fitness greater than the average local fitness $f(P_{av}^j)$ and replaces the remaining genes with the corresponding genes from the second parent. The average local fitness is defined by

$$f(P_{av}^j) = \frac{1}{K}\sum_{i=1}^{K} f(P_i^j) \tag{9.17}$$

Meanwhile, the mutation operator represents extraordinary random phenomena in the evolution process. Some useful genetic information may be lost from the selected population during the reproduction step. To compensate for this potential loss, the mutation operator is applied to introduce new genetic information into the gene pool (Michalewicz, 1994).

Morphological Operations

Morphological operations are performed on selected individuals prior to the crossover and mutation processes. In the crossover process, the probability that any given pair of chromosomes will exchange their bits (Figure 5a) is approximately 0.7. Crossover is performed by selecting a random gene along the length of the chromosomes and swapping all genes after that point ((Michalewicz, 1994).

In other words, In block selection, a large fraction of the weakest chromosomes in the population is thrown out, and the stronger chromosomes determine their location. The strength is measured according to the chosen optimization problem. Operationally, the optimization problem is represented by a ðtness function that maps, structures over to real numbers. The strongest structures are then those with the highest ðtness score. Block selection is controlled by one parameter, S, which speciðes that only the best fraction 1/S of the population is to be retained after the action of selection. Figure 5 shows the effects of selection with S = 2 on a population of size 8.

Figure 5. Crossover process: (a) Selected one random gene, and (b) Selected two random genes

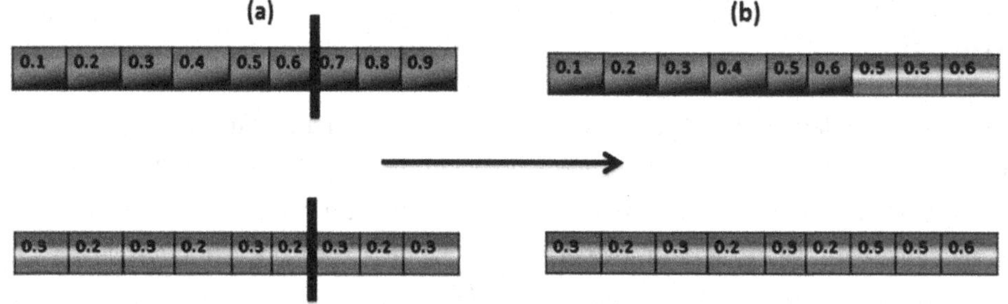

In the mutation process, the probability that the value of a given bit within a chromosome is flipped (0 becomes 1, 1 becomes 0) is usually very low for binarily encoded genes, perhaps 0.001. Every time chromosomes are selected from amongst the population, the algorithm first checks whether crossover has been applied, and the algorithm then iterates down the length of each chromosome, mutating bits if applicable (Davis, 1991). The probability of crossover was set to 1 while the mutation probability was set to 0. These values were set to encourage quick convergence to an optimum, either local or global. For instance, inserting mutation for permutations is required: (i) pick two allele values at random;(ii) move

Figure 6.

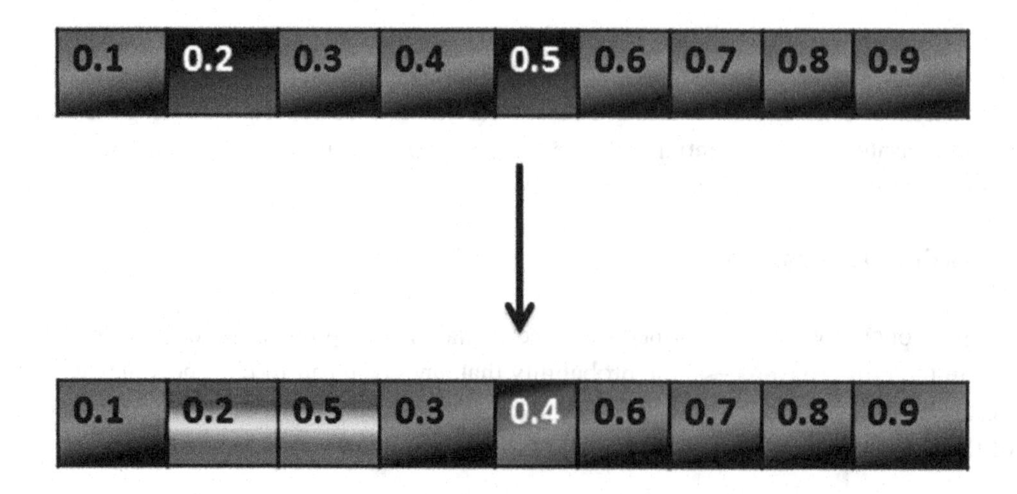

the second to follow the first, shifting the rest along to accommodate (Figure 7); and (iii) Note that this preserves most of the order and the adjacency information.

Figure 7. Spectral signature mutation procedures

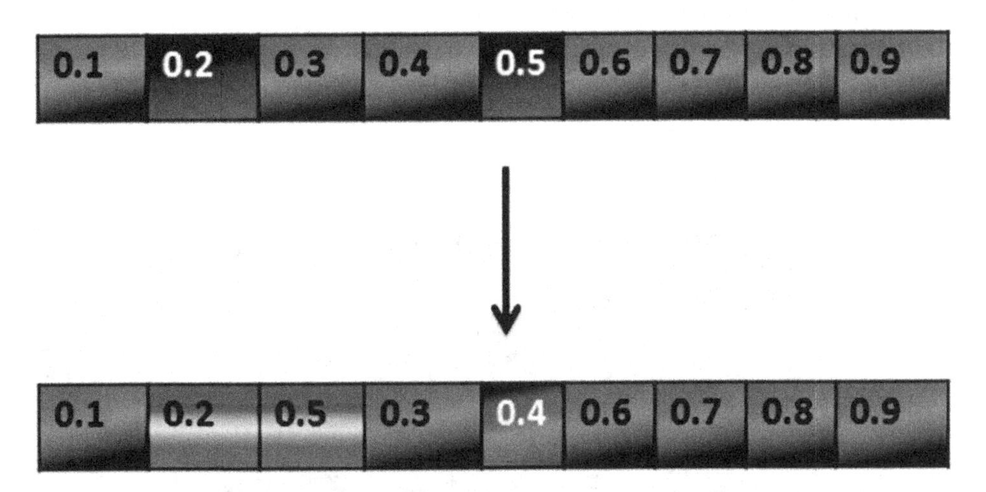

There are two major approaches to achieve the inversion mutation for permutations. The former is selecting two alleles at random and then invert the substring between them. The latter is preserving most adjacency information (Figure 8) (only breaks two links), disruptive of order information.

Figure 8. Spectral signature inversion mutation for permutations

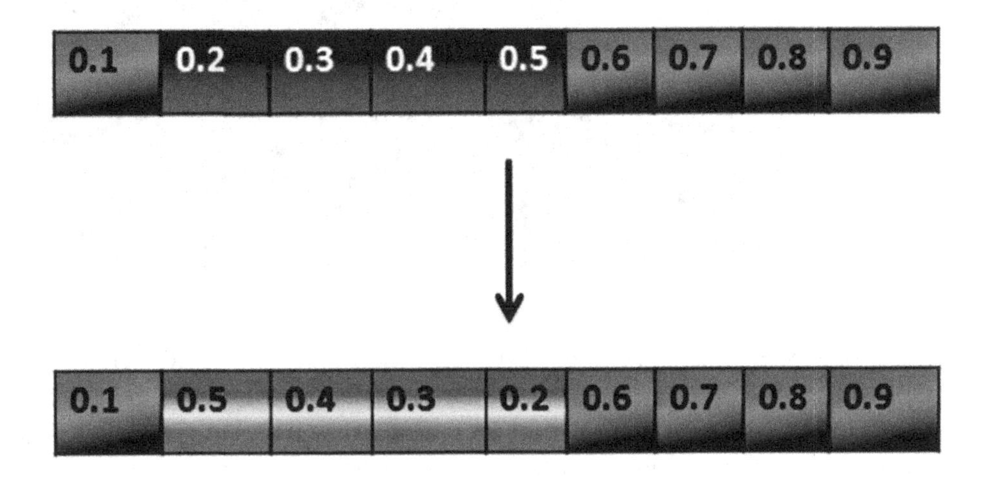

This procedure is designed to exploit the connectivity property of the different satellite data. The morphological operations are implemented during the reproduction step as follows: (i) closing, followed by (ii) opening. The accuracy of the dark patch segmentations achieved in this manner depends on the size and shape of the structuring element. Therefore, a structuring kernel with a square window size of-of 5 x 5 pixels in size is chosen to preserve the fine details of the suspected MH370 debris present in the satellite images (Mohanta and Sethi 2011).

Figure 9 represents the initial populations of 3000 for individual generations from WorldView-2 data panchromatic satellite data. The binary number between 0 and 1 is randomly generated and sorted as a string in the computer memory. It can also be noticed that they are not sorted in order, but randomly which represent only the row of the Gaofen (GF-1) panchromatic satellite data.

Figure 9. A genetic algorithm for an initial WorldView-2 data panchromatic satellite data generation

Consequently, the fitness values are changed between 0 and 1 (Figure 10) with iteration increments. The highest fitness value of 1.0 provides a sharp WorldView-2 data panchromatic satellite data (9.10c). As the fitness is gradually increased with iteration increments, the GF-1 panchromatic satellite data is being to be regenerated (9.10b). The highest fitness value indicates a clear GF-1 image.

Figure 9.11 provides an example of the crossover process with 10 individuals. Of these 10 individuals, the positive brighten patches represent suspected MH370, whereas the negative grey level patches represent the surrounding pixels. Accordingly, every cell is compared to the other corresponding cells to determine whether positive or negative. Consistent with Marghany (2013), in the GA procedures, a cell

Figure 10a. Fitness variations with iterations 0.1

has a positive value and should be propagated to subsequent generations when the cell in the intermediate prototype has a value larger than zero and greater than the threshold value. Such cells represent debris and ship evidence in GF-1 satellite data. By contrast, a cell that has a negative value represents a surrounding feature. In such a case, the cell in the intermediate prototype has a value of less than zero and below the threshold value and this cell's influence on subsequent generations should be diminished. The variation in the cell value (positive or negative) is a function of the cell's dissimilarity with comparable cells (Davis, 1991). This study confirms and extends the capabilities of the GA that were introduced by Kahlouche et al. (2002) and Marghany (2013).

NON-DOMINATED SORTING GENETIC ALGORITHM NSGA-II

This section presents a brief description of NSGA-II relevant to this study. NSGA-II is the second version of the famous "*Non-dominated Sorting Genetic Algorithm*" based on the work of Prof. Kalyanmoy Deb for solving *non-convex* and *non-smooth* single and multi-objective optimization problems. Its main features are: (i) A sorting non-dominated procedure where all the individual are sorted according to the level of non-domination; (ii) It implements elitism which stores all non-dominated solutions, and hence enhancing convergence properties; (iii) It adapts a suitable automatic mechanics based on the crowding

Figure 10b. 0.5

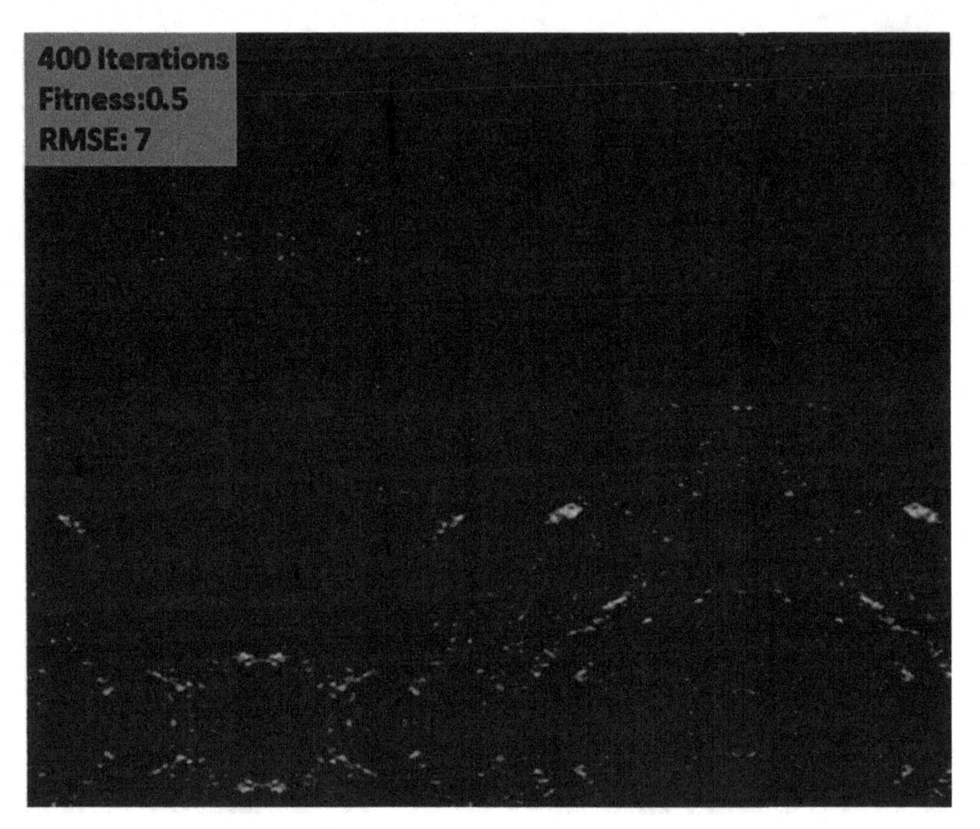

distance to guarantee diversity and spread of solutions, and (iv) Constraints are implemented using a modified definition of dominance without the use of penalty functions(Deb et al., 2000).

Perhaps, there is not exist one best solution in the case of multiple objectives. Therefore, there exists a set of solutions that are superior to the rest of the solution in the search space when all objectives are considered but are inferior to other solutions in the space in one or more objectives. These solutions are known as Pareto-optimal solutions or nondominated solutions (Zitzler et al., 2000).

The efficiency of NSGA lies in the way multiple objectives are reduced to dummy fitness function using nondominated sorting procedures. Consequently, NSGA can solve practically any number of objectives. In this regard, this algorithm can handle both minimization and maximization problems.

In order to sort a population of size N for $f(S_1,B_1),\ldots,f(S_N,B_N)$ according to the level of non-domination, each solution m must be compared with every other solution in the population to ðnd if it is dominated or nondominated. This requires comparisons $O(f(S_m,B_m))_N$ for each solution, where is m is the number of different pixels belong to debris, ships, and, wave whitecaps, and sea roughness (Zitzler et al., 2000). The initialized population N of $f(S_1,B_1),\ldots,f(S_N,B_N)$ is sorted based on the level of non-domination. Let S is each solution that must be compared to other every solution to determine the level of domination. In this regard, the fast sort algorithm was given by Deb et al., (2000) can be explored in the automatic detection of MH370 debris in satellite images (Figure 12).

For each individual $f(S_1,B_1)$ in the main population P do the following

Figure 10c. 0.7

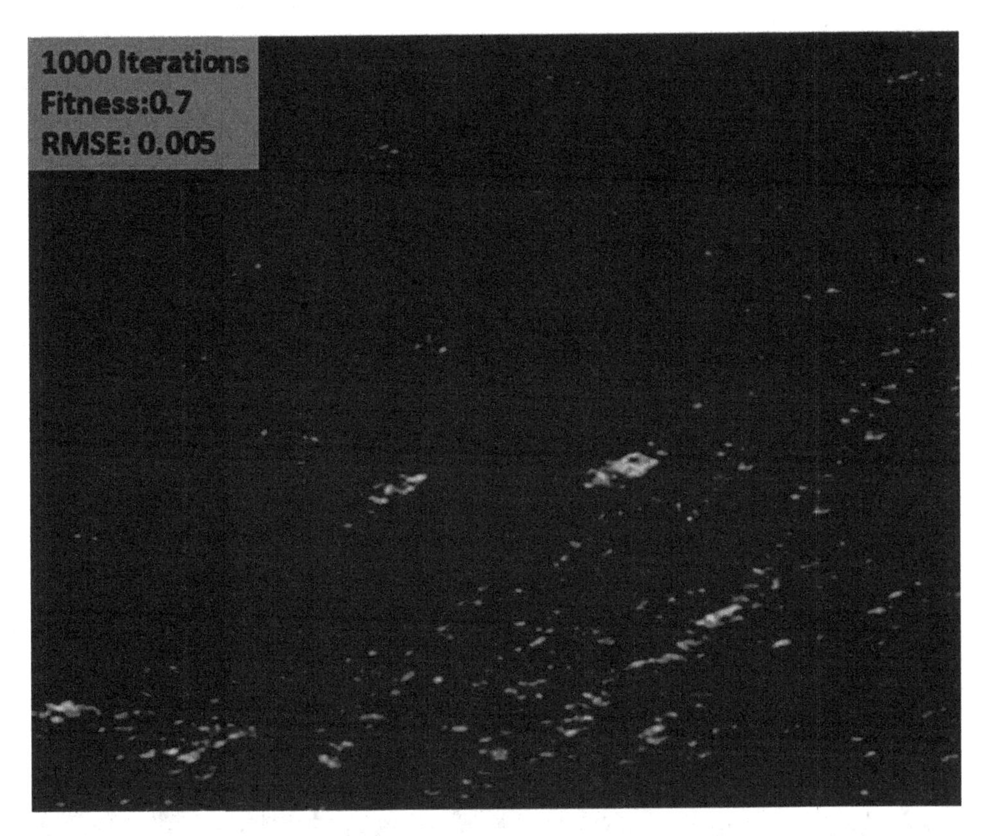

Initialize $s_{f(S_1,B_2)} = \Phi$. This set Φ would include all the individuals of $f(S_n,B_n)_N$ which is being dominated by $f(S_1,B_1)$.

Initialize $n_{f(S_1,B_1)} = 0$. This would be the number of individuals that dominate $f(S_1,B_1)$ i.e. no individuals dominate, then $f(S_1,B_1)$ belongs to the first front; set rank for an individual $f(S_1,B_1)$ to one i.e. $f(S_1,B_1)_{rank}=1$.

for each individual m in P

if $f(S_1,B_1)$ dominated m then

add m to the set Φ i.e. $\Phi = \Phi \cup \{m\}$

*else if m dominates $f(S_1,B_1)$ then

increment for domination counter for $f(S_1,B_1)$ i.e. $n_{f(S_1,B_1)} = n_{f(S_1,B_1)} + 1$

Let the first front set F_1 and then update by adding $f(S_1,B_1)$ to front 1 i.e. $F_1 = F_1 \cup \{f(S_1,B_1)\}$

Initialize the front counter to one. $i=1$

Then $F_i \neq \Phi$

Let $Q \neq \Phi$. The set for sorting the individuals for $(i+1)^{th}$ the front

of each individual $f(S_1,B_1)$ in front F_i

Figure 11a. Crossover procedures first individual

For every individual m in $s_{f(S_1,B_1)}$ and ($s_{f(S_1,B_1)}$) is the set of individuals dominated by ($f(S_1,B_1)$).

$n_{f(S_1,B_1)} = n_{f(S_1,B_1)} - 1$, decrement the domination count for individual m.
if $n_{f(S_1,B_1)}$=0 then none of the individuals in the subsequent fronts would dominate m. Hence set $f(S_1,B_1)$ $_{rank}$=i+1. Update the set Q with individual m i.e. $Q = Q \bigcup m$.
-increment the front by one.
-Now the set Q is the next front and hence F_i=Q.

Crowding Distance

Following Deb et al., (2000), the moment the non-dominated sort is achieved the crowding distance is designated. All the individuals in the population are assigned as the crowding distance value since the individuals are selected based on rank and crowding distance(Deb 2000). Crowding distance is assigned front wise and comparing the crowding distance between two individuals in different front is meaningless (Figure 13).

- For each front F_i, the number of individuals is represented by N.

Figure 11b. Resulting from an individual prior cancellation

- Reset the distance d_j to be zero for all the individuals of $f(S_1,B_1)$ i.e. $F_i(d_j)=0$, where j corresponds to j^{th} an individual of $f(S_j,B_j)$ in front F_i.
- For every objective function f
  *Sort the $f(S_j,B_j)$ in front F_i based on objective f i.e. $f(S_j,B_j)$=sort (F_i,f).
  *Assign infinite distance to boundary values for each individual $f(S_j,B_j)$ in F_i i.e.

$f(S_{d_1},B_{d_1})=\infty$ and $f(S_{d_n},B_{d_n})=\infty$

*for K= 2 to *(n-1)*

$$f(S_{d_K},B_{d_K})=f(S_{d_K},B_{d_K})+\frac{f(S,B)(K+1).q-f(S,B)(K-1).q}{f_q^{\max}-f_q^{\min}}$$

$E(S_{d_K},B_{d_K}).q$ is the value of qth an objective function of the Kth individual in $E(S_{d_K},B_{d_K})$,

Figure 11c. After cancellation

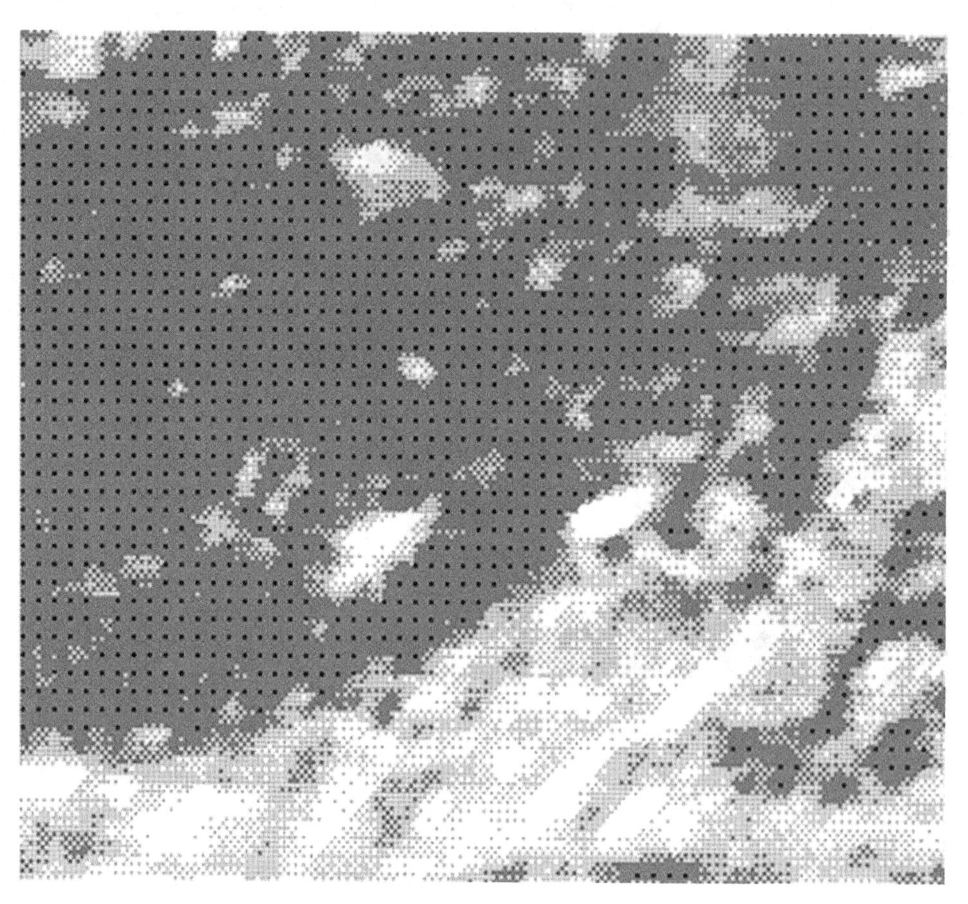

The main concept behind the crowing distance is estimating the Euclidian distance between each individual $f(S_j, B_j)$ in a front F_i which is based on their q objectives in the q dimensional hyperspace. The individuals $f(S_j, B_j)$ in the boundary are always selected since they have an infinite distance assignment (Deb, 2000 and Deb et al., 2000). In the Selection phase, once the individuals $f(S_j, B_j)$are sorted based on non-domination and with crowding distance $E(S_{d_K}, B_{d_K})$ assigned(Zamanifar et al., 2014), the selection is carried out using a crowded comparison operator $\prec_n$ which is based on(Guo et al., 2010 ; Fortin et al., 2013; Hu et al., 2016; Kamjoo et al., 2016):

(i) non-domination rank $f(S_1, B_1)_{rank}$ i.e. individuals $f(S_j, B_j)$ in front F_i will have their rank as $f(S_1, B_1)_{rank}$ =i.

(ii) crowding distance $f(S_{d_K}, B_{d_K})$

- $f(S_1, B_1) \prec_n m$

  - $f(S_1, B_1)_{rank} < m_{rank}$

- or if $f(S_1, B_1)$ and m belongs to the same front F_i, then $F_i(S_{d_K}, B_{d_K}) > F_i(d_m)$ i.e., the crowing distance should be more.

Figure 12. Pseudo Code of NSGA-II algorithm

For each individual $f(S_1, B_1)$ in the main population P do the following

Initialize $s_{f(S_1,B_2)} = \Phi$. This set Φ would include all the individuals of $f(S_n, B_n)_N$ which is being dominated by $f(S_1, B_1)$.

Initialize $n_{f(S_1,B_1)} = 0$. This would be the number of individuals that dominate $f(S_1, B_1)$ i.e. no individuals dominate, then $f(S_1, B_1)$ belongs to the first front; set rank for an individual $f(S_1, B_1)$ to one i.e. $f(S_1, B_1)_{rank} = 1$.

for each individual m in P

if $f(S_1, B_1)$ dominated m then

. add m to the set Φ i.e. $\Phi = \Phi \cup \{m\}$

*else if m dominates $f(S_1, B_1)$ then

. increment for domination counter for $f(S_1, B_1)$ i.e. $n_{f(S_1,B)} = n_{f(S_1,B)} + 1$

Let the first front set F_1 and then update by adding $f(S_1, B_1)$ to front 1 i.e. $F_1 = F_1 \cup \{f(S_1, B_1)\}$

Initialize the front counter to one. i=1

Then $F_i \neq \Phi$

Let $Q \neq \Phi$. The set for sorting the individuals for $(i+1)^{th}$ the front

of each individual $f(S_1, B_1)$ in front F_i

For every individual m in $s_{f(S_1,B)}$ and ($s_{f(S_1,B)}$) is the set of individuals dominated by $(f(S_1, B_1))$.

. $n_{f(S_1,B)} = n_{f(S_1,B)} - 1$, decrement the domination count for individual m.

. if $n_{f(S_1,B)}$=0 then none of the individuals in the subsequent fronts would dominate m. Hence set $f(S_1, B_1)_{rank} = i + 1$. Update the set Q with individual m i.e. $Q = Q \cup m$.

-increment the front by one.

-Now the set Q is the next front and hence $F_i = Q$.

Figure 13. The pseudo-code of the crowing distance stigmatization

- For each front F_i, the number of individuals is represented by N.
  - Reset the distance d_j to be zero for all the individuals of $f(S_1, B_1)$ i.e. $F_i(d_j) = 0$, where j corresponds to j^{th} an individual of $f(S_j, B_j)$ in front F_i.
  - For every objective function f

    *Sort the $f(S_j, B_j)$ in front F_i based on objective f i.e. $f(S_j, B_j)$=sort (F_i, f).

    *Assign infinite distance to boundary values for each individual $f(S_j, B_j)$ in F_i i.e.

    $f(S_{d_1}, B_{d_1}) = \infty$ and $f(S_{d_n}, B_{d_n}) = \infty$

*for K= 2 to (n-1)

$$f(S_{d_K}, B_{d_K}) = f(S_{d_K}, B_{d_K}) + \frac{f(S,B)(K+1).q - f(S,B)(K-1).q}{f_q^{max} - f_q^{min}}$$

. $E(S_{d_K}, B_{d_K}).q$ is the value of qth an objective function of the Kth individual in $E(S_{d_K}, B_{d_K})$.

The individuals $f(S_1,B_1)$ are chosen by exercising a binary contest selection with crowed comparison-operator $\prec_n$. Following Deb (2000), the point with a lower rank $f(S_1,B_1)_{rank} < m_{rank}$ is preferred between two solutions. Else the point that is included in the region with less number of $f(S_j,B_j)$ points is selected. Therefore, the diversity with non-dominated solutions is presented by using the crowding comparison procedure which is used in the tournament selection and during the population reduction phase. Since solutions compete with their crowding distance (Zamanifar et al., 2014).

Recombination and Selection

Following Bandyopadhyay and Bhattacharya (2014), the offspring population is merged with the current generation population and variety is completed to suit the individuals of the next generation. Elitism is confirmed, subsequently, all the best individuals are included in the population. In this context, the population is now sorted based on non-domination (Arora et al. 2018). Subsequently, the new generation is filled by each front until the population size surpasses the existing population size. For instance, the population exceeds N when adding all the individuals in front F_i, then the individuals in front F_i are chosen based on their crowding distance in the descending order until the population size is N (Moravej et al., 2015 and Arora et al. 2018).

DEBRIS SEGMENTATIONS FROM SATELLITE DATA

Panchromatic and multispectral satellite data are examined using the multi-objective evolutionary algorithm (MOEA). In this regard, genetic algorithm and non-dominated sorting genetic algorithm NSGA-II are implemented and compared.

Debris Investigations From Panchromatic Data Using MOEA

The panchromatic satellite data have claimed investigation of MH370 debris involved Worldview-2 PAN data. Briefly, Worldview-2 PAN has a wavelength, which is ranged between 0.45 and 0.80 μm. It has a ground resolution of 0.52 m. It was claimed that an object (yellow circle) of length 24 m detected in WorldView-2 data in 43° 58′ 34″S and 90° 58′ 34 ″ E in the southern Indian Ocean (Figure 9.14).

The genetic algorithm isolates the suspected MH370 bright debris from the surrounding ocean. However, GA cannot isolate the wave breaking and whitecaps from the suspected object. In this circumstance, the suspected object belongs to the Whitecaps (Figure 15). In this view, individual whitecaps are described by an illumination threshold, which is fluctuated to confirm that the results showed are not belong to MH370. The ocean is covered by the high density of Whitecaps due to the increment of wind speeds of 7 m/s to 14 m/s. Further, both wind stress and wave modulations cause a large spatial fluctuation of whitecaps of approximately 4 km in the WorldView-2 Pan data. The largest whitecaps have a maximum reflectance value of 0.72 in the PAN band of 0.45μm and disappear in a band of 0.70 μm to a band of 0.80 μm. This is another proof of the object that is not metal or man-made- object as the highest reflectance is located in the band of 0.45μm (Figure 16).

At a wavelength of 0.45 μm, the Whitecaps produce the highest reflectance than the wavelength of 0.80μm. On the contrary, the highest absorption mostly occurs with 0.80 μm (Minchin et al., 2017). The rapid rise in attenuation at longer wavelengths is because of absorption by water (Davis 1991).

Besides, on 16th March 2014, a PAN band of Worldview-2 spotted three bright objects which are located between the latitude of 45° 58′ 34″ S and 45° 58′ 30″ S and longitude of 90° 57′ 37″ E and 90° 57′ 40″ E (Figure 17a). On the contrary, the genetic algorithm suggests these objects are just whitecaps (Figure 17b). This confirms the results of Figure 15).

Figure 14. Suspected MH370 debris in WorldView-2 data

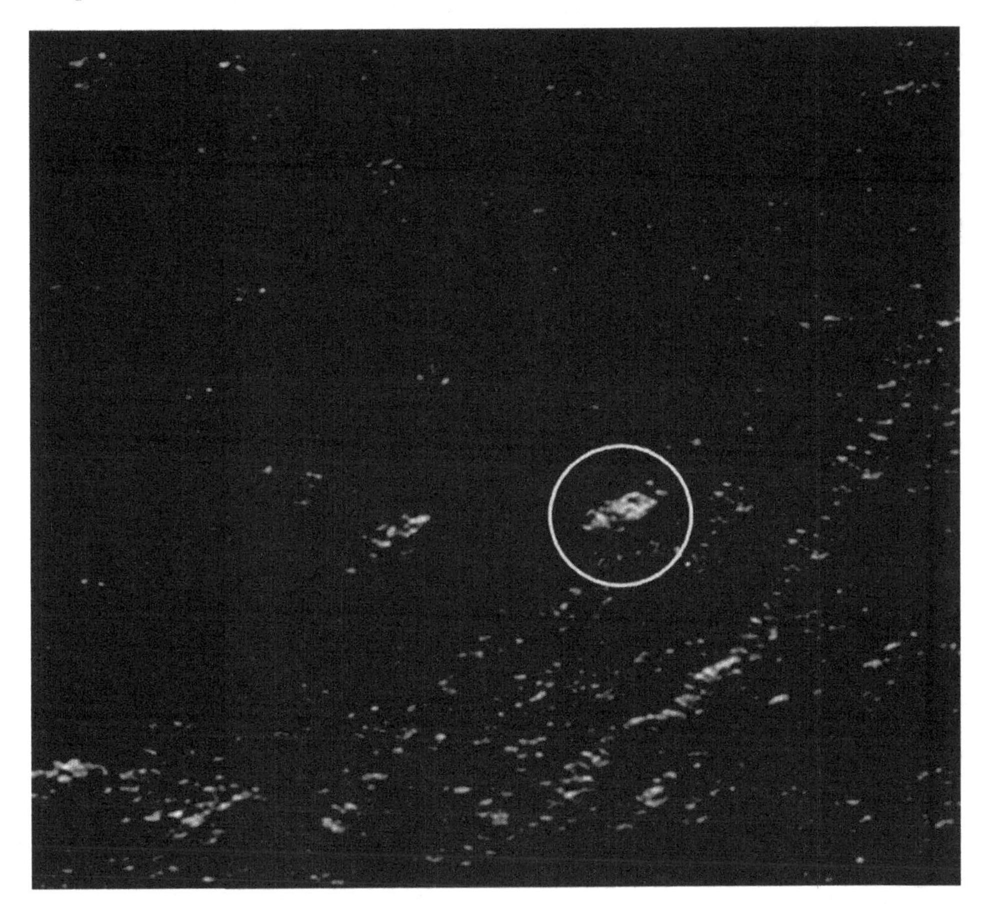

On 18 March 2014, China claimed that panchromatic GF-1 was able to track MH370 debris, which was located approximately 120 km southwest of the Perth, Australia (Figure 18a) between 44° 57′ 30″ S, and longitude 90° 13′ 45″E. On the other hand, a genetic algorithm indicates that the object belongs to an aircraft carrier (Figures 9.18b and 9.18c). The reflectance of the suspected object is close to 0.99 with a PAN band of 0.45μm. In other words, aircraft carrier (Figure 18b and 9.19) has the highest reflectance than Whitecaps (Figure 16). In the panchromatic band, the structure of the aircraft carrier reflects more radiation to the sensor than Whitecaps. Moreover, the spectral signature of the aircraft carrier and turbulent wake fluctuate while the spectral signature of the sea background is flat (Figure 19).

Lastly, the panchromatic Thaichote satellite spotted 300 objects on March 24, 2014, with 2 m resolution (Figure 20a). However, the genetic algorithm indicates that these 300 objects belong to low clouds (Figure 19.20b). Consequently, the reflectance and transmittance of clouds depend on the geometric thickness, the number density of the droplets, and their size distribution. At the wavelength of 0.45 to 2.5 μm, clouds have the spectral variant. In this sense, clouds may be either warmer or colder than the sea surface, and debris, so that one cannot reliably distinguish clouds from debris in the thermal wavelengths.

In the panchromatic band, clouds can be discerned by their textural properties, but not their spectral characteristics. In this context, the correlation between small bright objects and large clouds are similar,

Figure 15.

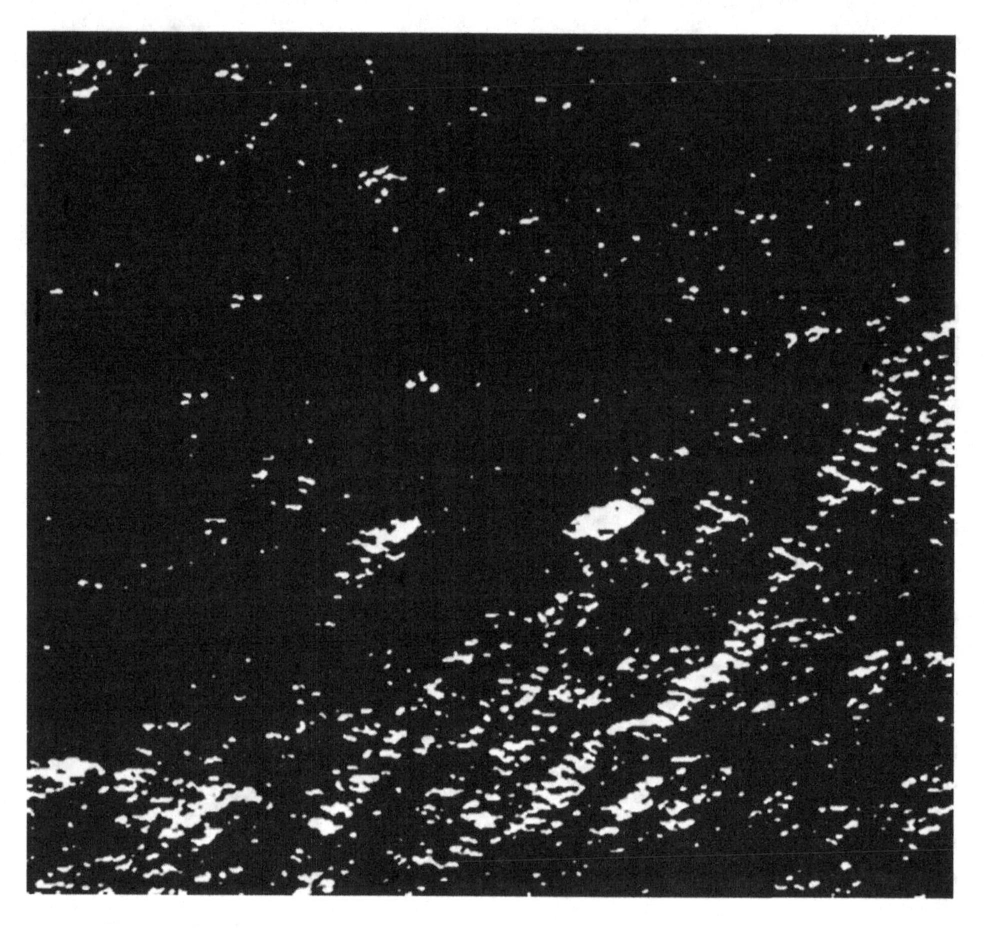

i.e., 1.0 with a standard error of 0.00003 (Table 2). This indicates that the claimed MH370 debris from the panchromatic Thaichote satellite is part of cloud variations.

Table 9.3 also proves that small objects are detected by genetic algorithm belong to small cloud covers. In fact, there is a significant relationship between large cloud covers and small cloud covers. Indeed, the F_{sig} of 0.00034 is smaller than F_{stat} where the p-value is less than 0.5 (Table 3).

Debris Investigations Using NSGA-II From Panchromatic Data

The proposed NSGA-II for the automatic identification of objects has been applied to generate spectral reflectance from three panchromatic satellite datasets (Figure 21). In the initial stages, the standard errors are increased with high population numbers of 55803, 34562 and 31006, respectively. Table 9.4 spectacles that the WorldView-2 image has the highest standard deviation of 0.75 compared to the other satellite data. In fact, it has extra features, for whitecaps, front, and wave crests. Feature generations using the NGSA-II algorithm required the highest numbers of populations. At the initial stage, there are, however, no distinct features in satellite data. Consequently, the random generation patterns are dissimilar among the satellite images due to various objects and variable spectral signatures of inconstant objects in each image(Kamjoo et al., 2016). This confirms the work of Marghany(2013).

Figure 16. The reflectance of the suspected object in the WorldView-2 panchromatic band

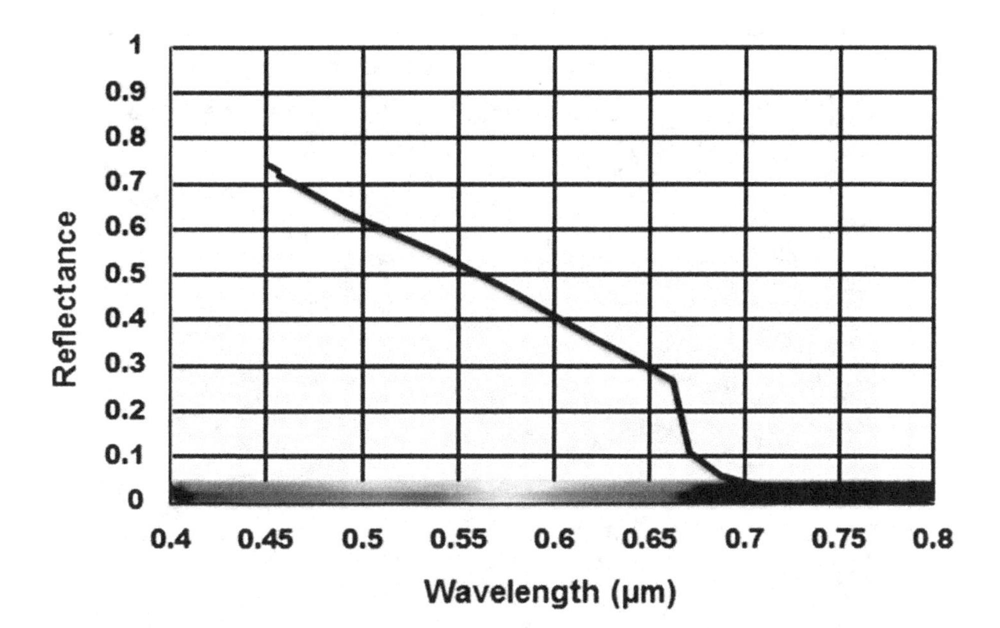

The distinct object features, consequently, are discriminated with the maximum fitness value of 100 and the lowest standard deviation of 0.34 (Table 9.5). In Figure 22, the fitness procedure isolated the objects with the strong identification of their edges in satellite images. The WorldView-2 (Figure 22a) and GF-1 images (Figure 22b) have been generated with a higher fitness value of 100 and the lowest standard error of 0.18, and 0.21, respectively. It is interesting to find that the highest fitness is accompanied by the lowest standard error; as the fitness values increase, the errors are reduced (Deb et al., 2000 ;Deb 2000; Marghany 2013). The NGSA-II confirms the results of a genetic algorithm as the objects existed in panchromatic Thaichote data belong to cloud covers. This proves with a fitness value of 98 and a standard error of 0.23 (Figure 22 c).

Debris Investigations in Multispectral Data Using NGSA-II

The multispectral satellite data of Pléiades-HR 1A data claimed to suspect MH370 debris in the southern Indian Ocean, which is away from Perth by 2245.57 km (Figure 23). This data was acquired on the 23rd of March 2014. This data contain four bands, which are blue, green, red and near-infrared. These contain also panchromatic-sharpened with a radiometric resolution of 12-bits per band, i.e., 4096 levels of intensity per band. Both genetic algorithm and NGSAII are implemented on these images to identify MH370 debris. On the contrary, the images contain mainly water, surface glint, cloud (of varying coherence and thickness), cloud shadows, shaded water and shaded glint (Figure 24).

NGSA-II explains that the three objects spotted in Pléiades-HR 1A (Figure 25a) are whitecaps (Figure 25c). This confirms also by segmentation delivers by genetic algorithm. Consequently, NGSA-II can determine the boundary features of the Whitecaps than a genetic algorithm. This proves a higher fitness value of 100 and less standard error of 0.023 (Table 6).

Figure 17a. Suspected Objects spotted in PAN Worldview-2

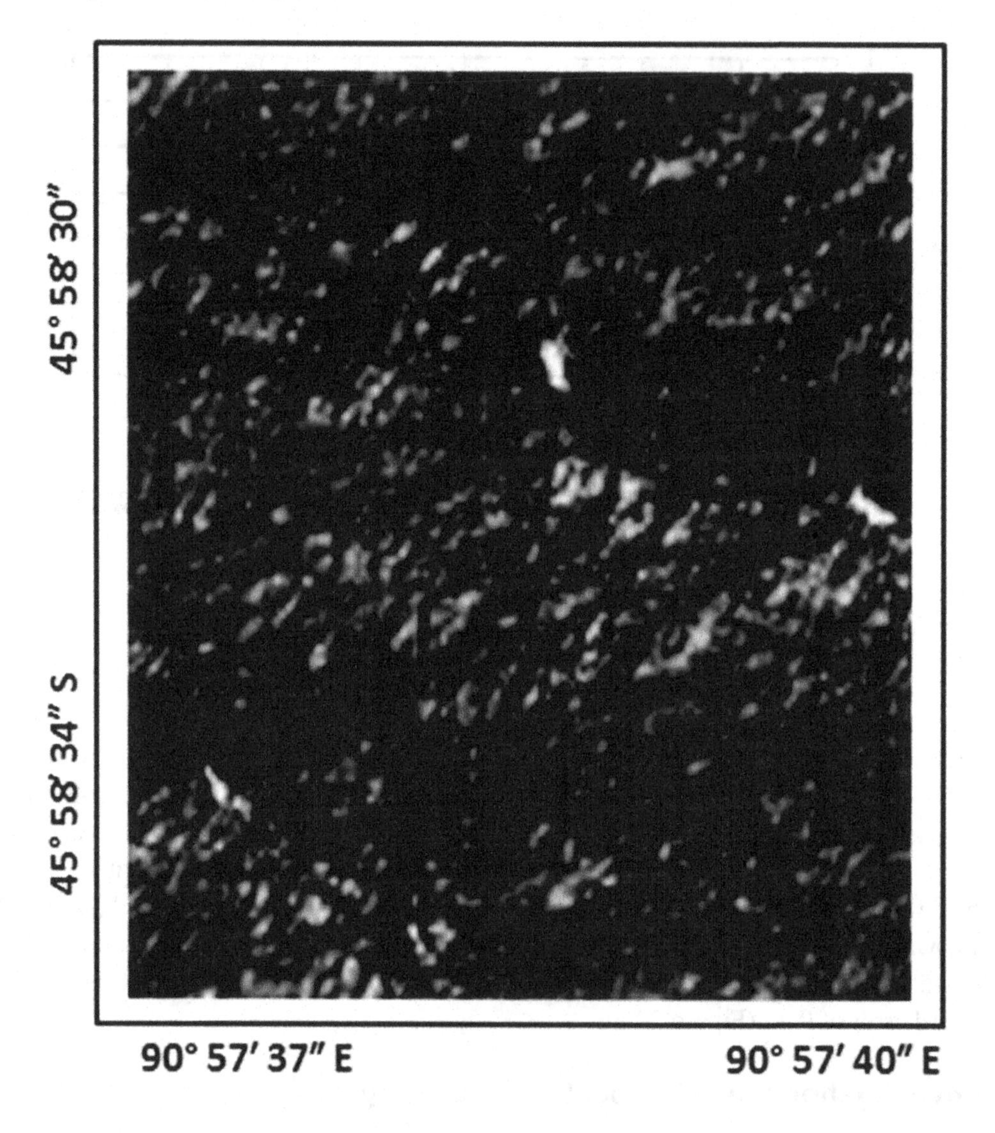

Notably, NGSA-II can define the Whitecaps and provides excellent discrimination of turbulent boundary pixels (Figure 26c). However, the genetic algorithm can segment the suspected objects as the Whitecaps and wave group movements (Figure 26b). On the contrary, NGSA-II distinguishes accurately the ships and their wakes in Pléiades-HR 1A images than a genetic algorithm. The NGSA-II has a higher fitness and less standard error than a genetic algorithm.

Pareto optimal curve confirms that the objects that exist in satellite data do not belong to MH370 debris. In other words, the identified objects are whitecaps, clouds, aircraft carrier, ships and their wakes (Figure 28). In this context, the objective function of the reflectance of the different objects and whitecaps are positioned entirely on the Pareto front curve. In other words, the major points of different object reflections on the Pareto front and are the imprecise points. They are correlated to the Pareto front curve because of the error reduction in the population's generations and fitness procedures (Deb et al., 2000 and Marghany 2013).

Figure 17b. Genetic algorithm

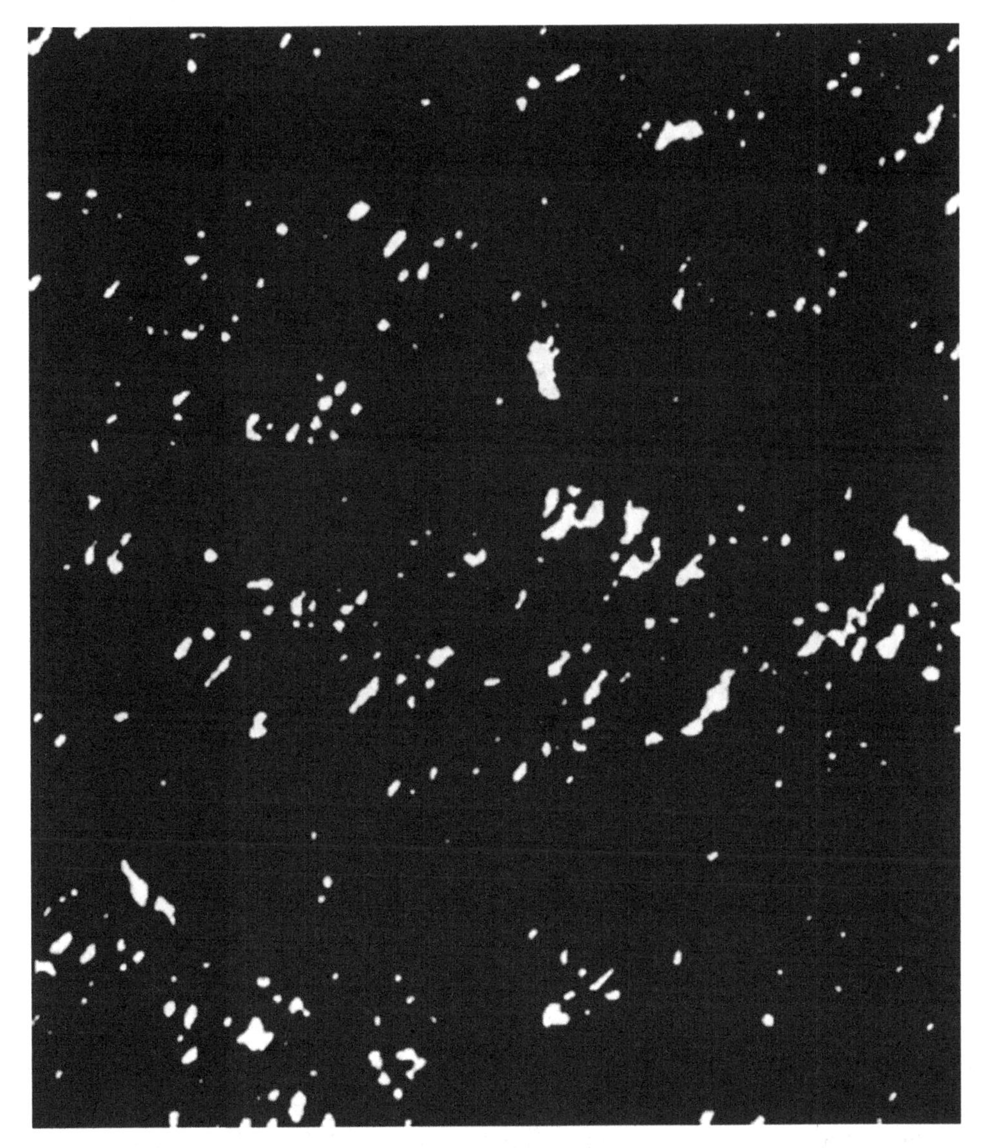

The Pareto set is estimated with excellent precision and effectiveness through adherence to the optimization algorithms. The selection of fewer functions in evaluations of MH370 debris is used to compute each Pareto solution precisely. Through the local approximation of the objective and constraint functions, many evaluations are saved because the approximated model is accurate close to the Pareto set and is left intentionally inaccurate far from it.

Indeed, the NGSA-II provides a set of compromised solutions called Pareto optimal solutions since no single solution can optimize each of the objectives separately. The decision-maker is provided with a set of Pareto optimal solutions to choose a solution based on the decision maker's criteria. This sort of NGSA-II solution technique is called an *a* posteriori method since the decision is taken after searching is finished. In this context, the Pareto-optimal approach does not require any a priori preference decision between the conflict of surrounding waters, clouds, and object pixels. Furthermore, Pareto-optimal points

Figure 18a. Debris investigation GF-1 satellite data

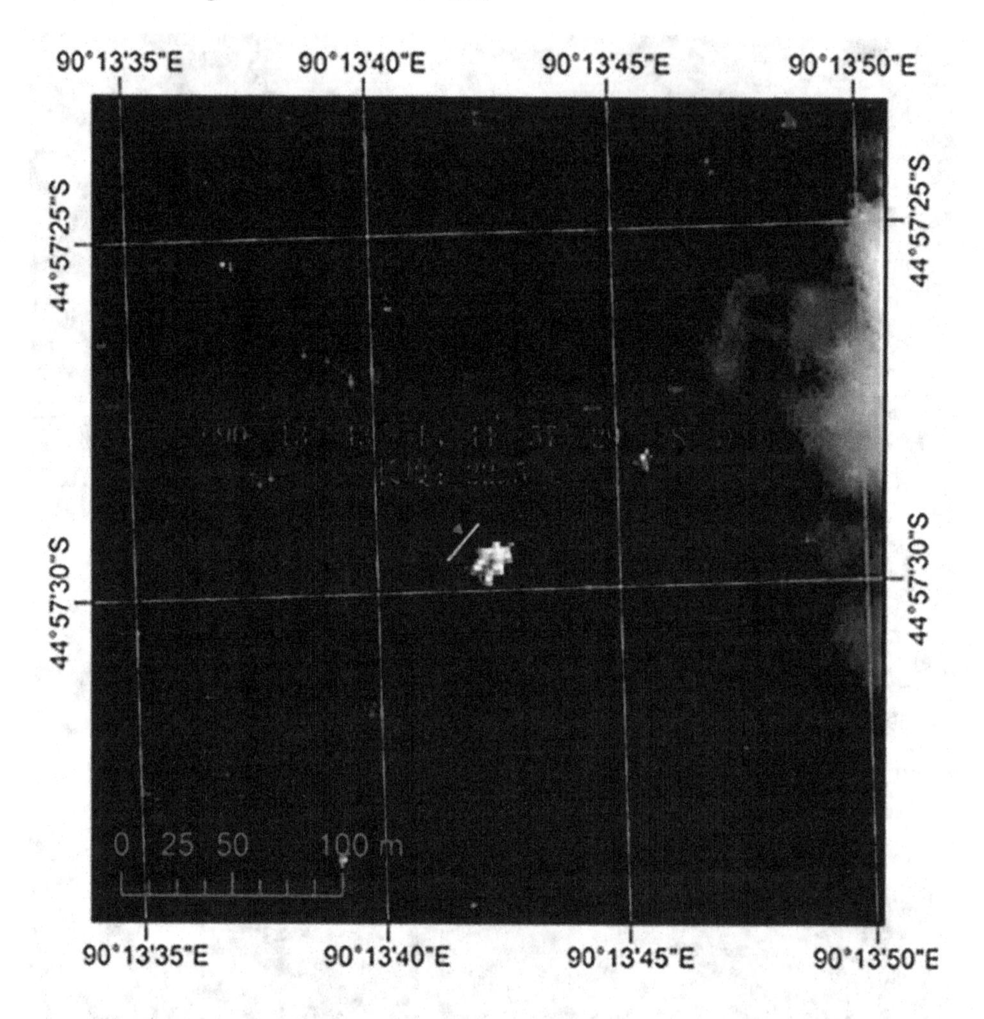

have formed a Pareto-front as shown in Fig. 28 in the multi-pixel objectives function of the satellite data space variables (Hu et al., 2016). NGSA-II, which is based on the Pareto optimal solutions provides excellent discrimination of whitecaps, clouds, aircraft carriers, and ships. This can be confirmed by the receiver–operator characteristics (ROC) curve (Figur 9.29). In this regard, the existence of a weighted sum of objective function converts a conflicting multi-objective problem of suspected MH370 debris and surrounding features into an objective one.

The Pareto-front contains the Pareto-optimal solutions, and in the case of a continuous front, it divides the pixels' objective function space into two parts, which are non-optimal solutions and infeasible solutions. In this regard, it improved the robustness of the pattern search and improved the convergence speed of NGSA-II (Yijie and Gongzhang 2008). In general, the NGSA-II algorithm can automatically identify suspected MH370 debris and confirms its absence in all satellite data along the search area in the Southern Indian Ocean. Furthermore, NGSA-II can identify these objects from the surrounding pixels without using a separate segmentation algorithm (Zamanifar et al., 2014). NGSA-II can identify man-made objects which are reported in previous studies as ships and their wakes (Bayat et al.,2014).

Figure 18b. Genetic algorithm s' result

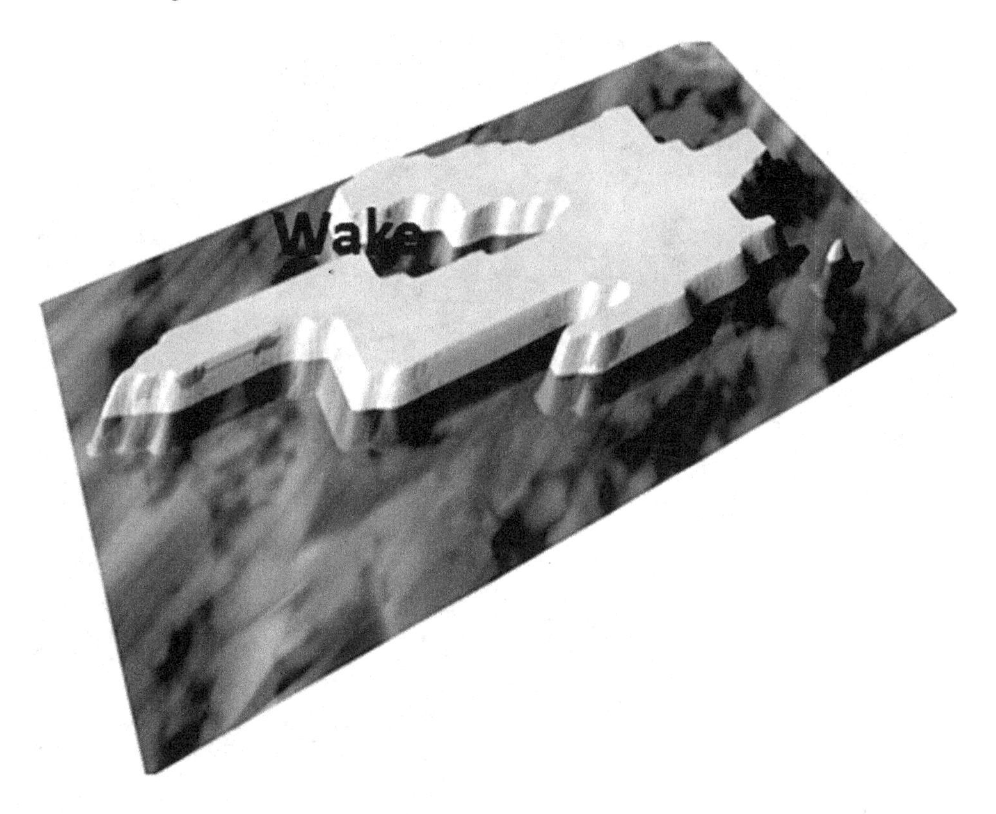

CONCLUSION

This chapter concludes that the objects that are spotted by different satellite data do not scientifically belong to the MH370 debris. In the early stage, the image processing tools are proposed such as PCA and supervised classifications also stated that the objects spotted are just man-made without accurate identifications. On the contrary, the multiobjective genetic algorithms can identify the dotted objects as whitecaps, clouds, ships, and an aircraft carrier. Consequently, where is the MH370 debris? Perhaps the next chapters can deliver more logical clues, which can assist to understand deeply what did occur to the MH370 debris.

Figure 18c. Aircraft carrier generated by the genetic algorithm.

Figure 19. The spectral signature of the aircraft carrier in panchromatic GF-1 data

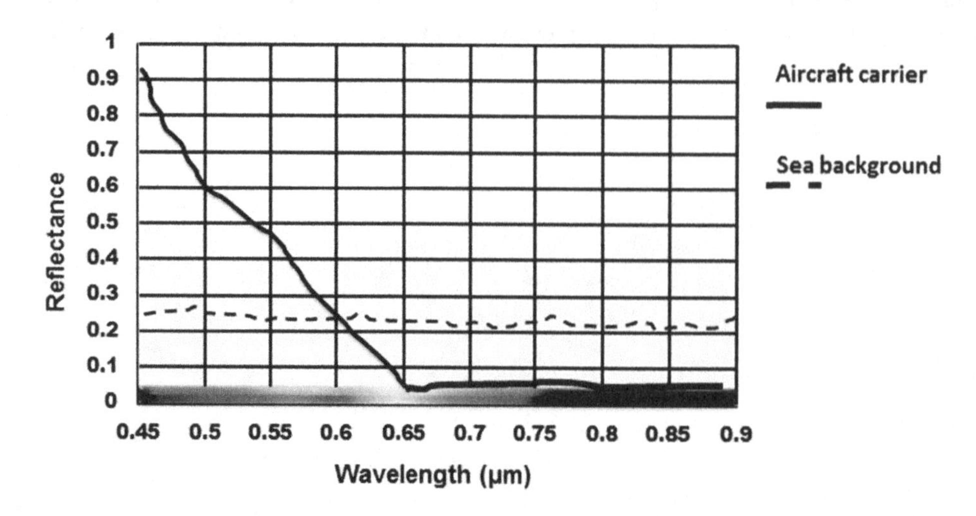

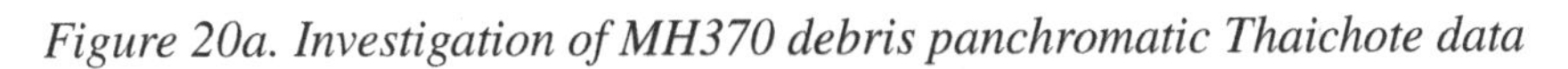

Figure 20a. Investigation of MH370 debris panchromatic Thaichote data

Figure 20b) Genetic algorithm results

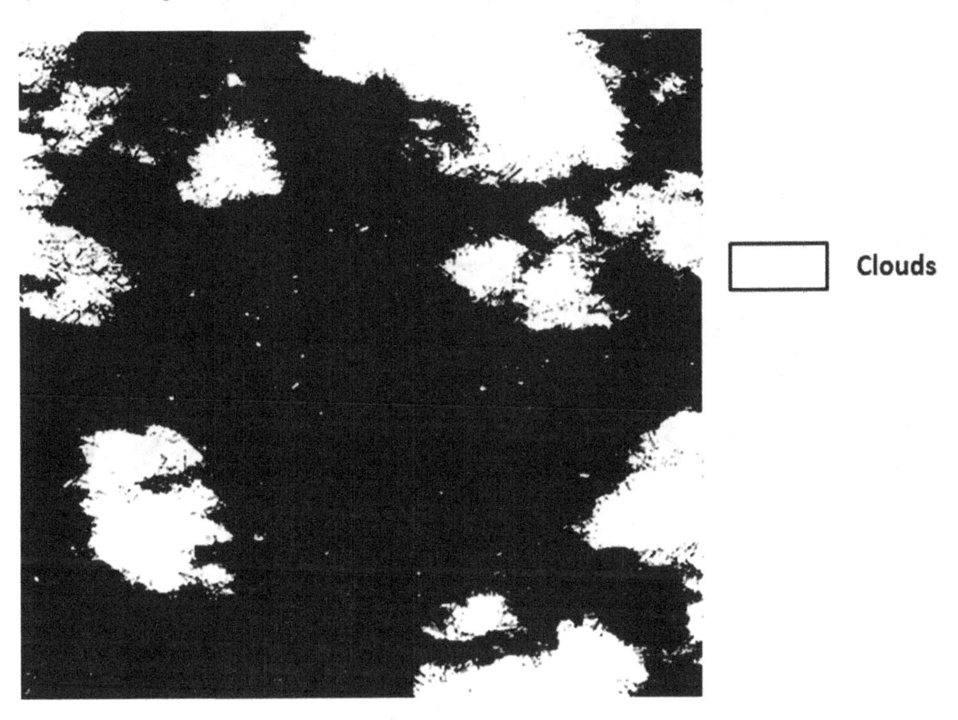

Table 2. Accuracy of cloud detection in panchromatic Thaichote data

Features	Correlation	Standard Error
Large clouds- Small Objects	1.0	0.00003

Table 3. The significant relationship between large and small cloud covers from panchromatic Thaichote data

Features	F_{sig}	F_{stat}	*p*-value	Significant
Large and small cloud covers	0.00034	765	0.00006	Significant

Figure 21a. Random generation of panchromatic satellite data for WorldView-2

Figure 21b. GF-1

Figure 21c. Thaichote data

Table 4. The standard deviation for the number of iterations for different satellite data

Satellite data	Number of iterations	Standard deviation
WorldView-2	55803	0.75
GF-1	34562	0.71
Thaichote data	31006	0.52

Figure 22a. The fitness of NGSA-II algorithm for automatic object identifications in WorldView-2

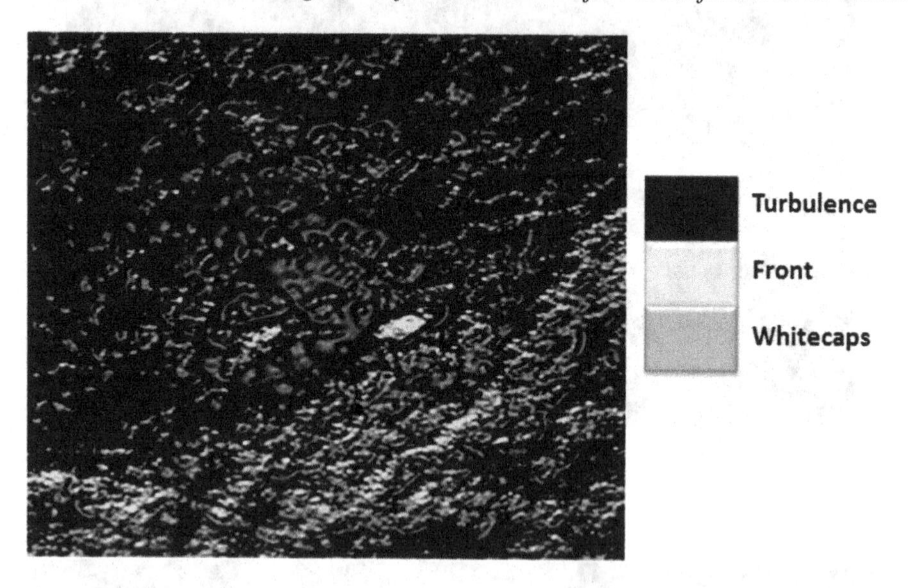

Figure 22b. GF-1

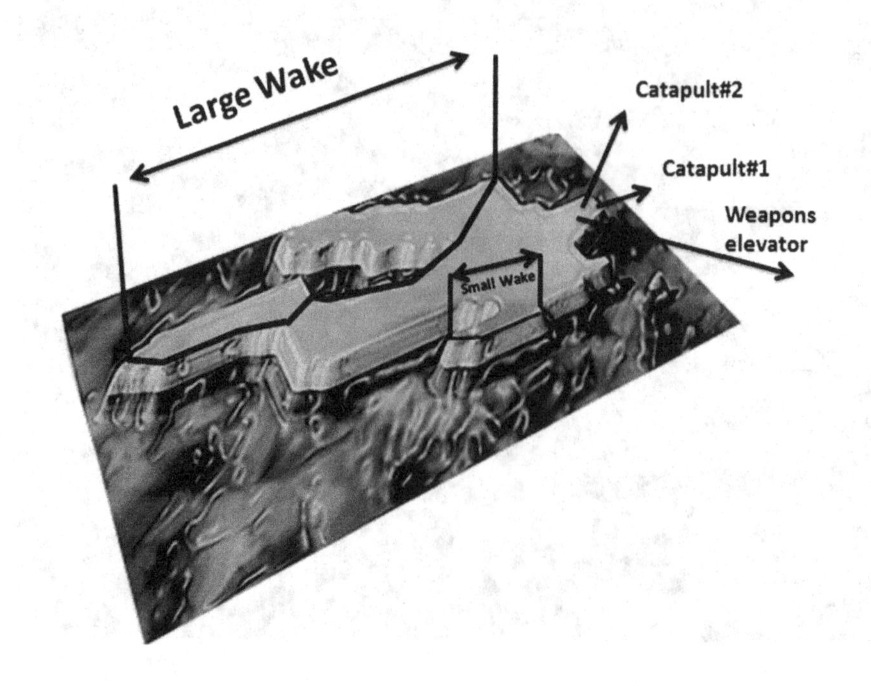

Figure 22c. Thaichote data

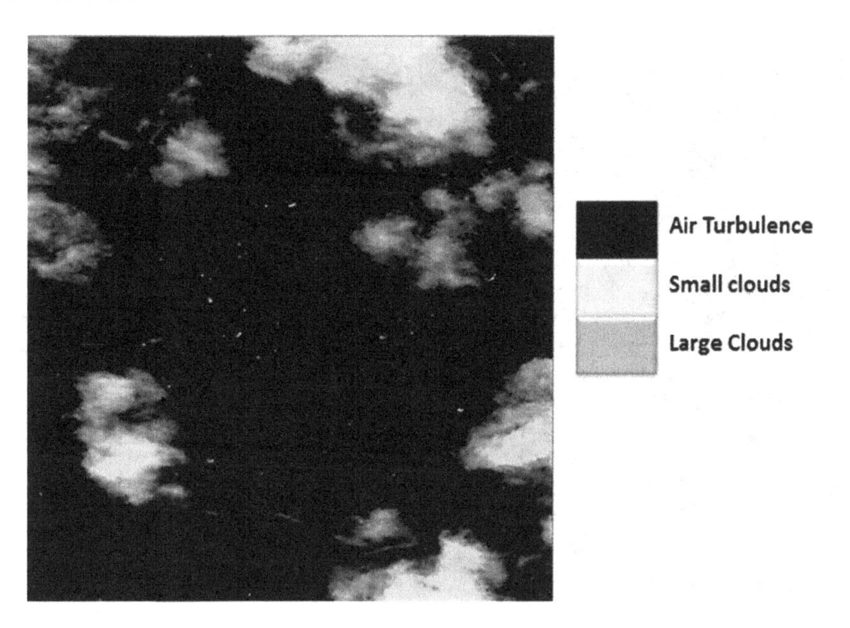

Table 5. Fitness and standard deviation

Satellite data	Features	Fitness	Standard error
WorldView-2	Whitecaps	100	0.18
GF-1	Aircraft carrier	100	0.21
Thaichote data	Clouds	98	0.23

Figure 23. The geographical coverage of Pléiades-HR 1A satellite data

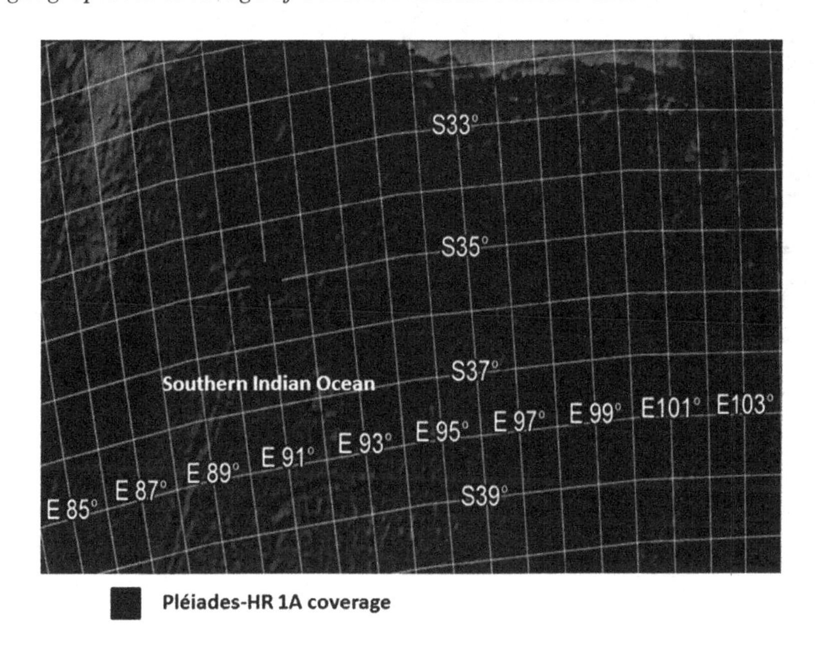

Figure 24. Cloud covers, surface glint and cloud shadows in Pléiades-HR 1A satellite data

Figure 25a. Object identifications from Pléiades-HR 1A

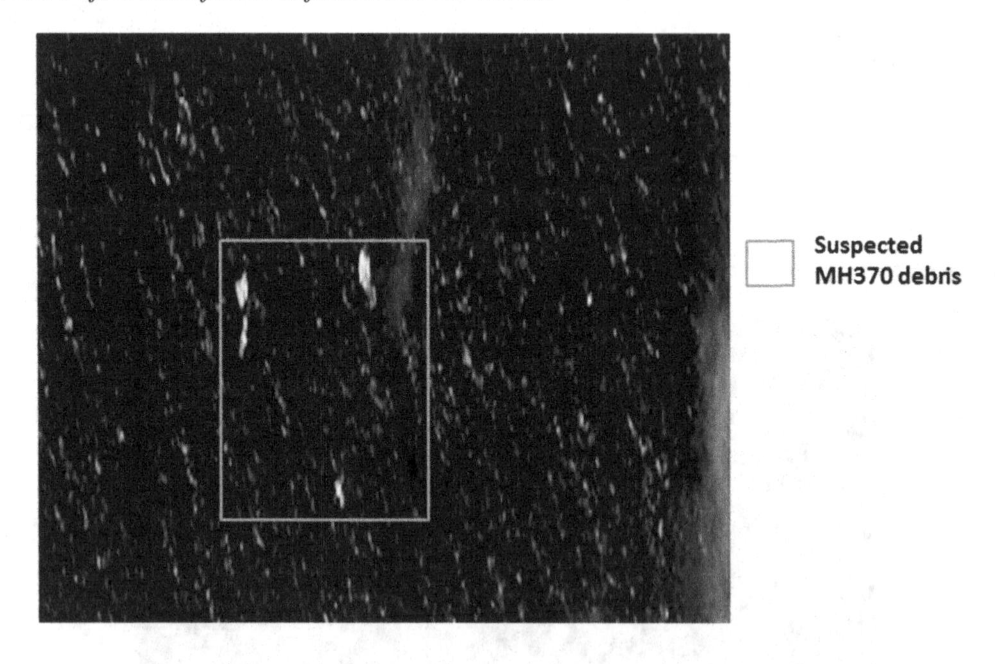

Figure 25b. Genetic algorithm

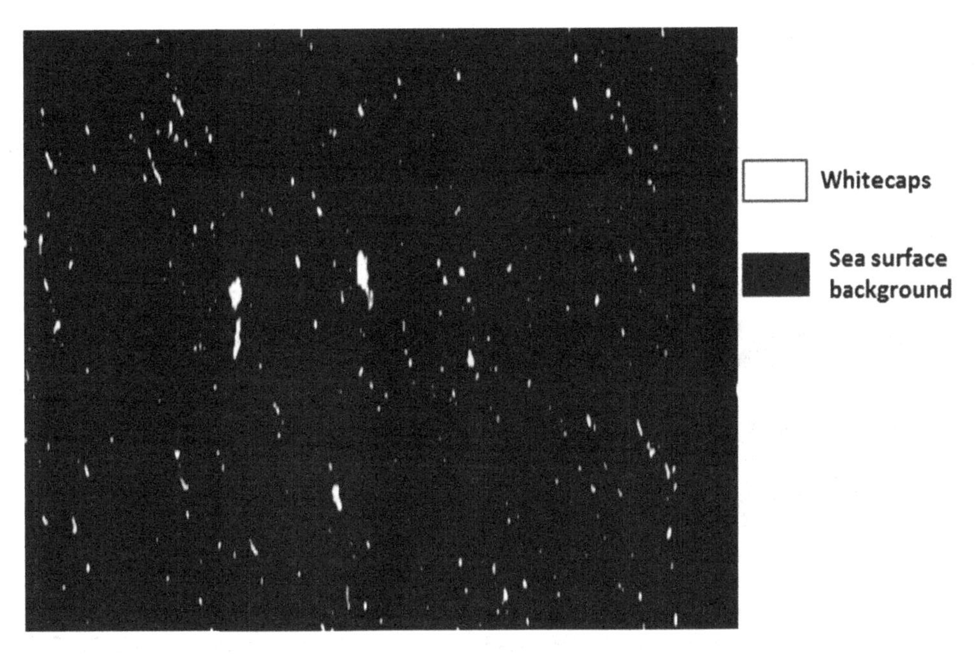

Figure 25c. NGSA-II algorithm

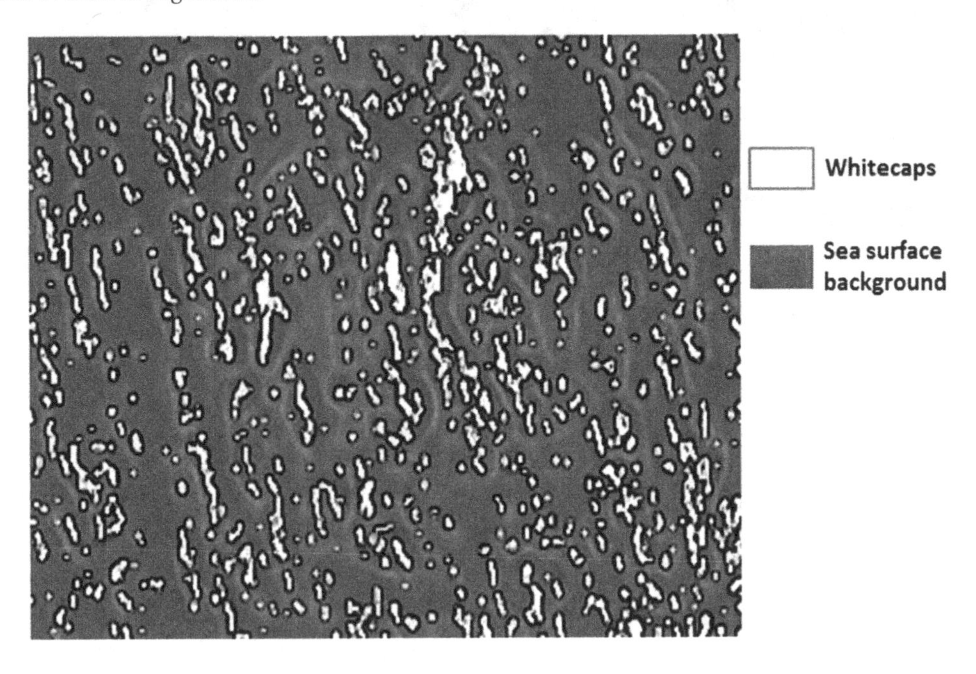

Table 6. Comparison between NGSA-II and Genetic Algorithm

Algorithm	Fitness	Standard error
NGSA-II	100	0.023
Genetic	96	0.15

Figure 27. Ships and their wakes identification in Pléiades-HR 1A satellite data

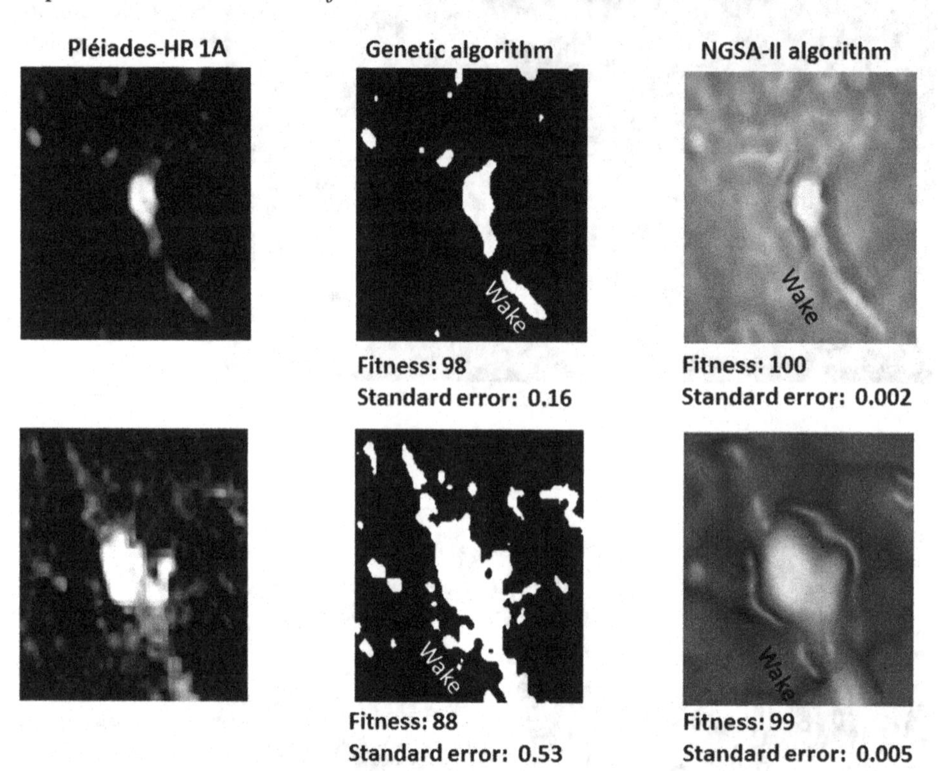

Figure 28. Pareto optimal solution for different objects

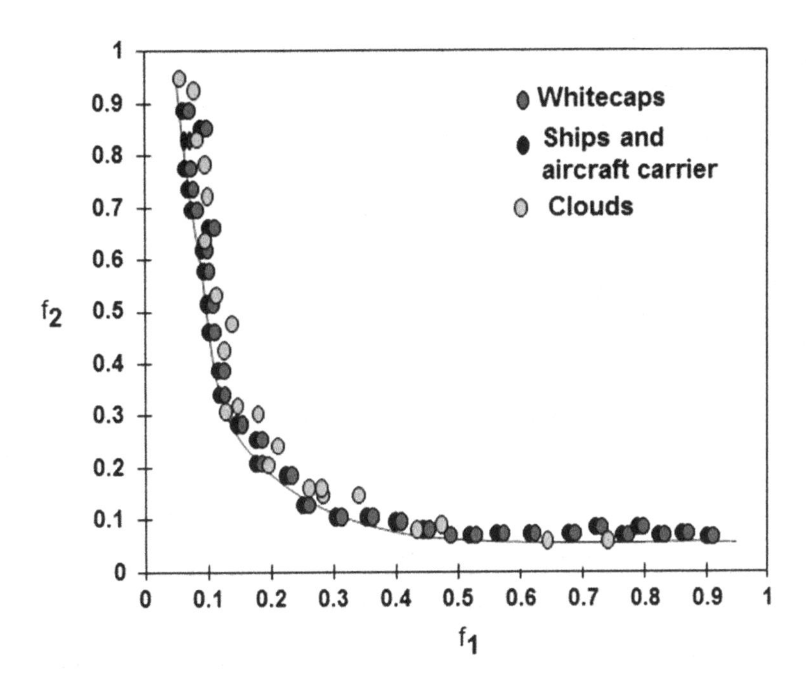

Figure 29. ROC for object identification using NGSAII

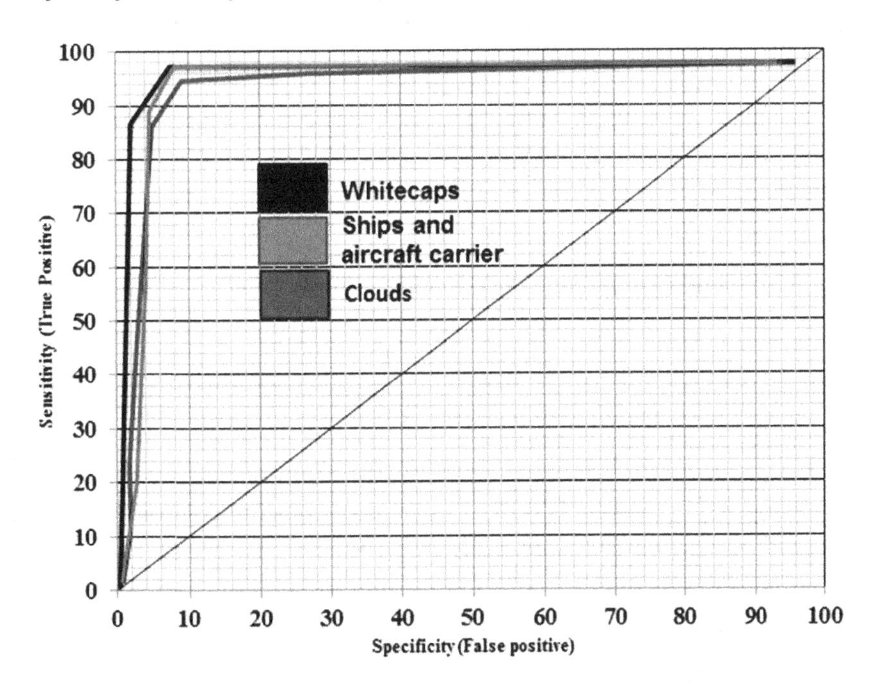

REFERENCES

Arora, R., Kaushik, S. C., Kumar, R., & Arora, R. (2016). Multi-objective thermo-economic optimization of solar parabolic dish Stirling heat engine with regenerative losses using NSGA-II and decision making. *International Journal of Electrical Power & Energy Systems, 74*, 25–35.

Bandyopadhyay, S., & Bhattacharya, R. (2014). Solving a tri-objective supply chain problem with modified NSGA-II algorithm. *Journal of Manufacturing Systems, 33*(1), 41–50.

Bayat, M., Dehghani, Z., & Rahimpour, M. R. (2014). Dynamic multi-objective optimization of industrial radial-flow fixed-bed reactor of heavy paraffin dehydrogenation in LAB plant using NSGA-II method. *Journal of the Taiwan Institute of Chemical Engineers, 45*(4), 1474–1484.

Buayai, K., Ongsakul, W., & Mithulananthan, N. (2012). Multi-objective micro-grid planning by NSGA-II in primary distribution system. *European Transactions on Electrical Power, 22*(2), 170–187.

Davis, L. (1991). *The Handbook of Genetic Algorithms*. Van Nostran Reingold.

Deb, K. (2000). An efficient constraint handling method for genetic algorithms. *Computer Methods in Applied Mechanics and Engineering, 186*(2-4), 311–338.

Deb, K., Agrawal, S., Pratap, A., & Meyarivan, T. (2000, September). A fast elitist non-dominated sorting genetic algorithm for multi-objective optimization: NSGA-II. In *International conference on parallel problem solving from nature* (pp. 849-858). Springer.

Dozier, J. (1989). Spectral signature of alpine snow cover from the Landsat Thematic Mapper. *Remote Sensing of Environment, 28*(1), 9–22.

Guo, D., Wang, J., Huang, J., Han, R., & Song, M. (2010, October). Chaotic-NSGA-II: an effective algorithm to solve multi-objective optimization problems. In *2010 International Conference on Intelligent Computing and Integrated Systems* (pp. 20-23). IEEE.

Fortin, F. A., & Parizeau, M. (2013, July). Revisiting the NSGA-II crowding-distance computation. In *Proceedings of the 15th annual conference on Genetic and evolutionary computation* (pp. 623-630). Academic Press.

Hu, Y., Bie, Z., Ding, T., & Lin, Y. (2016). An NSGA-II based multi-objective optimization for combined gas and electricity network expansion planning. *Applied Energy, 167*, 280–293.

Kamjoo, A., Maheri, A., Dizqah, A. M., & Putrus, G. A. (2016). Multi-objective design under uncertainties of hybrid renewable energy system using NSGA-II and chance constrained programming. *International Journal of Electrical Power & Energy Systems, 74*, 187–194.

Kahlouche, S., Achour, K., & Benkhelif, M. (2002, June). A new approach to image segmentation using genetic algorithm with mathematical morphology. In *Proc. 2002 WSEAS Int. Conf., 12-16 June 2002, Cadiz, Spain* (pp. 1-5). Academic Press.

Marghany, M. (2013, June). Genetic algorithm for oil spill automatic detection from ENVISAT satellite data. In *International Conference on Computational Science and Its Applications* (pp. 587-598). Springer.

Michalewicz, Z. (1994). *Genetic Algorithms+Data Structures*. Evolution Programs.

Minchin, S., Tran, M., Byrne, G., Lewis, A., & Mueller, N. (2017). *Summary of Imagery Analyses for Non-natural Objects in Support of the Search for Flight MH370: Results from the Analysis of Imagery from the PLEIADES 1A Satellite Undertaken by Geoscience Australia*. Geoscience Australia.

Mohanta, R. K., & Sethi, B. (2011). A Review of Genetic Algorithm application for Image Segmentation. *International Journal of Computer Technology & Applications.*, *3*(2), 720–723.

Moravej, Z., Adelnia, F., & Abbasi, F. (2015). Optimal coordination of directional overcurrent relays using NSGA-II. *Electric Power Systems Research*, *119*, 228–236.

Sivanandam, S. N., & Deepa, S. N. (2008). Genetic algorithm optimization problems. In *Introduction to Genetic Algorithms* (pp. 165–209). Springer.

Trinanes, J. A., Olascoaga, M. J., Goni, G. J., Maximenko, N. A., Griffin, D. A., & Hafner, J. (2016). Analysis of flight MH370 potential debris trajectories using ocean observations and numerical model results. *Journal of Operational Oceanography*, *9*(2), 126–138.

Weng, Q. (2011). *Advances in environmental remote sensing: sensors, algorithms, and applications*. CRC Press.

Yijie, S., & Gongzhang, S. (2008). Improved NSGA-II multi-objective genetic algorithm based on hybridization-encouraged mechanism. *Chinese Journal of Aeronautics*, *21*(6), 540–549.

Zamanifar, M., Fani, B., Golshan, M. E. H., & Karshenas, H. R. (2014). Dynamic modeling and optimal control of DFIG wind energy systems using DFT and NSGA-II. *Electric Power Systems Research*, *108*, 50–58.

Zitzler, E., Deb, K., & Thiele, L. (2000). Comparison of multiobjective evolutionary algorithms: Empirical results. *Evolutionary Computation*, *8*(2), 173–195. PMID:10843520

Chapter 10
Fundamentals of Altimeter Microwave Satellite Data

ABSTRACT

Large-scale oceans such as the Indian Ocean require a specific sensor to cover such a huge area. In the case of MH370 searching, ocean dynamic parameters over this extremely huge ocean are required to understand the trajectory movement of MH370 debris. The best sensor could assist is the altimeter satellite data. In fact, this data can deliver several ocean dynamic parameters such as wave height, Rosby wave pattern, sea level variability, and ocean surface current. This chapter aims at delivering a fundamental review of the altimeter satellite data. This chapter shows that there are two main components of radar altimeters: (1) frequency modulated continuous wave (FMCW) and (2) pulse altimeters, which are a function of used radar signals. Two sorts of FMCW altimeters are mainly implemented: broad-beamwidth types and narrow-beamwidth. Moreover, the chapter has listed the variety of altimeter satellite data.

INTRODUCTION

Up to date, there is no study that has implemented altimeter data to track the impact of ocean dynamic features, i.e., circulation, waves, etc., on the trajectory movement of MH370 debris. In this view, optical satellite sensors have used to track MH370 debris without success as discussed in Chapter 9. Prior to use altimeter satellite data, this chapter is addressing the principles of altimeter microwave data to complete thought of how these data can contribute to search MH370 debris.

Altimeter satellite data have an incredible achievement to apprehend the ocean circulation dynamic system. In this understanding, the trajectory movement of MH370 debris is usually a function of ocean circulation. In this view, the perfect empathetic of altimeter mechanics for monitoring, simulation, and tracing ocean circulation can assist to comprehend the mechanics of MH370 crashing in the Indian Ocean.

This chapter critically evaluates the existing altimeter sensors which monitored the Southern Indian Ocean circulation and wave dissipation to bridge the gap found between various remote sensing data recorded as a part of the MH370 crashing scenario. In this understanding, the wide range of ocean circulation is required to forecast and predict any MH370 debris trajectory movement across the Indian

DOI: 10.4018/978-1-7998-1920-2.ch010

Ocean. It is concluded that different comprehensive approaches are necessary to develop a forecasting tool for assessing and monitoring debris trajectory movement (Chelton et al., 2001).

An altimeter is referred to as an altitude meter. In this view, it is a device to compute the object's height above a stable point. In this regard, the estimation of altitude is known as an altimeter. The bathymetry, consequently, is associated with an altimeter which is the computation of depth beneath the sea surface. The dimension is usually measured from the altimeter platform, i.e. satellite or aircraft and the Earth's surface (Figure 1). In other words, altimeters are nadir-looking pulse-radars often. They transmit short microwave pulses and measure the round trip time delay to targets to determine their distance from the air- or spaceborne sensor.

Figure 1. The concept of the altimeter

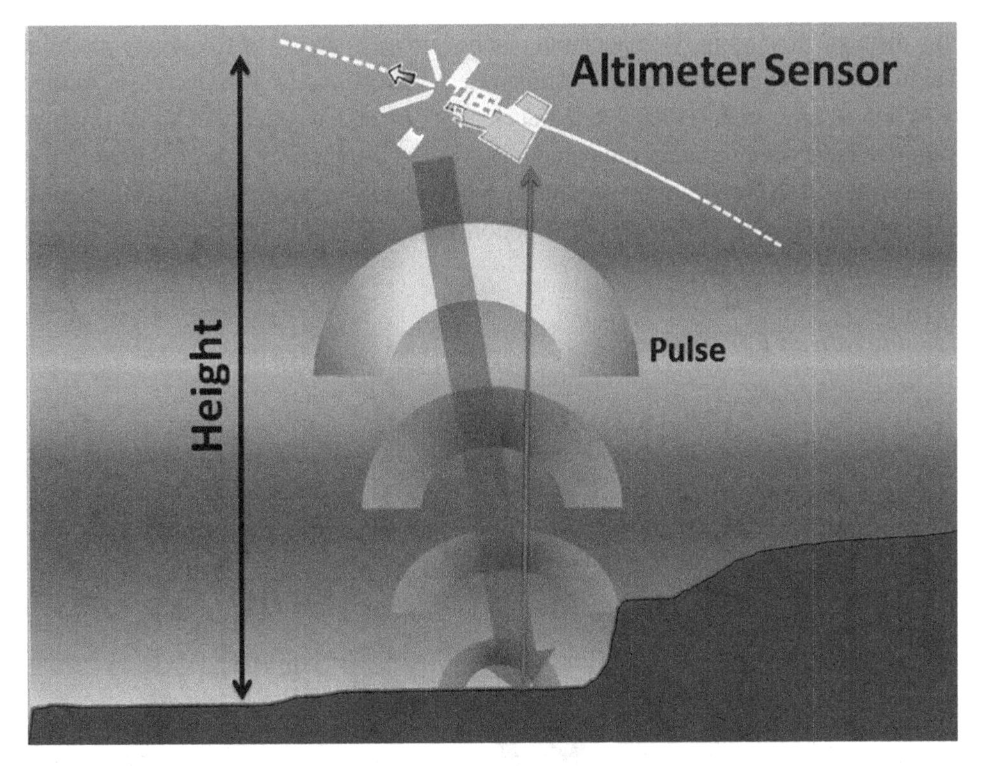

PRINCIPLES OF ALTIMETER

Like a synthetic aperture radar (SAR), altimeter emits a radar signal and then records the backscattered signal from the objects. Unlike SAR, the altimeter emits and receives radar waveform perpendicular to the object. This mechanism allows estimating the object height from the inverted backscatter signal. In this context, the ocean wave height can be inverted more easily than SAR as a function of perpendicular backscatter signal's amplitude (Cheney et al.,1987). The main signal bands used with altimeter are E-band, K_a-band, and S-band. Advanced sea-level retrieving parameters are easily made by S-band. In this regard, the reliable and precise ocean wave height is delivered by altimeter than SAR sensors.

Sort of Radar Altimeter

There are two main components of radar altimeters: (i) frequency modulated continuous wave (FMCW) and (ii) pulse altimeters which are a function of used radar signals. Two sorts of FMCW altimeters are mainly implemented (i) broad-beamwidth types and narrow-beamwidth. Both FMCW altimeters are a function of antenna beamwidth. In contrast, the pulse altimeters are well known as short-pulse altimeters or pulse-compressions which are a function of intrapulse modulation. Besides, an altimeter is also operating in optical bands, for instance, laser altimeters.

In other words, simple continuous wave radar devices without frequency modulation have the disadvantage that it cannot determine target range because it lacks the timing mark necessary to allow the system to time accurately the transmit and receive cycle and to convert this into range. Such a time reference for measuring the distance of stationary objects, but can be generated using of frequency modulation of the transmitted signal. In this method, a signal is transmitted, which increases or decreases in the frequency periodically. When an echo signal is received, that change of frequency gets a delay Δt (by runtime shift) like to as the pulse radar technique. In pulse radar, however, the runtime must be measured directly. In FMCW radar are measured the differences in phase or frequency between the actual transmitted and the received signal instead (Figure 2) (Witter and Chelton 1991).

Figure 2. Ranging with an FMCW system

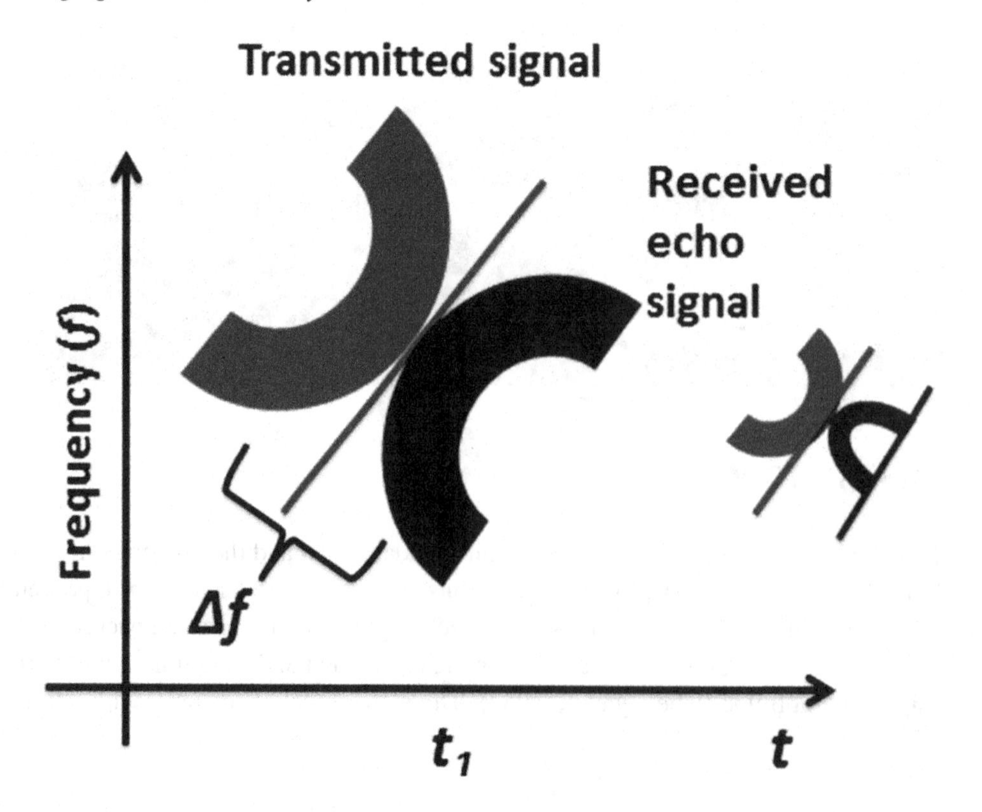

For lower height measurements a frequency modulated continuous wave radar (FMCW) can be used. This technology provides a better resolution but a lower maximum unambiguous height measurement at the comparable wiring effort. Often both technologies are used simultaneously (Cheney et al.,1987 and Chelton et al., 2001).

The Geoid

The geoid is termed as the shape of the ocean surface which it would have if it were enveloped with ocean surface at comparative relaxation to the Earth rotating. In this view, the geoid is the reference used to ensure the precise height measurements by altimeters (Figure 3). Therefore, mass concentration pulls the geoid away from a perfect sphere shape which is mainly affected by the Earth's rotation. In this context, the geoid is also considered as the total of the Earth rotation impact and gravity (Hayne, 1980).

Figure 3. The principle of altimeter measurements

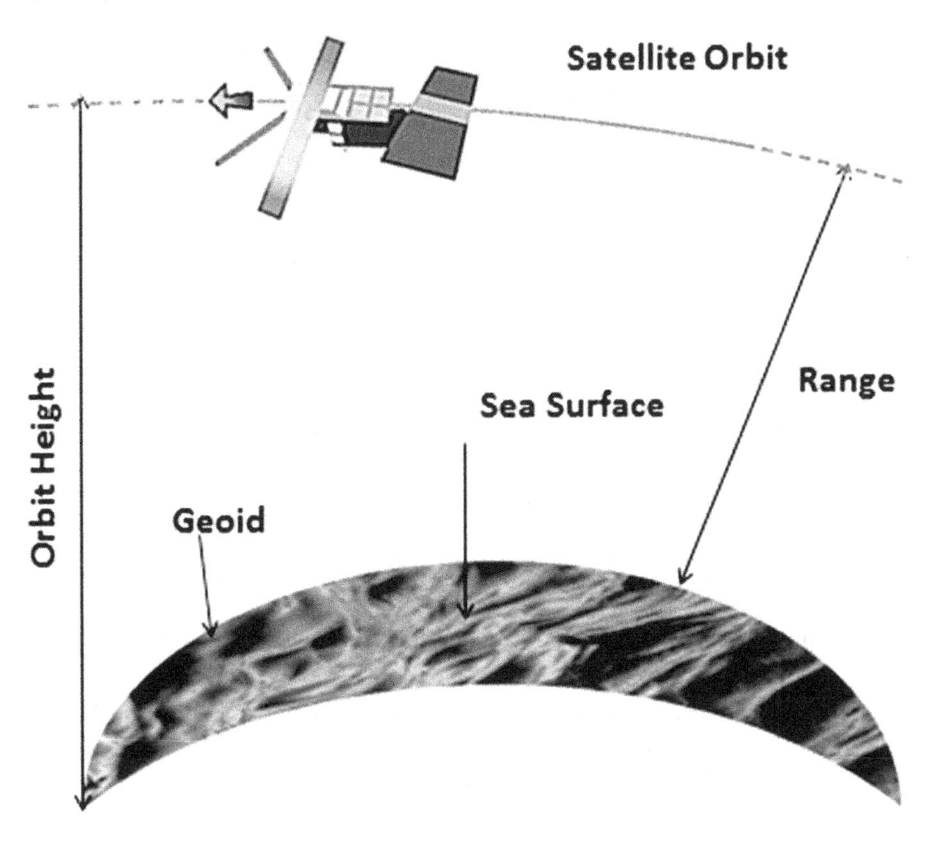

Geoid heights, moreover, are obtained relative to a reference ellipsoid. Indeed, the reference ellipsoid is ultimately suitability. In other words, using a reference ellipsoid is reducing the implantation of massive numbers, which allows for precise computations.

Reference Ellipsoid

It is the best fitting ellipsoid to the geoid. An ellipsoid is essentially a sphere with a bulge at the equator. To first order, this accounts for over 90% of the geoid (Figure 4). Consequently, sea surface peak measurements from the center of the Earth are in the order of approximately 6000 km. With the aid of putting off a reference surface, the heights corresponding to the ellipsoid are about 100 m.

Figure 4. Surface reference ellipsoid

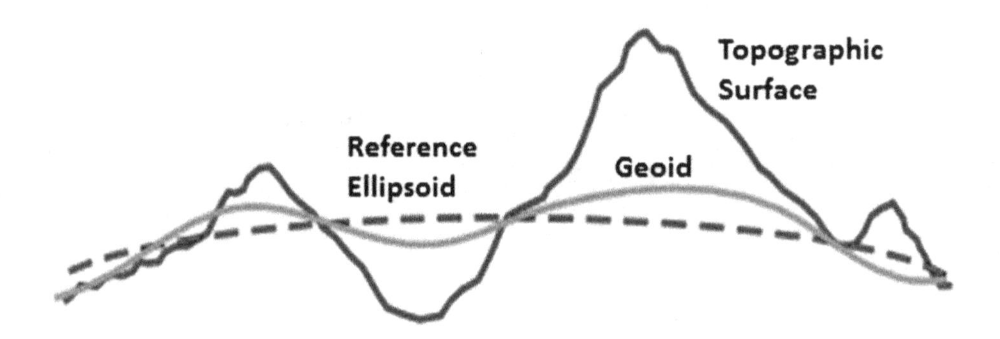

In reality, any reference surface can be used. A sphere could be paintings; nevertheless, sea surface peak differences can be as huge as 20 km. In this regard, an ellipsoid can create less accurate information since the geoid isn't widely recognized regionally. In this context, altimeters are commonly flown in orbits which have a precisely repeating ground track, every 9.9156 days. By subtracting sea-surface height from one traverse of the ground track from height measured on a later traverse, changes in topography can be observed without knowing the geoid. Modifications in topography, therefore, can be revealed without identifying the geoid.

Generally, the geoid is steady in time, and the deduction eliminates the geoid, revealing modifications because of exchanging currents, along with mesoscale inconsistency, supposing tides being eliminated from the facts. Mesoscale changeability contains eddies with diameters more or less than 20 and 500 km. The tremendous accuracy and precision of Topex/Poseidon's altimetric device approve the measurements of the oceanic topography over ocean basins with an accuracy of ± 5 cm.

Range and Azimuth Resolutions

Like SAR, range R measurement is included in altimeter devices. In time t, altimeter devices transmit more than 1700 pulses per second as signals to the Earth's surface and then receive the backscatter signal in the waveform. This mechanism is a function of a range R. Altimeter range resolution is mathematically estimated by:

$$R_r = \tau\ c/2\cos(\gamma) \quad (10.1)$$

where τ is pulse length γ and c is the speed of light which converts pulse length from units of time to distance. Far-field targets are seen at high resolution than near-field targets! The travel time is longer, so a shorter distance can be determined at a specified pulse length. When the travel time is short, either the targets must be spaced out or the pulse must be short. In this regard, Range resolution grows with decreasing the pulse length.

On the contrary, the nearer range resolution is higher than the far range resolution in the azimuth direction. To be settled, targets must be more apart than the beam width. The function of the wavelength of the pulse (λ), slant range distance (S), and antenna length (D) are given by:

$$R_a = (0.7)(S)(\lambda)/D \quad (10.2)$$

here $S = \text{Altitude}/\cos(90-\gamma)$ and γ is depression angle.

Nevertheless, as electromagnetic waves propagate through the atmosphere, they may be decelerated by using water vapour or ionisation. Once those phenomena had been corrected, the final range can be expected with high accuracy. The ultimate aim is to compute the surface height. This requires independent measurements of the satellite's orbital trajectory, i.e., genuine latitude, longitude and latitude coordinates (Robinson, 2004).

Satellite Altitude

The important orbital parameters for satellite altimeter missions are altitude, inclination and duration period of the transmitted signal. The altitude of a satellite altimeter is the satellite's distance with respect to an arbitrary reference (e.g., the reference ellipsoid, a difficult approximation of the earth's surface). The altitude of altimeter relies upon some of the constraints (e.g., inclination, atmospheric drag, gravity forces performing on the satellite, area of the sector to be mapped, and so on) (Marghany, 2018). The period or 'repeat orbit' is the time wished for the satellite altimeter bypass over the equal function on the ground, uniformly sampling the earth's surface. Inclination delivers the highest range at which the satellite can perform the measurements (Chelton et al., 2001).

HOW DOES A RADIO ALTIMETER WORK?

Radio altimeters are simpler and work in a similar way to radar (the system planes, ships, and other vehicles use to navigate). In other words, they just radiate a beam of radio waves down from the satellite and plane and wait for the reflections to return. Since radio waves travel at the speed of light (300,000 km or 186,000 miles each second), it takes only a few hundredths of a second for a radio beam to make the 20,000-meter or so round trip to Earth's surface and back. The plane times the beam and calculates its altitude in kilometres by multiplying the time in seconds by 150,000 (that's 300,000 divided by two). In this understanding, the beam has travelled twice as far as its own altitude going to the ground and back again). Radio altimeters are much quicker and more precise than pressure instruments and are widely used in high-speed aeroplanes or ones that need to fly at particularly low altitudes, such as jet fighters.

In general, the speed of light is about a million times faster than the cruising speed of a typical plane (v), so a radio signal bounced to the ground and back travels a distance of about twice the plane's altitude ($2h$) (Figure 5). In this regard, the altitude must multiply the time of the signal, which radiates from the transmitter and reflects the receiver by half the speed of light. In theory, the faster the plane travels, the less accurate the measurement, because the radio beam has further to travel; in practice, the speed of light is so much faster than the speed of the plane that any error is minimal.

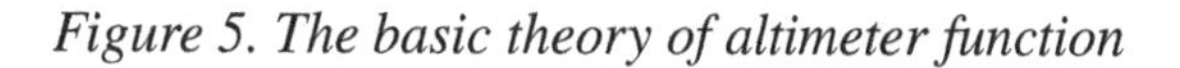

Figure 5. The basic theory of altimeter function

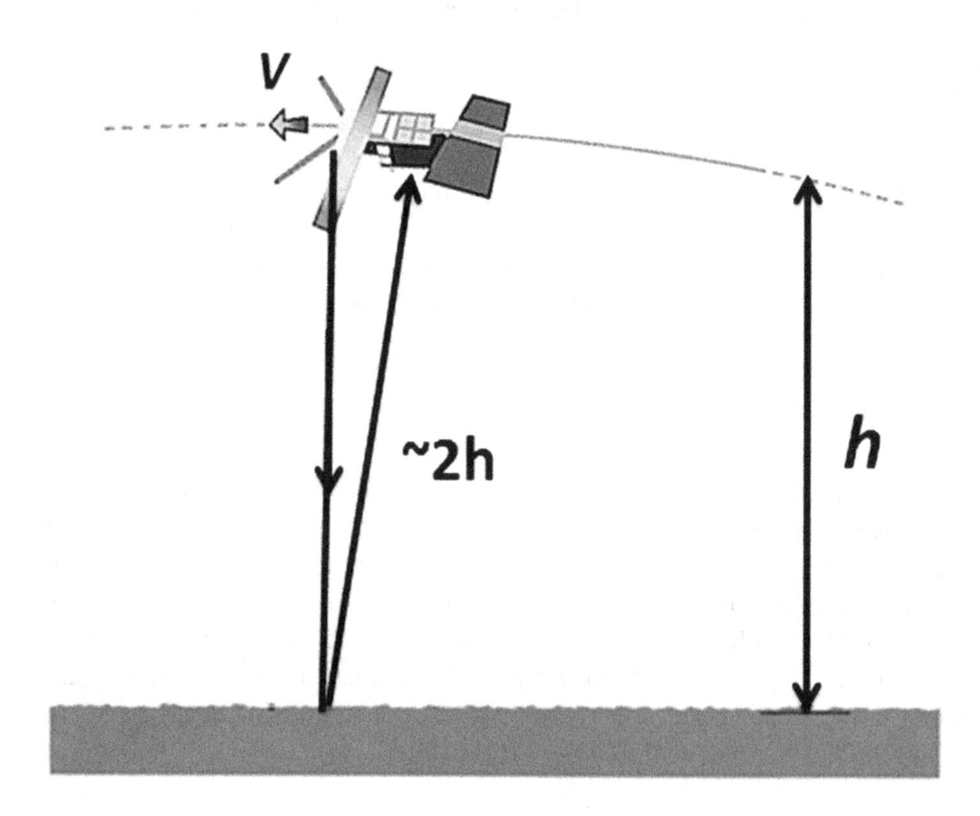

HOW DOES SURFACE HEIGHT ESTIMATE FROM RADIO ALTIMETER?

The transmitted pulse reaches the surface first on a small point. As the pulse advances, the illuminated area grows rapidly from a point to a disk, as does the returned power. The reflecting area depends on the beamwidth of the antenna. The energy from the centre of the main beam has to travel the shorter path than the energy from the edges. An annulus is formed and the geometry is such that the annulus area remains constant as the diameter increases. The returned signal strength, which depends on the reflecting area, grows rapidly until the annulus is formed, remains constant until the growing annulus reaches the edge of the radar beam, where it starts to diminish. The simple mathematical formula to identify the surface height (H) is given by:

$$\text{(Corrected) height} = \text{altitude} - \text{(Corrected) variety} \tag{10.3}$$

In case of the sea surface height, there are several circumstances must be considered. These include (i) ocean surface height, ocean circulation, and other physical parameters, for instance, wind speed, eddies and seasonal variations. Ocean surface height is determined without referring to other physical parameters which are associated with the sea surface, for instance, tide, wind speed, and currents. In fact, the geoid is governed by the sea surface due to the impact of gravity distribution over the world. Under this circumstance, the geoid is fluctuated due to changing in the water masses and densities. In other words, a hill at the geoid is noticeable as a seafloor has a denser rock zone at the seafloor that would distort sea level by tens of meters. Furthermore, dynamic topography which is known as ocean circulation is a function of the Earth's rotation. It is derived impact of about 1 m (Rosmorduc et al., 2016). By removing the geoid from sea surface height, the dynamic topography is then computed. In practice, the mean sea level is deducted to yield the variable component (sea degree anomalies) of the ocean signal (Raney, 1998).

Figure 6. Radar Pulse

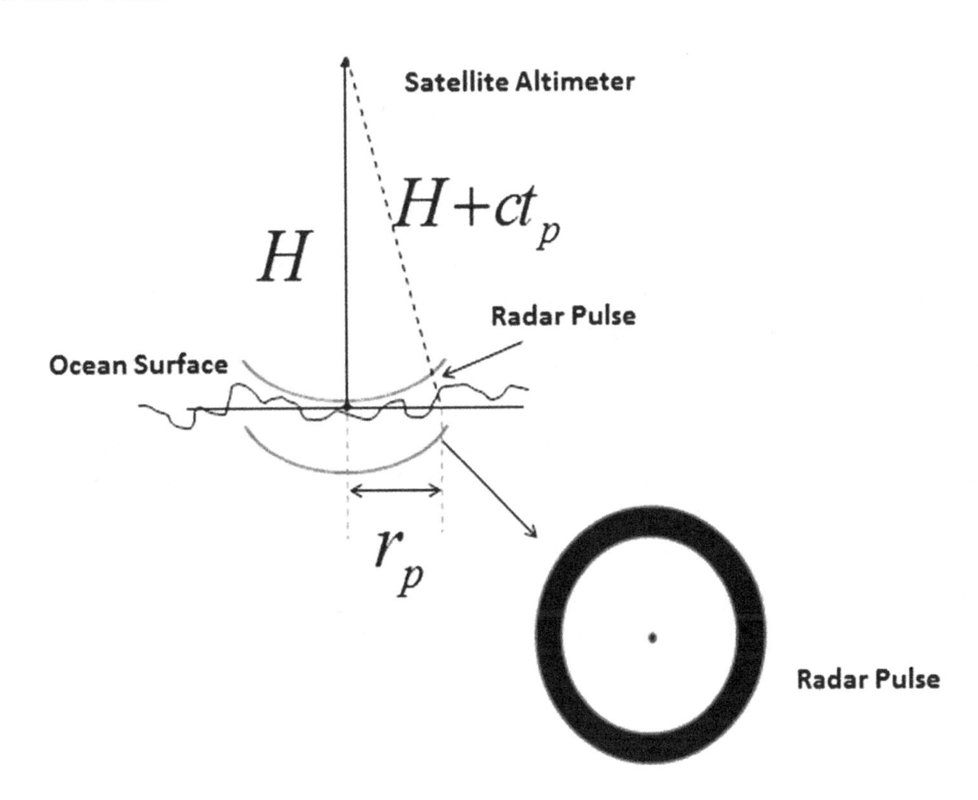

PULSE-LIMITED ALTIMETRY

Consider a radar pulse emanating from a radar beacon propagating downwards and interacting with a flat ocean surface. Figure 6 shows an illustration of the vertical cross-section and top-down view of the radar pulse.

Implementing the Pythagorean theory, the leading edge r_p of the pulse is casted as:

$$r_p = \sqrt{Hct_p} \tag{10.4}$$

where H is satellite Height, c is the speed of light and t_p is pulse time. Equation 10.4 demonstrates that the pulse time fluctuation of the backscatter signal of the ocean or land surfaces are identified as (i) the period before the pulse arrives,(ii) the period after the pulse arrives and before the tail of pulse has been received by the antenna, and after the tail, the pulse has been received by the antenna (Robinson, 2004 and Rosmorduc et al., 2016).

To determine the power signal of the delay-Doppler radar as a function of time, we'll need to assume that the footprint of the pulsed radar is small enough to be considered two rectangles of width W. This can be expressed by:

$$P(W) = \begin{cases} 0 & \mathrm{t} < \mathrm{t}_0 \\ 2W_r(t) & \mathrm{t}_0 < \ \mathrm{t} < \mathrm{t}_0 + t_p \\ 2W_r[r(t) - r(t - t_p)] & \mathrm{t} > \mathrm{t}_0 + t_p \end{cases} \tag{10.5}$$

The altimeter radiates a pulse towards the Earth's surface. The time, which intervenes from the transmission of a pulse to the reception of its backscatter of the Earth's surface, is proportional to the satellite's altitude. Some theoretical details of the principle of radar are applied to altimeter which can assist a great understanding of the different behaviours and characteristics of the pulse in function of irregularities on the surface encountered. In this regard, the magnitude and shape of the echoes (or waveforms) also contain information about the characteristics of the surface which caused the backscatter. The greatest results are acquired over the ocean, which is spatially homogeneous and has a surface which conforms with known statistics. In contrast, land surfaces which are not homogeneous and contain discontinuities or significant slopes, make precise analysis further challenging. Even in the best case (the ocean), the pulse should last no longer than 70 picoseconds to achieve an accuracy of a few centimetres. Technically, this means that the emission power should be greater than 200 kW and that the altimeter would have to switch every few nanoseconds (Rosmorduc et al., 2016 and Marghany 2018).

These problems are solved by the full deramp technique, making it possible to use only 5 W for emission. The range resolution of the altimeter is about half a metre (3.125 ns) but the range measurement performance over the ocean is about one order of magnitude greater than this. This is achieved by fitting the shape of the sampled echo waveform to a model function which represents the form of the echo.

MAXIMUM RANGE AND RANGE RESOLUTION

By suitable choice of the frequency deviation per unit of time can be determined by the radar resolution, and by choice of the duration of the increasing of the frequency can be determined by the maximum non-ambiguous range. The maximum frequency shift and steepness of the edge can be varied depending on the capabilities of the technology implemented circuit. For the range resolution of FMCW radar, the bandwidth *BW* of the transmitted signal is decisive (as in so-called *chirp radar*). However, the technical possibilities of Fast Fourier Transformation are limited in time (i.e. by the duration of the sawtooth *T*). The resolution of the FMCW radar is determined by the frequency change that occurs within this time limit (Robinson, 2004 and Marghany 2018).

$$\Delta f_{FFT} = \frac{\partial(f)}{\partial(t).(f_{up} - f_{down})} \tag{10.6}$$

Equation 10.6 demonstrates that Δf_{FFT} is the smallest measurable frequency difference, $\delta(f)/\delta(t)$ is steepness of the frequency deviation. Consequently, f_{up} is upper frequency (end of the sawtooth) while f_{dwn} is lower frequency (start of the sawtooth).

The reciprocal of the duration of the sawtooth, pulse leads to the smallest possible detectable frequency. This can be expressed in the equation (1) as $|\Delta f|$ and results in a range resolution capability of the FMCW radar.

The signal bandwidth of FMCW-Radar can be from 1 MHz up to 390 MHz. (Its upper border is mostly limited by legal reasons. For instance, the most used for FMCW-applications European ISM-radio band is defined from 24,000 MHz to 24,250 MHz with a given bandwidth of 250 MHz.) As the bandwidth increases, the achievable range resolution is decreasing and this means the monitored objects can be seen more accurately. The maximum detected range becomes smaller when the bandwidth increases (Table 1).

Table 1. Bandwidth different characteristics

Bandwidth	Range Resolution	Maximum Range	Required Power
400 kHz	4 000 m	120 km	1,4 kW
50 -500 kHz	1 500 -100 m	15 -250 km	30 W
1 MHz	150 m	75 km	1,4 ... 4 kW
2 MHz	75 m	37,5 km	
10 MHz	5 m	7,500 m	
50 MHz	3 m	500 m	4 mW
65 MHz	2.5 m	1 200 m	100 mW
250 MHz	0.6 m	500 m	4 mW
8 GHz	3.5 cm	9 m	4 mW

MODULATION PATTERN

The altimeter signals have several modulation patterns, which can allow altimeter for the different sort of measurements. These patterns are sawtooth modulation (Figure 7), triangular modulation, square-wave modulation, stepped modulation (staircase voltage), and sinusoidal modulation. The former is sawtooth modulation, which is depleted in a reasonably great range (maximum distance) fused with a trivial influence of Doppler frequency (Raney, 1998).

Figure 7. Sawtooth modulation

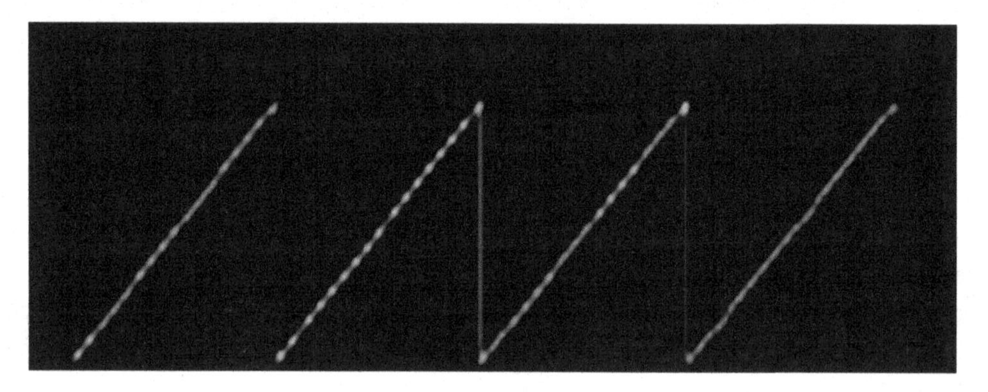

Therefore, triangular modulation (Figure 8) permits laid-back separation of the difference frequency Δf of the Doppler frequency f_D .

Figure 8. Triangular modulation

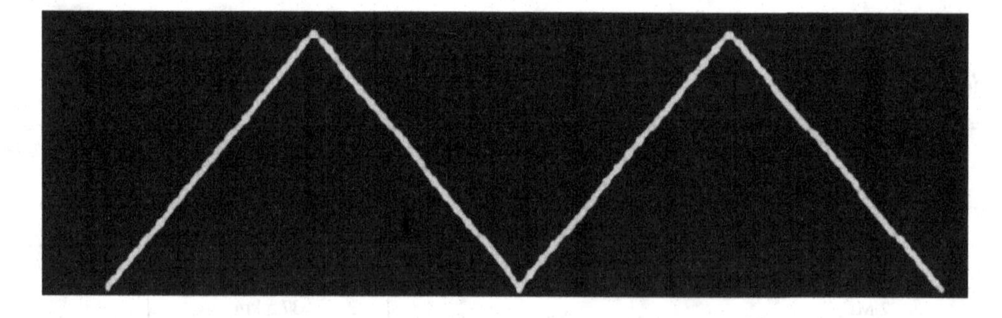

Square-wave modulation is known as a simple frequency-shift keying (FSK) (Figure 9). Consequently, square-wave modulation is implemented to measure a very precise distance at close range. To this end, the phase comparison of the two echo signal frequencies is used. However,it has the disadvantage, that the echo signals from several targets cannot be separated from each other, and that this process enables only a small unambiguous measuring range (Marghany, 2018).

Figure 9. Square-wave modulation

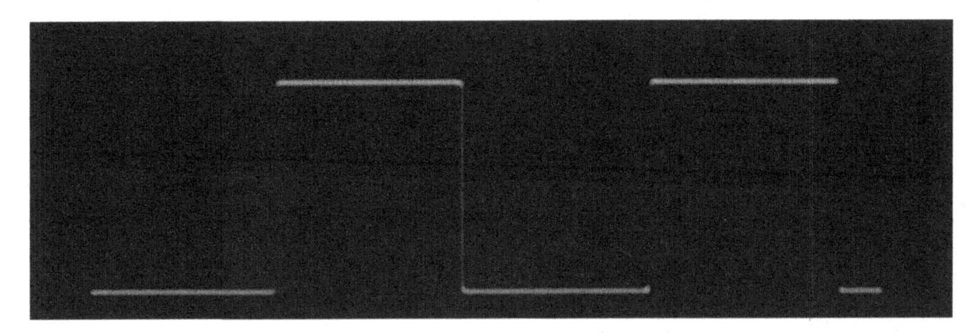

On the contrary, stepped modulation (staircase voltage) (Figure 10) is used for interferometric measurements and expands the unambiguous measuring range. The latter is the sinusoidal modulation forms (Figure 11), which have been used in the past. These could be easily realized by a motor turned a capacitor plate in the resonance chamber of the transmitter oscillator. The radar, then depleted only the relatively linear part of the sine function near the zero crossing (Rosmorduc et al., 2016).

Figure 10. Stepped modulation form

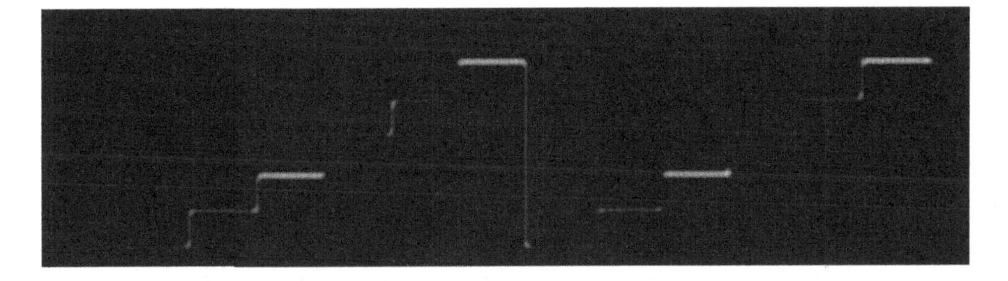

Figure 11. Sinusoidal modulation form

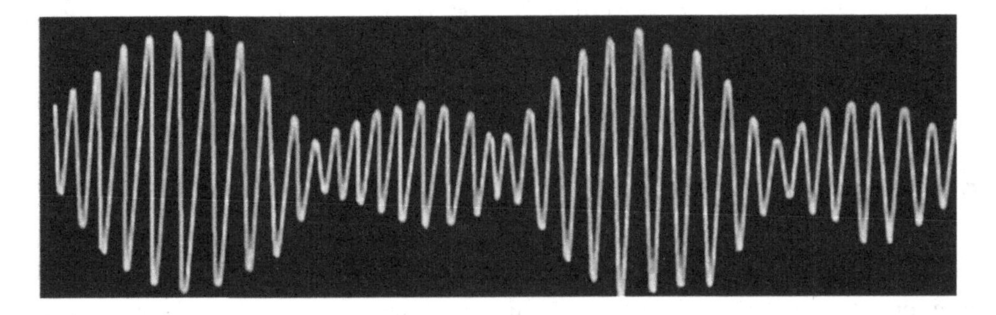

FREQUENCIES USED AND THEIR IMPACTS

Numerous altered frequencies are exploited for radar altimeters. The choice is determined by regulations, mission objectives and constraints, technical possibilities — and impracticalities, for each frequency band, has its advantages and disadvantages.

ku band with 13.6 GHz is the utmost regularly- operating frequency for Topex/Poseidon, Jason-1, Envisat, ERS, etc. It is sensible to atmospheric alarms and disconnection of the effect of ionospheric electrons (Figure 12).

Figure 12. Electromagnetic waves for altimeter bands

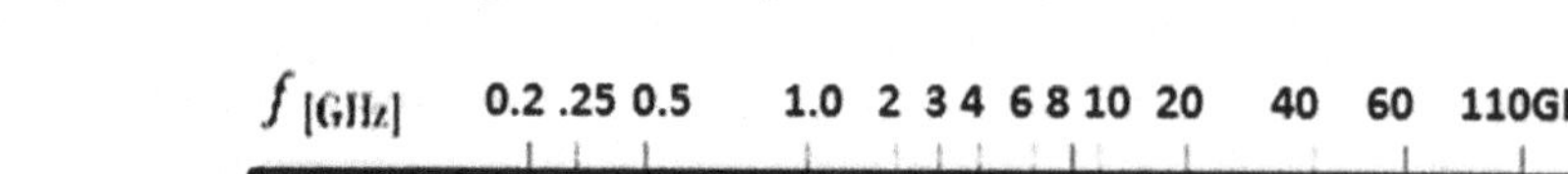

Further, C band with 5.3 GHz is known to be more sensitive than Ku to ionospheric perturbation, and less sensitive to the effects of atmospheric liquid water. Its main function is to enable correction of the ionospheric delay in combination with the Ku-band measurements. To obtain the best results, an auxiliary band like this must also be as far as possible from the main one (Rosmorduc et al., 2016 and Marghany 2018). Consequently, S-band with 3.2 GHz is also used in combination with the Ku-band measurements for the same reasons as the C band. Finally, signal frequencies in the Ka-band with 35 GHz enable better observation of ice, rain, coastal zones, land masses (forests, etc.) and wave heights. Due to international regulations governing the use of electromagnetic wave bandwidth, a larger bandwidth is available than for other frequencies, thus enabling higher resolution, especially near the coast. It is also better reflected in the ice. Nevertheless, attenuation owing to water or water vapour in the troposphere is high, meaning that no measurements are produced when the rain rate is higher than 1.5 mm/h (Vaijayanthi and Vanitha, 2015).

ALTIMETRIC MEASUREMENTS OVER THE OCEAN

The echoes or reflected waves received by altimeter are a function of the sea surface situations which creates variations in echo powers over a time. In this regard, the echo amplitude increases since the leading pulse strike the flat surface. Nonetheless, if the surface is extremely rough, it causes the reflected waves to increase gradually. Ocean wave height estimation is a function of the reflected waves since the slope of the curve representing its amplitude over time is proportional to wave height (Figure 13).

Figure 13 is a keystone to retrieve six parameters from the shape of reflected waves (Wingham et al., 2006). Epoch at mid-height delivers the delay time of the reflected pulse which is a function of range.

Figure 13. The concept of ocean surface measurements by Altimeter

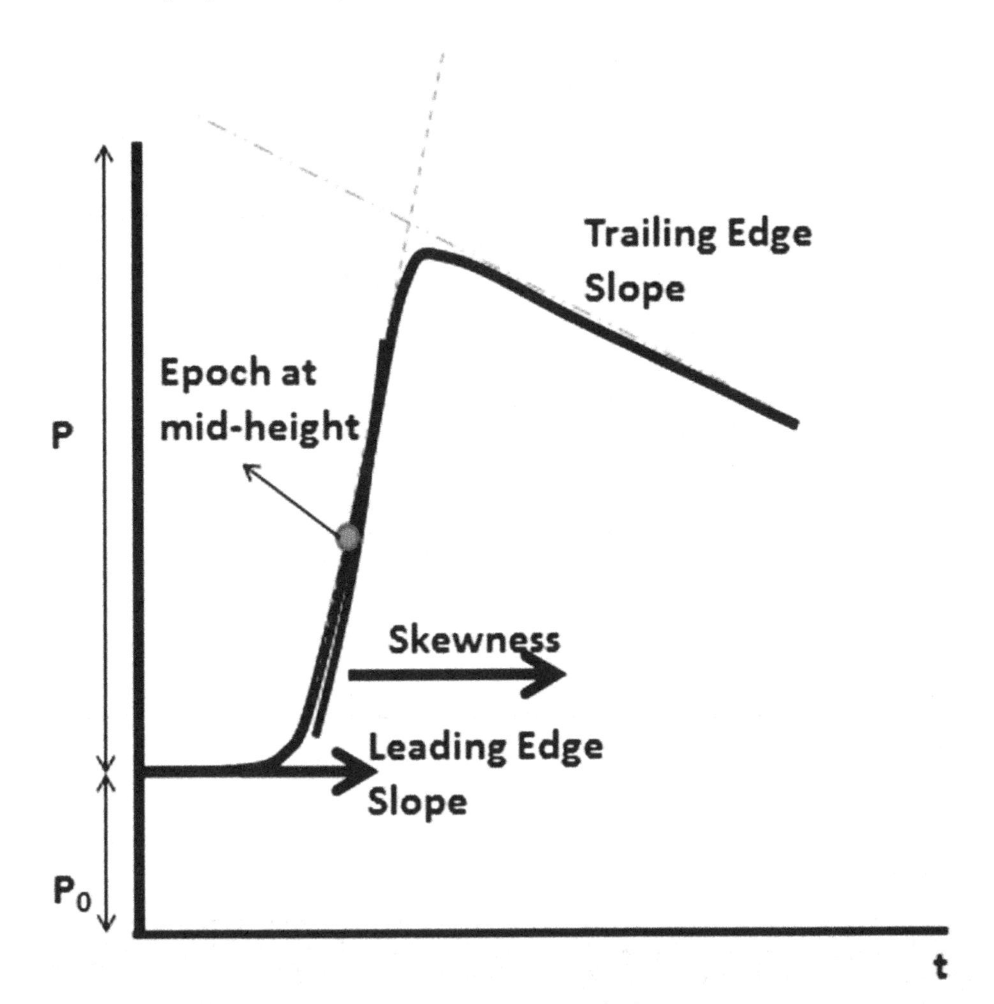

The amplitude of the radar signal power reveals the backscatter coefficient and sigma0. Therefore, significant wave height can be estimated from the leading edge slope. The rate of leading curvature can be acquired by skewness. The mispointing of the radar antenna is associated with a trailing edge slope. In other words, it means any deviation from the radar altimeter nadir of the radar pointing. Furthermore, the leading edge slope must occur above the thermal noise P_0 (Marghany 2018).

The altimeter irradiates a circle of the beam over the ocean or land surfaces with a 3 to 5 km wide diameters, which depend on the sea condition, the wave height or the ridged land. A rough sea surface or land delivers a wider footprint of approximately 10 km (Figure 14). In contrast, the calm sea surface or flat land provides a narrow footprint of about 2km (Witter and Chelton,1991).

ALTIMETER SENSORS

Figure 14. Reflected pulses from different ocean surface conditions

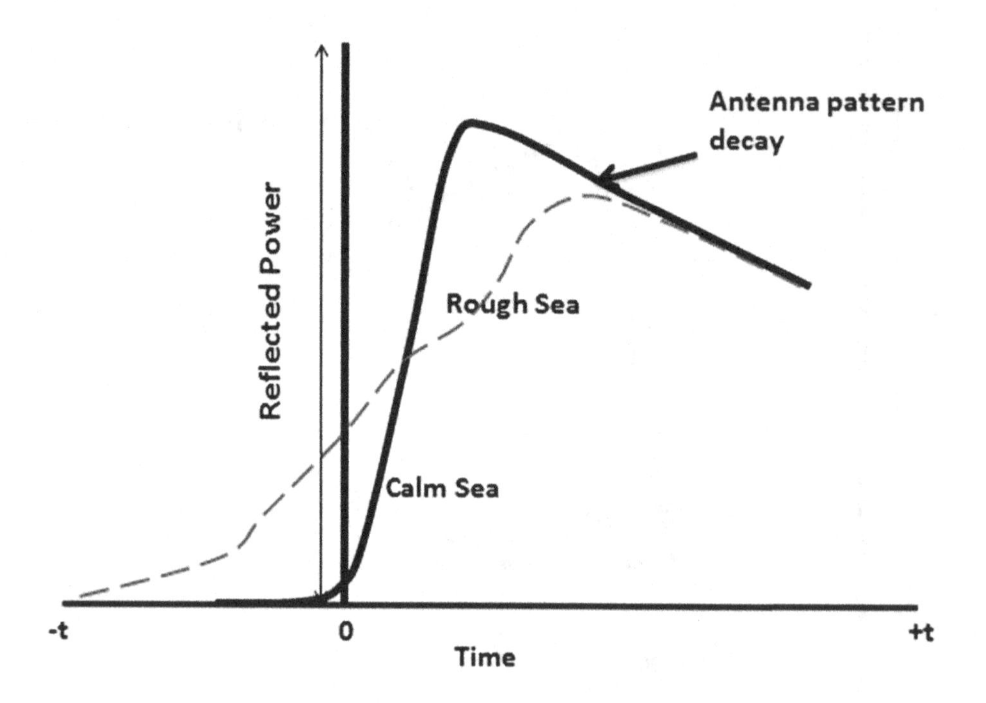

Figure 15. Different altimeter satellites

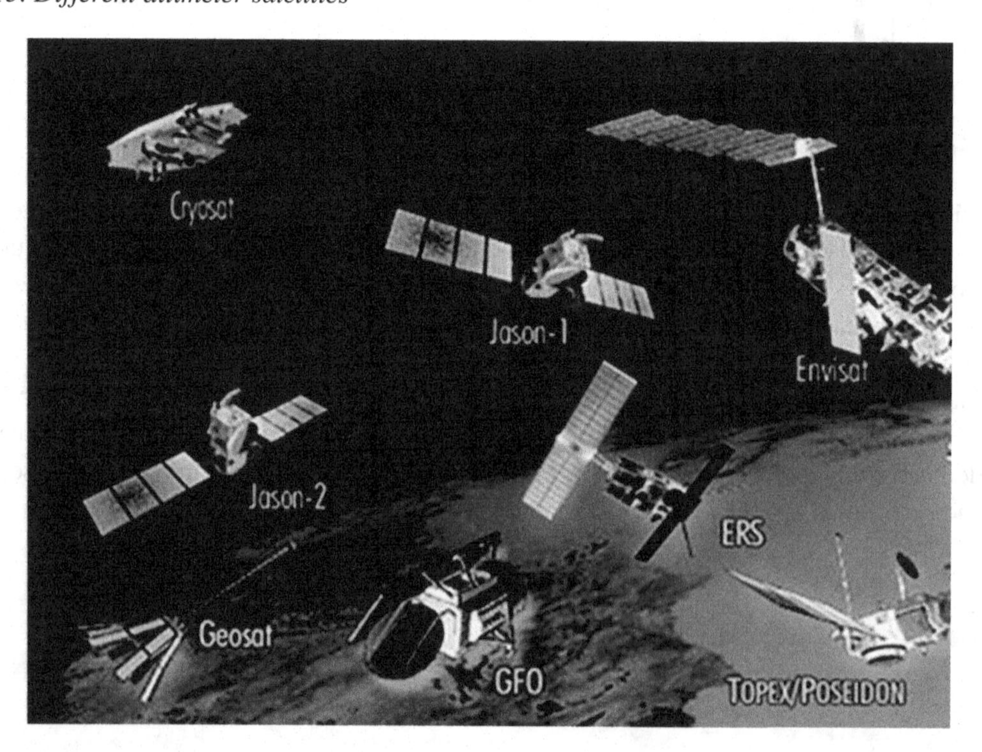

The detailed concurrent measurements of the sea surface height (SSH) and radar backscattering strength at a nadir in deep water are available. These measurements were made with microwave radars on board the Jason-1, Topex/Poseidon, Envisat, and Geosat Follow-on (GFO) altimetric satellites (Figure 15) (Table 2) (Chelton et al., 2001).

Table 2. List of altimeter satellite sensors

Sensors	Band	Radar frequency (GHz)
Geosat	ku	13.5
ERS-1	ku	13.8
ERS-2	ku	13.8
TOPEX	Ku c	13.6 5.3
Poseidon	ku	13.65
GFO	ku	13.5
Jason-1	ku c	13.6 5.3
ENVISAT	ku s	13.6 3.2
Jason-2	ku c	13.6 5.3
Jason-3	ku c	13.6 5.3
SENTINEL-3	ku c	13.575 5.41
Cryosat-2	ku	13.575

The details of the coastal wave interaction, for the most part, do not account for in the propagation models that simulate open ocean wave dynamics. Therefore, coastal measurements are difficult to use for validation and comparison of propagation model data. All four satellites provided unambiguous deep ocean measurements of the wave, to be used in propagation models (Vaijayanthi S and Vanitha 2015).

The majority of altimeter sensors are listed in Table 10.2 can't be used for monitoring and forecasting trajectory movements of the MH370 debris due to the fact them missions have ended. These sensors are Geosat, ERS-1/2, Topex/Poseidon, GFO, Envisat and Jason-1. On the contrary, few altimeter satellites can be implemented to simulate and forecast trajectory movements of the MH370. These altimeter sensors involve Jason-2 and Cryosat-2. However, Sentinel-3 and Jason-3 are launched post the MH370 event in 2016.

RETRIEVING WAVE HEIGHT AND PROPAGATION FROM ALTIMETER DATA

Three methods have been used to study physical properties of wave propagation: (i) multi-satellite time-spatial interpolation;(ii) investigation of tsunami wave height; and (iii) estimation of Background

Level (Marghany 2018). Multi-satellite time-spatial interpolation is performed to define the reference heights, which must be estimated at sea level anomaly (SLA) ''the SLA under the assumption of wave movements.'' The reference heights are defined by the weighted mean as:

$$SLA_{ref}(\phi,\vartheta,t)=\sum(e^{(r_i^2/R^2-t_i^2/T^2)}\bullet SLA_{obs,i})/\sum w_i \qquad (10.7)$$

where r_i is the distance between the location of the ith datum and the wave searching point; ϕ,ϑ, and t is the latitude, longitude, and date of a wave searching point; t_i is the time difference of observations between a wave observation point (t) and the ith datum; and R and T are scale parameters (Rosmorduc et al., 2016 and Marghany 2018). Then the wave height (h_t) at any sampling point is derived from

$$h(\phi,\vartheta,t)=SLA_{obs}(\phi,\vartheta,t)-SLA_{ref}(\phi,\vartheta,t) \qquad (10.8)$$

where SLA_{obs} is an anomaly in an observed sea surface height, and SLA_{ref} is the reference height defined by equations 10.7. Finally, the background level at any sampling point (ϕ and ϑ) is defined as the root mean square (RMS) of the residual error calculated by

$$r_k(\phi,\vartheta,t)=SLA_{obs}(\phi,\vartheta,t+kc)-SLA_{obs}(\phi,\vartheta,t+(k-1)c) \qquad (10.9)$$

where k =-5,-4,-3,-2,-1, 1, 2, 3, 4, 5, and c is the recurrence cycle, for instance Jason-1 has 9.9156 cycles, day, TOPEX/POSEIDON is 9.9156 cycles, day, GFO is 17.0506 cycles, day and ENVISAT is 35.0000 cycles, day (Wingham et al., 2006).

The inverse of the sea level variance is expended to simulate the Indian surface current patterns. This information is the keystone to reveal the trajectory movements of the MH370 debris throughout the Southern Indian Ocean. The subsequent chapter will tackle this issue evidently.

CONCLUSION

This chapter has demonstrated the principles of altimeter satellite data. In this view, it is explained how the altimeter signals can be used to retrieve components of the ocean dynamic features, for instance, sea level variation, bathymetry and ocean wave height. Moreover the chapter has listed the variety of altimeter sensors. The next chapter will implement principles and some altimeter sensors to model the impact of the ocean wave pattern on the MH370 debris across the Indian Ocean.

REFERENCES

Chelton, D. B., Ries, J. C., Haines, B. J., Fu, L. L., & Callahan, P. S. (2001). Satellite altimetry. In *International geophysics* (Vol. 69, pp. 1–ii). Academic Press.

Cheney, R. E., Douglas, B. C., Agreen, R. W., Miller, L., & Porter, D. L. (1987). *Geosat altimeter geophysical data record user handbook.* NASA STI/Recon Technical Report N, 88.

Hayne, G. S. (1980). Radar Altimeter Mean Return Waveforms from Near-Normal-Incidence Ocean surface scattering. *IEEE Transactions on Antennas and Propagation*, *28*(5), 687–692. doi:10.1109/TAP.1980.1142398

Marghany, M. (2018). *Advanced remote sensing technology for Tsunami modelling and forecasting.* CRC Press. doi:10.1201/9781351175548

Raney, R. K. (1998). The delay/Doppler radar altimeter. *IEEE Transactions on Geoscience and Remote Sensing*, *36*(5), 1578–1588. doi:10.1109/36.718861

Robinson, I. S. (2004). *Measuring the oceans from space: the principles and methods of satellite oceanography*. Springer Science & Business Media.

Rosmorduc, V., Benveniste, J., Bronner, E., Dinardo, S., Lauret, O., Maheu, C., & Ambrozio, A. (2016). Radar altimetry tutorial. In J. Benveniste & N. Picot (Eds.), ESA & CNES. Academic Press.

Vaijayanthi, S., & Vanitha, N. (2015). Aircraft Identification in high-resolution remote sensing images using shape analysis. *International Journal of Innovative Research in Computer and Communication Engineering*, *3*(11), 11203–11209.

Wingham, D. J., Francis, C. R., Baker, S., Bouzinac, C., Brockley, D., Cullen, R., ... Phalippou, L. (2006). CryoSat: A mission to determine the fluctuations in Earth's land and marine ice fields. *Advances in Space Research*, *37*(4), 841–871. doi:10.1016/j.asr.2005.07.027

Witter, D. L., & Chelton, D. B. (1991). A Geosat altimeter wind speed algorithm and a method for altimeter wind speed algorithm development. *Journal of Geophysical Research. Oceans*, *96*(C5), 8853–8860. doi:10.1029/91JC00414

Chapter 11
Advanced Altimeter Interferometry for Modelling the Wave Pattern Impacts of MH370 Flaperon

ABSTRACT

Previous studies investigated the Indian Ocean's currents' impacts on the trajectory movement of MH370 debris. This chapter introduces the novel approach of investigating the wave pattern variations in the Indian Ocean on the MH370 debris. The novel approach based on the altimeter interferometry technique is utilized in this chapter. To this end, dual SIRAL instruments on-board of CryoSat-2 are applied to obtain the annual cycle of significant wave height across the Indian Ocean. In this chapter, in a one-year significant wave height cycle, the swell remains propagating from the Southwest to the Northeast from January to March 2015 with a maximum significant wave height of 5 m in the Northeast Offshore Australian Shelf and 7 m significant wave height Southwest of Australian Shelf. In this circumstance, the Pareto algorithm proves that the flaperon would submerge to a water depth less than 300 m on account of the impact of wave power of 22000 KJ/m/wave. It can be said that the flaperon would be submerged further to a water depth of 1000 m because of the wave power of 30000 KJ/m/wave.

INTRODUCTION

Ocean dynamic does not only involve current movements as considered by studies concerned about MH370 debris trajectory movements. Researchers and scientists only implemented current movements to investigate the trajectory movements of MH370 debris and ignored other forces such as wave propagation.

Waves play a tremendous role in ocean dynamics. In fact, MH370 debris is governed by ocean dynamics. In this view, the ocean waves are required precise technology to be investigated and model. Altimeter satellite data have the potential to determine ocean waves pattern over the huge scales of the oceans. In this regard, satellite altimetry has modernized sea-level quantities due to the fact it affords

DOI: 10.4018/978-1-7998-1920-2.ch011

measurements of sea-surface height with world coverage and a revisit time of various days. For instance, Jason-1 and -2, which function on a 10-day repeat cycle. The European Space Agency (ESA) satellites ERS-1 and ERS-2 have operated in distinctive repeat cycles (3, 168, and frequently 35 days); the currently operating Envisat offers information on a 35-day repeat cycle. In contrast, Cryosat-2 has a long 369 days with the 30-day sub-repeat cycle, which adequately to investigate the wave pattern along the Indian Ocean (Wingham et al., 2006).

Consequently, the significant question, which arises is how the interferometry technique can implement with altimeter satellite to derive significant wave height? Besides, how significant wave heights across the Southern Indian Ocean impact the MH370 debris and flaperon stabilities through the surface water? This chapter, therefore, devotes to answer a critical question of did flaperon reach the coastal waters of Réunion Island?

PRINCIPLES OF SYNTHETIC APERTURE RADAR ALTIMETER INTERFEROMETRY

The Synthetic Aperture (SAR) Interferometer Radar Altimeter SIRAL-2 design is based on existing equipment, but with several major enhancements designed to overcome difficulties associated with measuring ice surfaces. It works by bouncing a radar pulse off the ground and studying the echoes from the Earth's surface. By knowing the position of the spacecraft - achieved with an on-board ranging instrument called DORIS (Doppler Orbitography and Radiopositioning Integrated by Satellite) - the signal return time will reveal the surface altitude. Correct antenna orientation is vital for this and is maintained using a trio of star trackers. Moreover, SIRAL is the primary instrument of the mission, designed and developed for ESA by Thales Alenia Space (formerly Alcatel Alenia Space), France. SIRAL is of Poseidon-2 heritage flown on the Jason-1 mission. The objective is to observe ice sheet interiors, the ice sheet margins, for sea ice and other topography.

Since the launch of Cryosat-2 in 2010, a new technology of altimeters expending Doppler and interferometric competencies have emerged and will most probably become the fashion for the upcoming altimeters, at least the Doppler one as it is already the case for the Sentinel-3 altimeter launched in 2016. In this context, the Delay-Doppler Altimeter (DDA) notion (also acknowledged as SAR altimeter) used to be first proposed with the aid of Raney (Galin et al., 2012).

Delay-Doppler altimeters have an excessive pulse repetition frequency (PRF) to make sure pulse-to-pulse coherence, leading to a practicable along-track resolution approximately 300 meters, improved signal-to-noise ratio and improved altimeter ranging performance (Martin-Puig et al., 2008). For instance, the Cryosat-2 Synthetic Aperture Interferometric Radar Altimeter (SIRAL) exploits the SAR mode over ocean zones.

The SAR interferometric mode (SARin) is CryoSat's most advanced mode, particularly used approximately over the ocean surface margins At this juncture, the altimeter performs synthetic aperture processing and exploits the second antenna as an interferometer to detect the across-track angle to the earliest radar returns. The SARin mode offers thus the precise surface location being measured when the surface is sloping and can be used to find out about extra contrasted sea surface slopes. Over most of the rough ocean, SIRAL operates in the preferred Low Rate Mode (LRM) that is the conventional pulse-limited radar altimeter mode. In this mode, the data rate is rapidly lower than the different dimension modes. The SIRAL data provides a very excellent possibility to investigate the competencies and deserves three distinctive altimeter function modes for the detection and estimation of small debris characteristics (free-board and surface).

ALTIMETER INTERFEROMETRY TECHNIQUE

The concepts of interferometric altimeter has been proposed first by Jensen, (1999) and lead to the improvement of the Cryosat mission. A detailed description of the concepts and processing of the Cryosat SARin information is given in Wingham et al. (2006). The primary (left) antenna transmits the radar signal and the two antennas measure the backscattered echo waveform (Figure 1.). The primary complex waveform is multiplied with the complex conjugate of the second antenna waveform. The phase of the ensuing cross-channel waveform is then defined as the interferometric phase difference, which consequences from the slight range distinction of an off-nadir scatterer for the two antennas. The normalized modulus of the conjugate product offers an estimate of the signal coherence. The stacked SAR echoes for each antenna are computed using the SAR mode processing. The SAR echoes, phase, and coherence are furnished with ESA Level-1B products.

In SARin mode the waveform evaluation window is elevated to 512 bins (240 m) to better sample sloping terrains. In the Baseline-C data, merchandise used. The use of zero-padding prior to FFT processing, besides, increase the variety of range bins in 1024 except altering the range window. Each bin corresponds consequently to 1.565 ns or 0.23 m along with the range.

Let us assume that the two SAR antennas S_1 and S_2 mounted on a satellite platform are separated by a baseline B with a baseline angle α, and observe the complex response at the point δ with a slant range H_I (Figure 1.). The mathematical relationships between the interferometric phase difference, Δψ, and the off-nadir angle, α, can be given by (Raney, 1998):

$$\Delta\psi = \frac{2\pi B}{\lambda}\sin(\alpha) \tag{11.1}$$

where λ is the radar wavelength and B is an interferometer baseline (distance between the two antennas). Under the small-angle approximation, the off-nadir angle α is expressed by:

$$\alpha = \frac{\lambda\Delta\psi}{2\pi B} \tag{11.2}$$

Galin et al., (2012) are estimated the across-track distance to nadir, d_0 by:

$$d_0 = \frac{ct_i}{2}\bar{\alpha} = H_i\bar{\alpha} \tag{11.3}$$

Equation 11.3 demonstrates that range H_i equals half of the multiplication of pulse time t_i by the speed of light c then multiply by angle scaling $\bar{\alpha}$. Consequently, the freeboard δ due to the earth curvature is defined as:

Figure 1. Geometry of altimeter interferometry

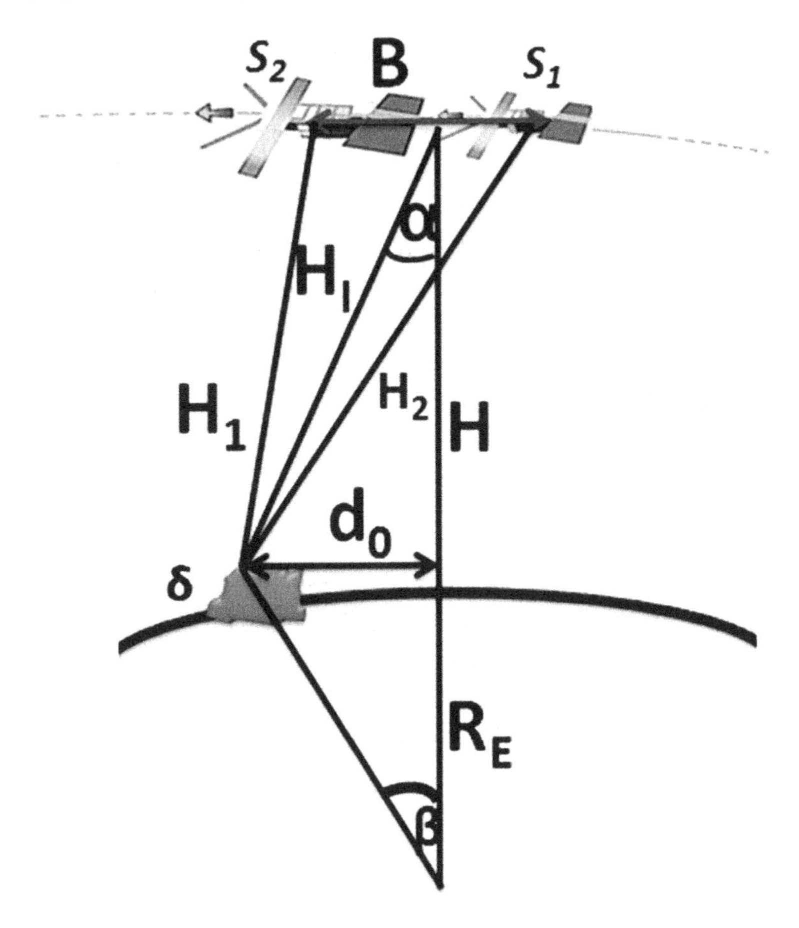

$$\delta = (H - H_i \cos\alpha_1 + R_E(1 - \cos\beta))\cos\beta \tag{11.4}$$

where $\beta = H/R_E\alpha_I$ and $H_i = ct_I/2$. The SARin echoes are similar to the SAR ones, except that the number of range bins in the echoes thermal noise part (TNP) is significantly larger (125×2 vs 50). Altimeters over open waters have a detectable signature in the thermal noise part (TNP), which is accounted as above the sea surface of high resolution (HR) waveforms of pulse-limited altimeters that can be easily detected (Goda, 2010).

The swath over which sea level can be detected which is of the order of 6 km is thus significantly increased to 12 km. The SAR detection algorithm can be applied to the SARin waveforms without modification. However, in the echoes, TNP the signals received by the two antennas are by nature random noise and thus incoherent (Rey et al., 2001 and Tournadre et al., 2012).

Precision Procedures Of InSAR Altimeter Scheme

InSAR altimeter accurate analysis is a function of the accuracies of satellite altitude, slant range, baseline length, baseline angle and phase difference which are denoted as σ_H, σ_B, σ_{H_1}, σ_α and σ_β respectively. In these circumstances, these parameters are autonomous. These independent parameters can be expressed mathematically as:

$$\sigma_{R_E} = \sigma_H \left(\frac{\partial R_E}{\partial H} \right) + \sigma_B \left(\frac{\partial R_E}{\partial B} \right) + \sigma_{H_1} \left(\frac{\partial R_E}{\partial H_1} \right) + \sigma_\alpha \left(\frac{\partial R_E}{\partial \alpha} \right) + \sigma_\beta \left(\frac{\partial R_E}{\partial \beta} \right) \tag{11.5}$$

Equation 11.5 demonstrates that the height approximation accuracy is a function of a decreasing of the look angle α. In this circumstance, α equals 0.5° to obtain accurate height estimation. On the contrary, decreasing α causes decreasing of the range resolution. Also, low leads sea surface to be presented as a quasi-specular scattering. In this regard, the echo-tracking model has differences in backscattering echoes from the sea surface (Raney, 1998).

Echo-Tracking Algorithm

The echo-tracking algorithm is implemented in the slant range enhancement post the phase and geometry computations to deliver a precise slant range estimation. More precisely, Figure 2 demonstrates that the location of the sea surface at the point δ α is the radar look angle and *r* is a function of α. In other words, the echoes waveform is a function of the plane surface impulse response $\delta_{FS}(\tau,r)$, system point target response $S_r(\tau,r)$, and height probability density function of the sea surface scattering elements $Q(\tau)$. In this view, the mathematical expression of the sea surface echoes waveform is based on the convolution (Martin-Puig et al., 2008), which is given by:

$$\delta_r(\tau,r) = \delta_{FS}(\tau,r) x S_r(\tau,r) x Q(\tau) \tag{11.6}$$

$$\delta_r(\tau,r) = \int_{SeaSurfaceArea} \frac{\lambda^2 \delta(\tau)\sigma_0(r)}{(4\pi)^3 L_p \|r_1\|^4} \|G(r)\|^2 \, dA \tag{11.7}$$

Equation 11.7 is considered as a presentation of the radar equation. In this context, δ(τ) is the transferred waveform in the duration period τ with wavelength λ. At the position *r* over the changing area *dA*, the sea surface backscatter is σ_0 and L_p is the double loss of the atmosphere. Further, *G* is the antenna gain. Consequently, system point target response $S_r(\tau,r)$ is formulated as:

$$S_r(\tau,r) = C^2 \sin c^2 \left(\frac{\tau_a}{2\pi} \left(2\pi f_a - 2\pi f_d \left(1 - \frac{k_r t_0}{f_0} \right) \right) \right) \sin c^2 \left(\tau_c \left(\frac{k_r}{cH} \|r\|^2 - \tau \right) \right) \tag{11.8}$$

Equation 11.8 indicates a filter length of along-track and cross-track τ_a and τ_c, respectively. Consequently, f_a, f_d and f_0 are azimuth, Doppler and wave frequencies, respectively. Therefore, the linear fre-

Figure 2. SAR geometry

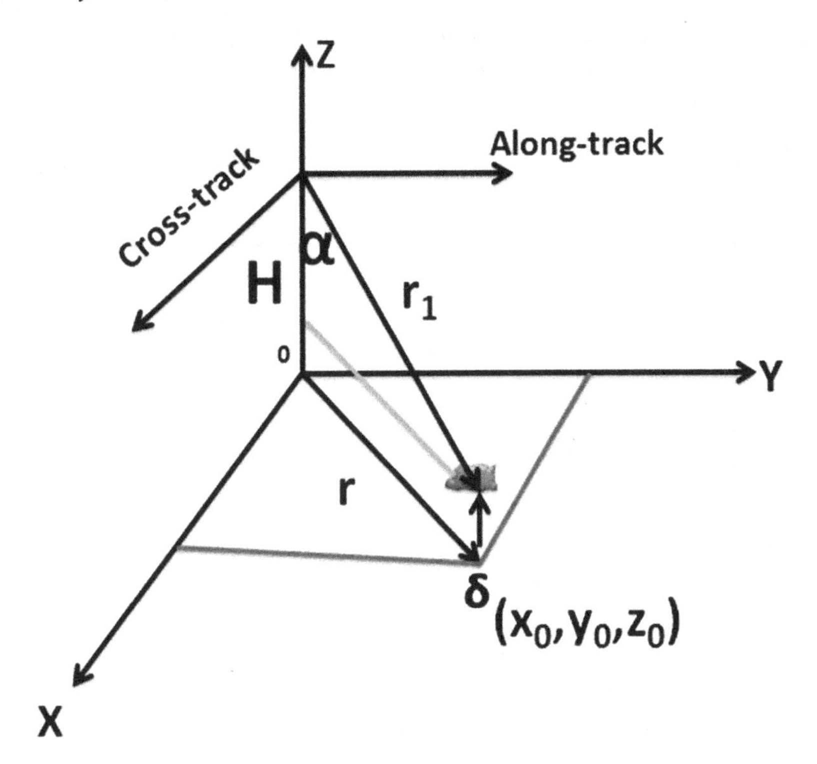

quency modulation rate k_r equals the ratio of baseline B and period T i.e. $k_r = \frac{B}{T}$. Finally, C is a function of SAR along-track and cross-track filter length.

The height probability density function of the sea surface scattering elements $Q(\tau)$ is determined from:

$$Q(\tau) = \frac{1}{\sqrt{2\pi\sigma_s}}\left[1 + \frac{\lambda}{6}\left(\frac{\tau^3}{\sigma_s^3} - 3\frac{\tau}{\sigma_s}\right)\right]e^{\left(-\frac{\tau^2}{2\sigma_s^2}\right)} \quad (11.9)$$

Equation 11.9 is the keystone to estimate the significant wave height H_s from the backscatter quantity σ_s, which indicates the root-mean-square wave height as follows:

$$\sigma_s = \frac{1}{2C}H_s \quad (11.10)$$

Figure 3 explains the delay time of the echo waveform model of SAR observation for instance, at the position of $\delta(\tau)$ (10000,0,1) with H equals 800 km and H_s at about 2 m, which produces the maximum normalized epoch power of 1 with a minimum delay time less than 0.008 s. It is clear that maximum delay time 0.04 s coincides with low normalized echo power closes in 0.

The time delay of the emitting signals derives from the mid-power epoch of the echo under the circumstance of fitting the backscatter signal to the echo waveform model. The prices slant range is obtained by multiplying by the light speed (Gomes et al., 2010).

Figure 3. The epoch waveform of near-nadir SAR observation with a certain position

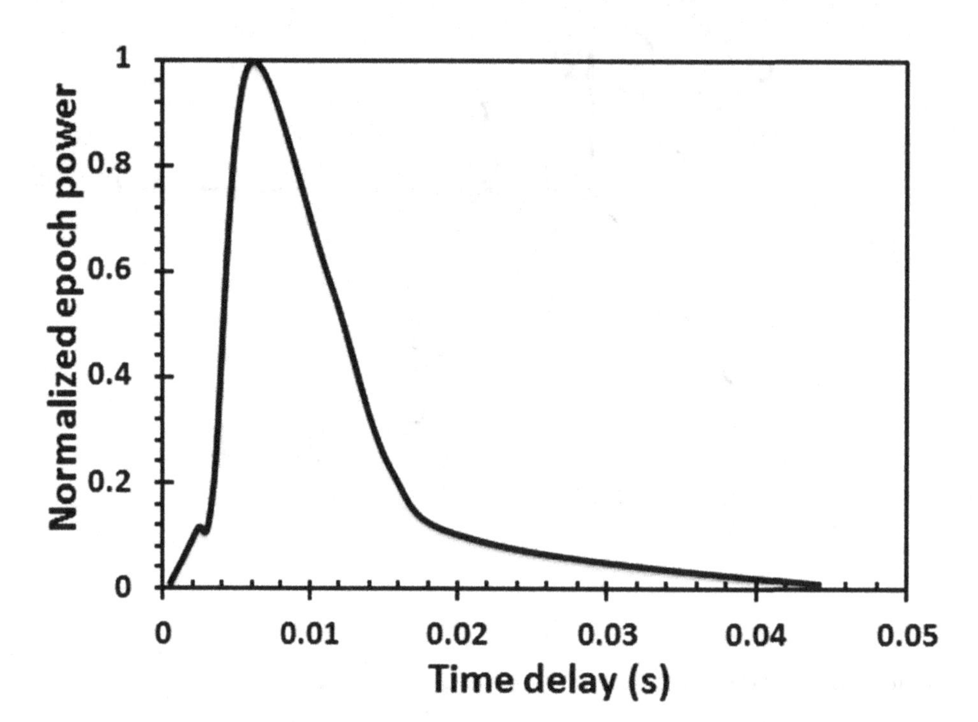

DELAY-DOPPLER ALTIMETER

Conventional radar altimeters emit pulses with a long interval period of approximately 500 μs. SIRAL, however, emits a burst of pulses with an interval of only 50 μs between them. The receiving echoes are thus interrelated, and by considering the complete burst of pulses in one campaign, the data processor can discrete the echo into the set of strips across the track by manipulating the slight frequency shifts. In this context, the Doppler effect causes a frequency shift in the forward- and aft-looking parts of the beam. In this view, the noise reduction is achieved by superimposing and averaged of strips on each other.

In interferometric mode (SARIn), however, a second receiving antenna is triggered to determine the receiving angle. SARIn, therefore, qualifies to sense approximately radar backscatter signals receiving from a target not precisely positioned beneath the altimeter sensor. Consequently, the variance in the path-length time of the radar echoes is tiny between radar echoes on the track and radar echoes out of the track. In this circumstance, α must be accurate.

The Delay-Doppler Altimeter alters from a common radar altimeter theory in which it utilizes coherent dispensation of sets of transmitting pulses. In this view, the full Doppler bandwidth is demoralized to achieve the most efficient use of the backscatter intensity, which is reflected from the Earth's surface (Figure 4). On the contrary, the conventional altimeter system is to determine the distance between the satellite and the mean ocean surface. In this context, the Delay-Doppler Altimeter method differs from conventional altimeter instruments by two techniques. The primary technique involves pulse-to-pulse coherence and full Doppler processing, which permit for the quantity of the along-track spot of the range measurement. The latter technique includes the implementation of two antennas and two receiver channels, which allow for the dimension of the across-track angle of the range quantity.

Figure 4. Delay-Doppler Altimeter interferometry

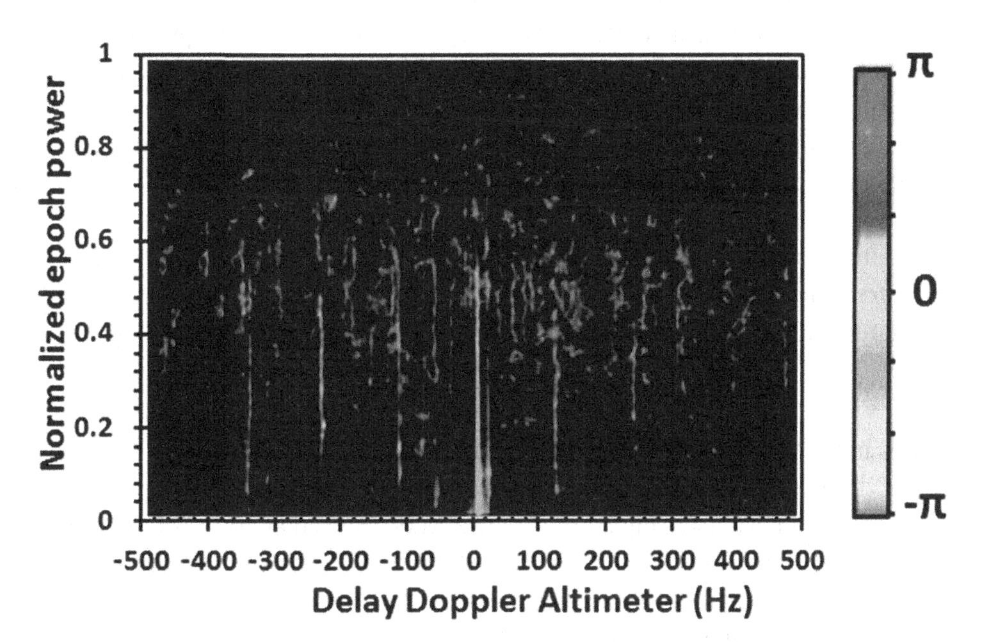

The Delay-Doppler Altimeter technology delivers numerous compensations than a conventional altimeter. The sea surface height precision offered from Delay-Doppler Altimeter technology is approximately twice that of the conventional altimeter. Simulations of the associated signal processing concepts have produced 0.5 cm precision in a calm sea, with precision remaining better than 1.0 cm even in significant wave heights as great as 4 m. The Delay-Doppler Altimeter technique reduces the uncertainties caused by ocean waves (Harries et al. 2006).

For instance, both Delay-Doppler Altimeter and conventional altimeter experience comparable levels of random noise due to a calm sea. Nevertheless, as the waves grow, a conventional altimeter suffers an intense noise level growth. Consequently, the coherent processing of the Delay-Doppler Altimeter experiences only a trivial intensification in random noise with wave height. In this circumstance, the Delay-Doppler Altimeter predominantly thrives appropriate for geodetic uses. The random error owing to ocean waves is the dominant error source. Consequently, Wind speed and wave height retrievals from the Delay-Doppler Altimeter has twice the precision of current sensors(Hemer et al., 2008).

DATA ACQUISITIONS

In this investigation, CRYOSAT-2 SIRAL data are implemented to verify the existence of the MH370 and its trajectory movements. CryoSat-2 revolves on a non-sun-synchronous polar orbit with an inclination angle of 92° inclination at an altitude of 713 km. Consequently, the repeat cycle for CryoSat is 369 days and 5344 orbits. This cycle provides full coverage of the earth and the cycle number is provided in the CryoSat product headers. The repeat cycle is made up of approximately 30-day sub-cycles, defined and used primarily for statistic and quality reporting (Sui et al., 2017).

There are two instruments attached to a board of CryoSat-2's mission. The primary two instruments aboard CryoSat-2 are SIRAL-2. Two SIRAL instruments serve as a backup in case the other flops.

A second instrument is Doppler Orbit and Radio Positioning Integration through Satellite, or DORIS, which is depleted to compute precisely the spacecraft's orbit. An array of retroreflectors is additionally conducted aboard the spacecraft and permit quantities to be made from the ground to insist on the orbital data supplied through DORIS (Sui et al., 2017).

In LRM the altimeter achieves as a conventional pulse narrow altimeter. This mode functions at a pulse repetition frequency (PRF) of 1970 Hz. This PRF is low enough to ensure that the echoes are decorrelated (Martin-Puig et al., 2008b and Sui et al., 2017) . Consequently, the echoes obtained may be incoherently extra to decrease speckle noise by $\frac{1}{\sqrt{M}}$ a factor, where M is the number of being an average of pulses in the selected time interval (Martin-Puig et al., 2008a).

In SAR mode, the pulses are emitted in bursts. The correlation between echoes is anticipated. Thus, in SAR mode, the PRF within a burst is higher than the LRM one and equal to 17.8 KHz (Sui et al., 2017). At this PRF, every burst emits 64 pulses. Every time, the pulses are transmitted, the altimeter delays for the returns and transmit again for the succeeding burst. Hence, there is not merely an 'intra-burst' PRF then similarly a burst repetition frequency (BRF), which, according to SIRAL measurements is 85.7 Hz (Martin-Puig et al., 2008b).

Both SIRAL LRM and SAR modes transmit pulses of equivalent pulse length. The principal distinction between these modes is the PRF and its related effects. In SAR mode, pulse to pulse correlation is a consequence of its excessive PRF, whilst pulse to pulse correlation is not existent in LRM, nor preferred. In this regard, the decorrelation PRF is obtained

by dividing the spacecraft velocity V by the decorrelation distance and correcting for the curvature of the earth R_e (Raney, 1998). as given by:

$$PRF = \frac{V}{0.31\lambda H(r^{-1})\left(\frac{R_e + H}{R_e}\right)} \tag{11.11}$$

here λ is the radar altimeter wavelength,H is the altimeter height and r is the radius of the circular uniformly illuminated area. The mathematical relationship between r and significant wave height can be expressed as:

$$r = \sqrt{\frac{(ct + 2H_s)H}{1 + \frac{H}{R_e}}} \tag{11.12}$$

being t is a lag time between the leading and trailing edges of the pulse. Consequently, Equation 11.12 demonstrates that the increase of PRF is due to the extreme growth of significant wave height H_s (Figure 5).

Figure 5. PRF as a function of significant wave height H_s

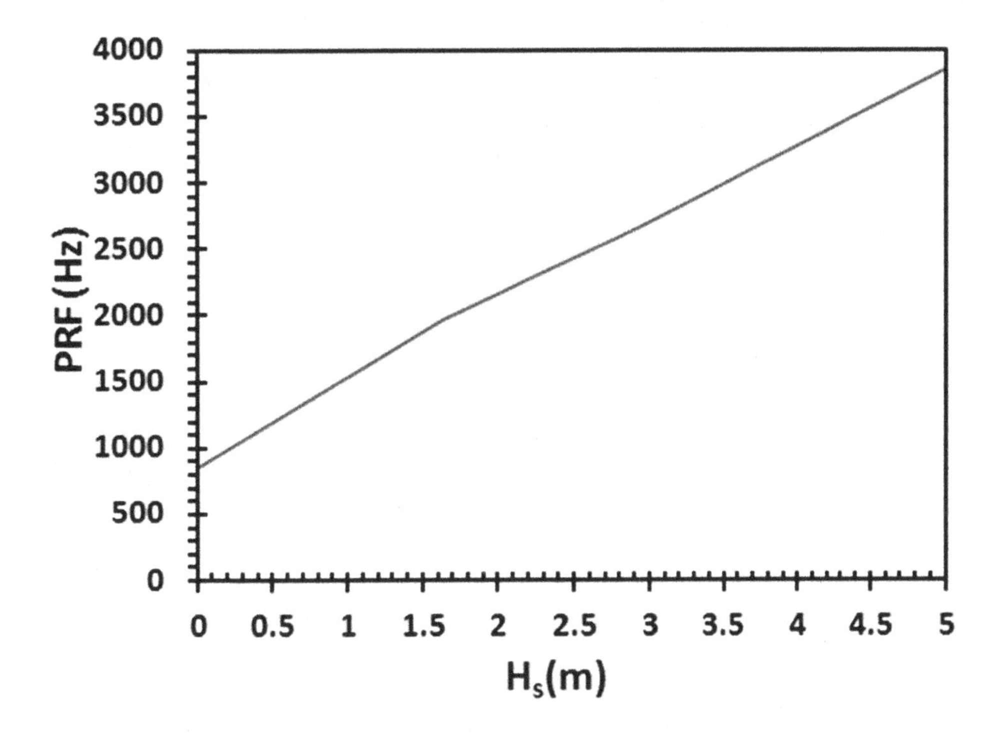

Figure 6 shows the simulated H_s with an area of 300 km x 300 km, with a 500 m wavelength and 4 m H_s. The centre of this area is the location of the suspected MH370 flaperon between 23°57′26.98″S and 101°18′18.76′E.

MULTIOBJECTIVE ALGORITHM FOR MODELLING SIGNIFICANT WAVE HEIGHT ON MOVEMENT OF MH370 DEBRIS

The mathematical components of the optimization process embraces two accurate optimization ranges, every with the exclusive objectives. Initially and concerning the first stage, the target of the technique is to identify a set of the most reliable options that are non dominated with recognizing to each other,

Figure 6. Simulation of H_s using SIRAL LRM data

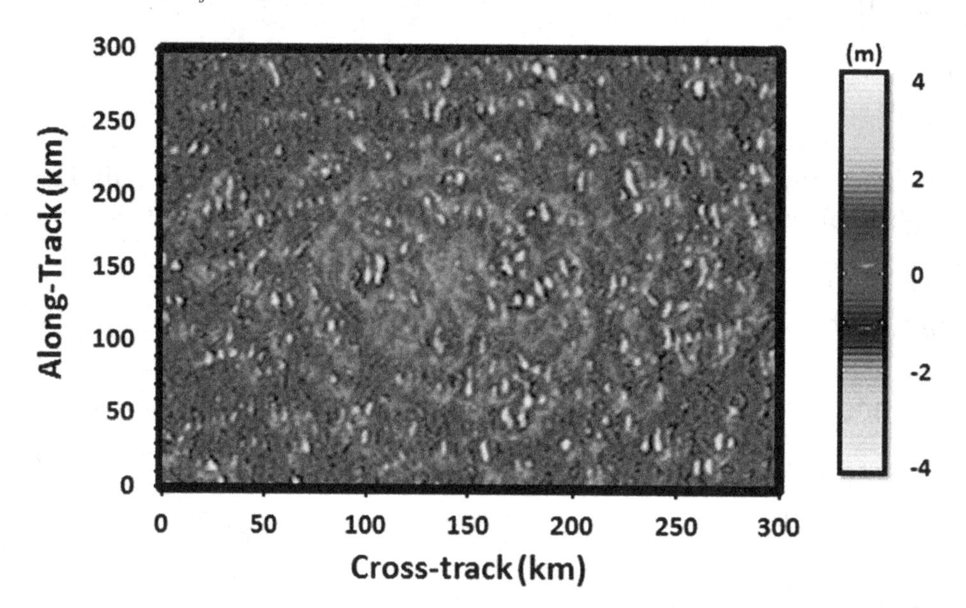

through the adoption of genetic algorithms. A genetic algorithm is a stochastic global explore an approach that mimics the metaphor of expected biological evolution.

Genetic algorithms operate on a population of feasible solutions relating the precept of survival of the fittest one to produce restored and improved approximations to a solution. At every generation, a new set of the population is created using the technique of choosing individuals according to their stage of fitness in the hassle domain and breeding them together using operators, crossover and mutation, borrowed from natural genetics. This technique leads to the evolution of populations of folks that are higher appropriate to their environment than the individuals they have been created simply as in real adaptation, and eventually for the case of multi-objective optimization issues lead to a set of ideal solutions (Louis and Radok,1975; Sanderson and Eliot,1999; Harries et al., 2006; Hemer et al., 2008).

Let assume that the flaperon is placed on an area of the ocean surface *A* under the action of the incident regular waves of varying wave frequency and/or under the action of irregular waves. The significant wave heights have delivered by CryoSat-2 are used to compute the irregular wave power spectra by using the Pierson–Moskowitz (P-M) spectrum (Steedman, 1993), which is given by:

$$S_I = \frac{H_s^2 T_z}{8\pi^2}\left(T_z \frac{\omega}{2\pi}\omega\right)^{-5} e^{\left[-\frac{1}{\pi}\left(T_z \frac{\omega}{2\pi}\right)^{-4}\right]} \tag{11.13}$$

Here T_z is the zero up-crossing periods and ω is wave frequency rad/s. In this regard, ω varies between 0.6 rad/s to 7.6 rad/s. Then the averaged produced power in irregular waves is computed as follows (Rey et al., 2001and Goda, 2010):

$$S_P(\omega) = \bar{P}(\omega) S_I(\omega) \tag{11.14}$$

In this context, $\overline{P}$ is the total averaged power implemented on the flaperon.

To investigate the impact of the wave fields in the flaperon, a mathematical formulation must be developed based on the Pareto optimal solution. Consequently, the principal quantities of flaperon involve length, L_f, the width, B, the height, H_f and the draft, d_r, which are presented in Cartesian coordinate x,y, and z. Moreover, the flaperon is considered as a rigid body with six degrees of freedom, (i.e. ζ_1, ζ_2, ζ_3, ζ_4, ζ_5, and ζ_6) (Figure 7).

Figure 7. Basic quantities of flaperon floating on the ocean surface

The scientific explanation of the potential wave effects Ψ on the flaperon can mathematically be written as:

$$\Psi = A^{-1}\sum_{a=1}^{A}\frac{\zeta(x_a, y_a)}{\eta} \tag{11.15}$$

where A expresses the total number of flaperon area exposed to the potential wave effects, ζ is free surface elevation, and η is the incident wave amplitude at any point behind and on the body of the flaperon.

In this domain, the flaperon is impacted by wave elevation oscillation H_s, and the total averaged power, which is related to relative movements and oscillations of the flaperon. Moreover, internal loads stress T_s on the rigid body. In this sense, the flaperon exposed to other control parameters which are (a) the rotational stiffness of the debris R_s;(b) the translational stiffness T_s; (c) the damping d_c and the angle between an incident wave and the longitudinal axis of the flaperon θ_i and (e) the grid type configuration of the modules of the flaperon *C*. The control force Φ on the flaperon can be mathematically expressed as:

$$\Phi = [R_s, T_s, d_c, \theta_i, C] \in \Omega \tag{11.16}$$

here Ω denotes the constrained tolerable scheme space in which the force variables fluctuate through the application of the optimization process. The optimal set of the non dominated options offer a more profound thoughtful of the physical problem in concurrence with the consequences of the design variables that each solution has. This set of elucidations is well known as the Pareto set of solutions. The initial phase of the optimization procedure embodies a constrained multiobjective optimization problem for which the variation of the values, Φ does not imply simultaneous enhancement of the performance criteria. The multi objective nature of the problem effects of the existence of multi different objectives that have to be optimized as stated in equation 11.16.

The multi-objective functions do no longer have an express mathematical shape and are defined following the following equations:

$$f_1(\Phi) = T_s(\Phi) \tag{11.17}$$

$$f_2(\Phi) = \Psi(\Phi) \tag{11.18}$$

$$f_3(\Phi) = d_c(\Phi) \tag{11.19}$$

$$f_4(\Phi) = \bar{P}(\Phi) \tag{11.20}$$

The submerged flaperon Ψ_s takes a place under the constrain circumstances of:

$$\max \Psi_s = \max \sum_{i=1}^{k} \left| F_i(\Phi) - z_i \right|^p \tag{11.21}$$

$$\text{Subject to } \Psi_s \in \Phi \in \Omega \tag{11.22}$$

$$g(p_i) > k > 0 \quad i=1,2,3,\dots\dots\dots k \tag{11.23}$$

here p is the duration of Φ impacts on the flaperon, k is the number of the objective functions $F_i(\Phi)$ and n is several impact durations which can be ranged from a day to a few months. Moreover, z_i is the orientation point. It should be strained that the orientation point could be the ideal point or a user-defined point with predefined values for each ith objective function. The ideal point is an infeasible point in which all the objective function components have their optimum value of the problem considering each objective separately (Gomes et al., 2010).

Wave Pattern Impacts On Flaperon

The CryoSat-2 coverage cycles coincide with the MH370 search area, in addition, it covers the location where the flaperon fell down in the Southern Indian Ocean between 23°57′26.98″S and 101°18′18.76′E (Figure 8). In other words, CryoSat-2 satellite data can deliver the significant wave height cycles in the Southern Indian Ocean.

Figure 8. CryoSat-2 coverage cycles over MH370 suspected crash zone

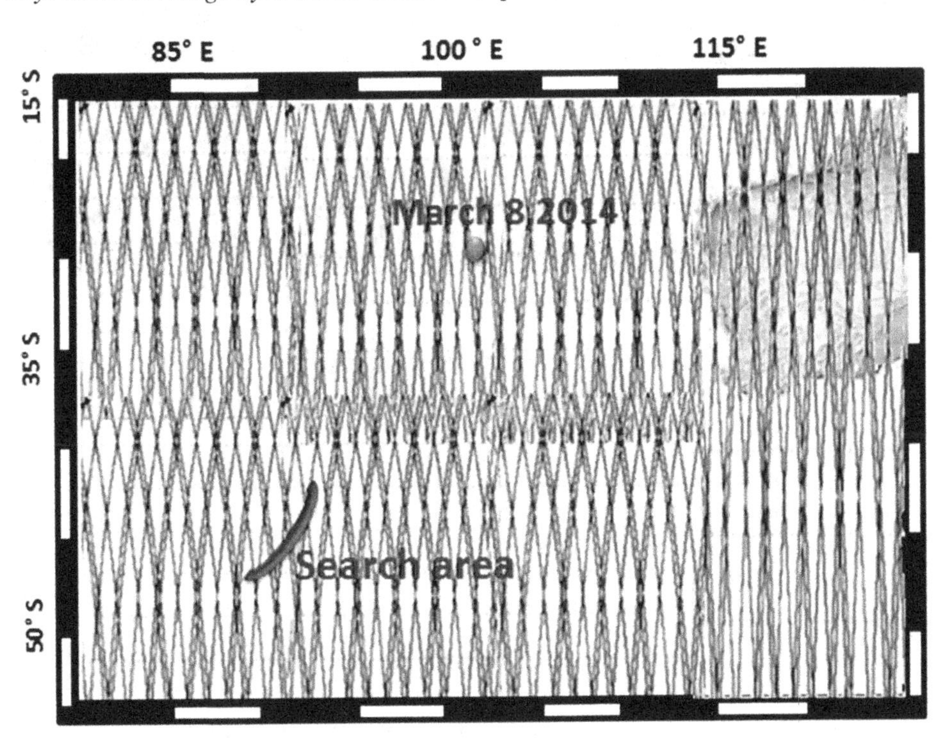

Offshore regions of the South West Australian Shelf are dominated by oceanic swell and seas with an average significant wave height about 4 m and the period from 5 to 20 s. The search area was dominated by a maximum significant wave height of 6 m from March to May. Further north of the search area where the flaperon fell down, the significant wave height was 3 m. In other words, there was a swell pool occupied the offshore regions of the South West of Australia. The southwest significant wave height of 172° propagated towards the northeast of 83° (Figure 9).

From June to August 2014, the maximum significant wave height of 7 m approached the South West Australian Shelf. In the northern West Australian shelf, the significant wave height fluctuated from 1m to 3 m (Figure 1.0). Consequently, the swell propagated from 174 ° to 62°. On the contrary, September

Figure 9. Significant wave height spatial variations from March to May 2014

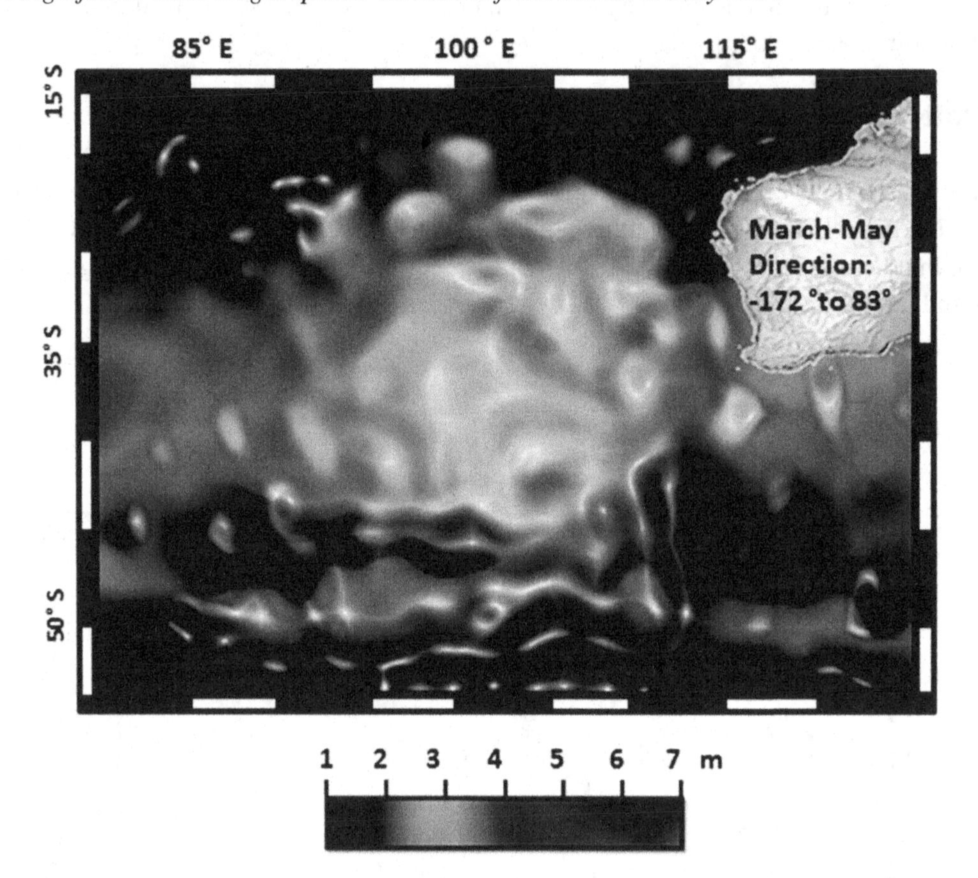

to October 2014 was dominated by maximum significant wave height 6 m, in north-west Australian Shelf with the dominated the direction of 74°. On the contrary, the South West Australian Shelf had a swell pool of 7 m significant wave height, which propagated from the southwest with the direction of 171°(Figure 11).

The significant wave height pattern increased dramatically to 7 m, in both the South West and North East of the Australian Shelf (Figure 12) from November to December 2014. However, the swell remains traveling from southwest to northeast and turned to the north-west of the offshore of Australian Shelf.

For a one-year significant wave height cycle, the swell remains propagating from the southwest to the northeast from January to March 2015 with a maximum significant wave height of 5 m in the northeast offshore of Australian Shelf and 7 m significant wave height southwest of Australian Shelf (Figure 13). On the contrary, inshore, the wave climate responds rapidly to the afternoon onset of the sea breeze. Wind waves associated with the sea breeze have a significantly shorter period and length than the swell and are less affected by the inshore bathymetry.

In this view, most of the year the waves arrive from the southwest, through winter storms often result in waves from the west and north-west bringing high energy conditions to the coast for short periods. Inshore, much of the coastline is sheltered from the direct impact of the swell-wave activity by an extensive chain of reefs, which cause significant attenuation of the ocean waves (Steedman, 1993 and Michailides and Angelides (2015). In this understanding, the search area dominated by the maximum wave power

Figure 10. Significant wave height spatial variations from June to August 2014

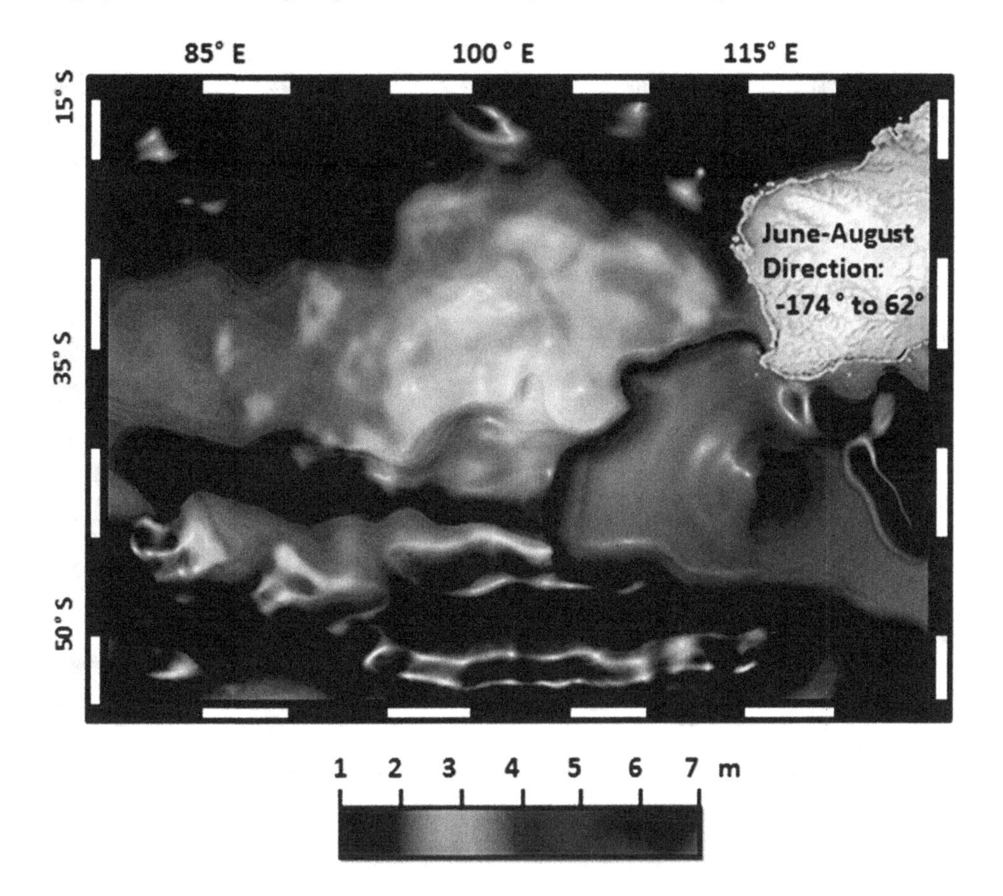

of 30000 kJ/m/wave which is located southwest of the Australian Shelf. However, the North East and West of the Australian shelf dominated by the maximum wave power of 20 kJ/m/wave. This quantity fluctuated from month to month (Figure 14 to Figure 17). The first three months post MH370 crashing i.e., March to May 2014, there was the highest wave power pool southwest of the Australian Shelf with 30000 KJ/m/wave (Figure 14). From June to August 2014, there was a slight reduction of the wave power of 25000 KJ/m/wave (Figure 15). The northwest sector offshore of Australian Shelf dominated by wave power of fluctuation between 5000 to 25000 KJ/m/wave. From September to October southwest of Australian Shelf dominated by maximum wave power of 30000 KJ/m/wave, while north offshore of Australian Shelf dominated by wave power of 15000 KJ/m/wave(Figure 16). This wave power of 30000 KJ/m/wave remained in the southwest of the Australian Shelf and decreased to 20000 KJ/m/wave in northeast Australian Shelf from November to December 2014(Figure 17). From January to March 2015, the wave power increased rapidly in the northeast to 25000 KJ/m/wave (Figure 18).

In general, Around Australia, annual mean wave power is greater in the southern waters than in the northern waters (Louis and Radok1975 and Sanderson and Eliot, 1999). The largest and longest-period powerful) waves occur off the west coast of Western Australia. Low mean heights and shorter periods occur on the Northwest Shelf. This pattern of wave climate is generally consistent with previous studies of waves on the continental shelf around Australia (Louis and Radok1975) Maximum wave power

Figure 11. Significant wave height spatial variations from September to October 2014

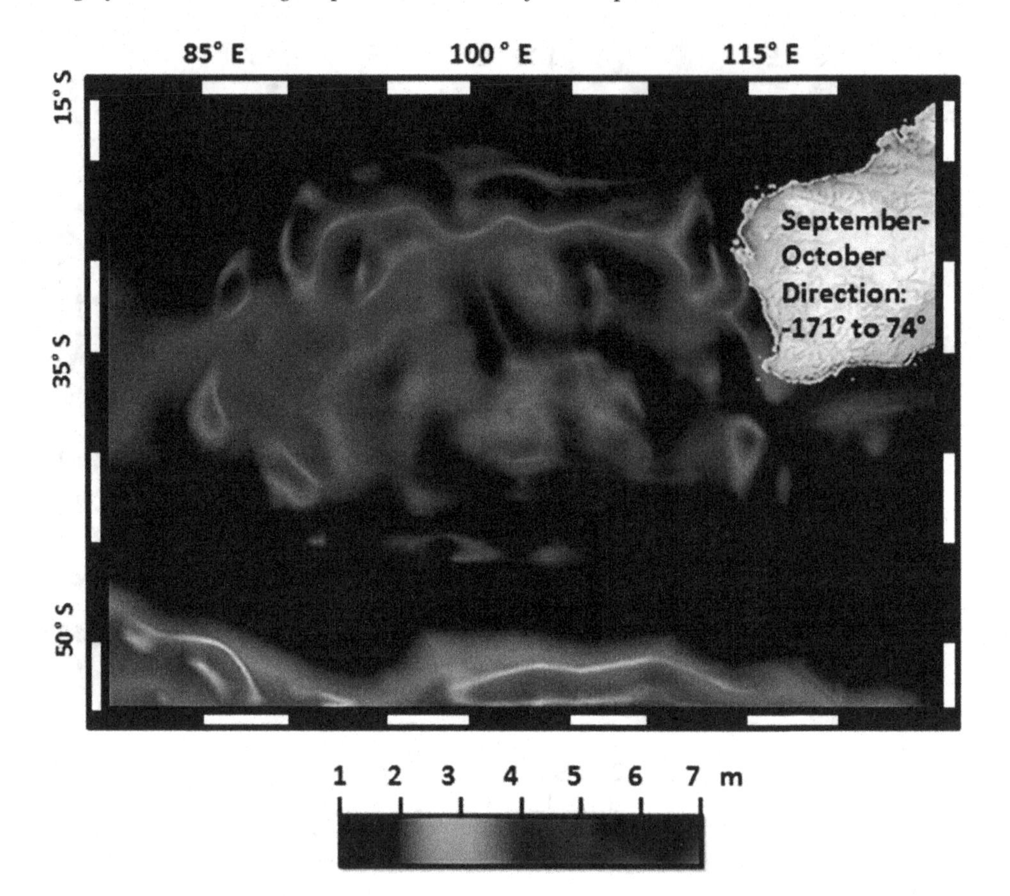

peaks in the south-west of Australian Shelf are associated with waves generated by Tropical cyclone conditions (Hemer et al., 2008).

Furthermore, the offshore wave climate of Australia is dominated by a persistent moderate-energy wave regime with waves of 2 to 4 meters in height. It is characterized by south to south-westerly swells along the south and west coasts, and low-energy waves from the west and north-west along the northern coast. The east coast of Australia is characterized by a strong south-easterly swell, interrupted only briefly by a locally generated choppy north-easterly following the passage of an anticyclone. The inshore wave energy along the northern coasts of the continent is generally low since it is dissipated across a broad continental shelf. In late summer and autumn, tropical cyclones may generate high-energy waves during intense but short-lived storms. Consequently, the swells of the west, south and east coasts are generated in the storm belt of the Southern Ocean, between 50° and 60°S (Louis and Radok1975 ; Sanderson and Eliot, 1999; Hemer et al., 2008).

Despite the strong to moderate offshore wave energy of south-western Australia, the inshore wave energy is considerably less due to dissipation via refraction and diffraction processes around reefs and headlands. This effect is particularly apparent on the west coast, where an extensive reef chain parallels the coast and may attenuate wave energy by up to 50%. Hence, the shoreline of south-western Australia experiences modally low wave energies. Isolated reefs and offshore islands offer some degree of local

Figure 12. Significant wave height spatial variations from November to December 2014 to May 2014

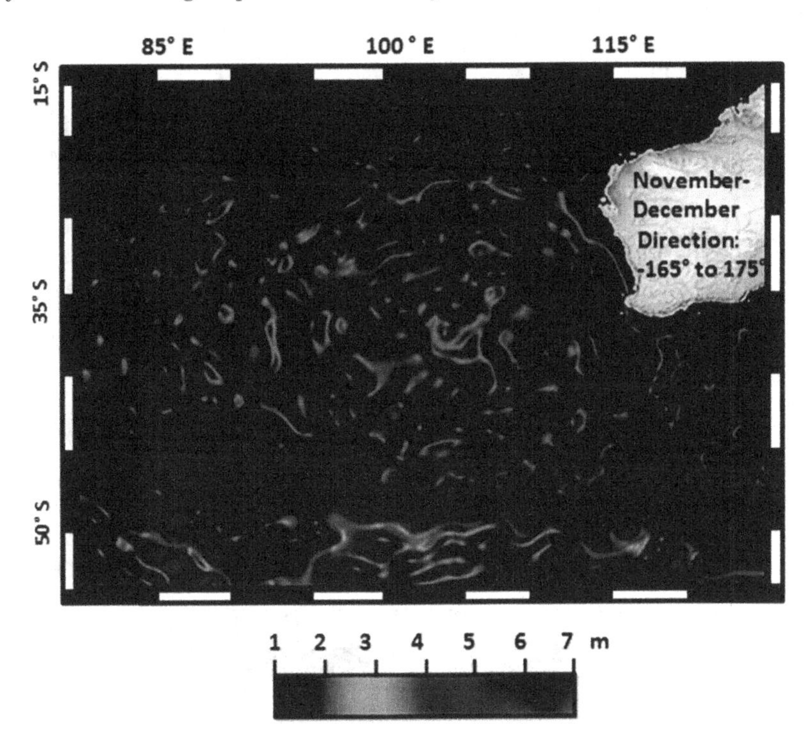

Figure 13. Significant wave height spatial variations from January to March 2015

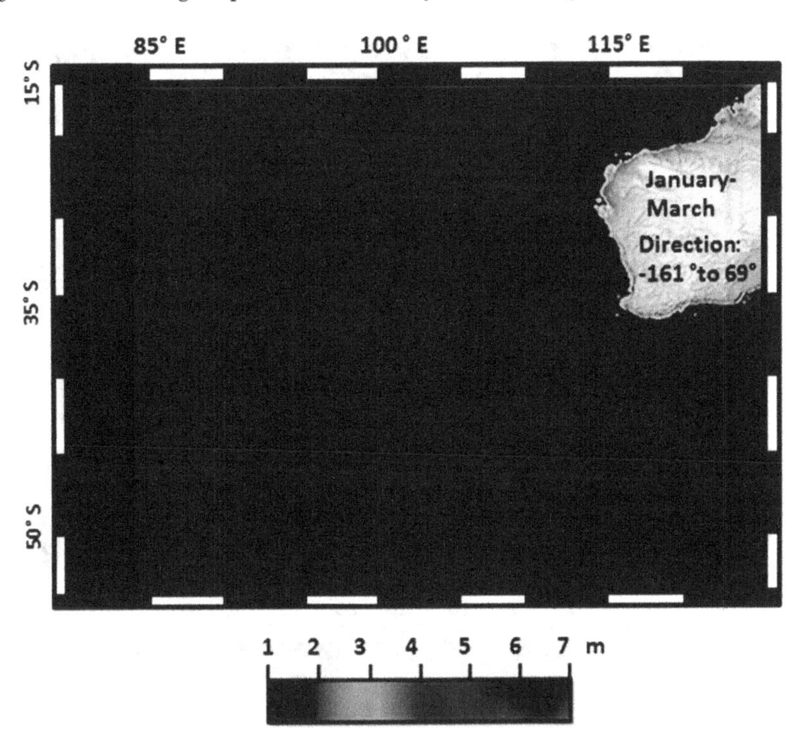

Figure 14. Wave power from March to May 2014

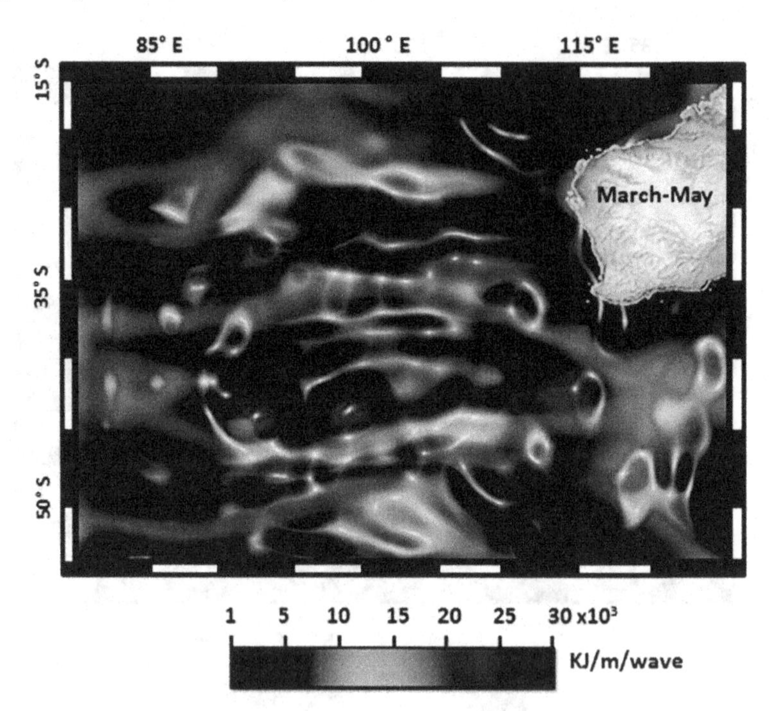

Figure 15. Wave power from June to August 2014

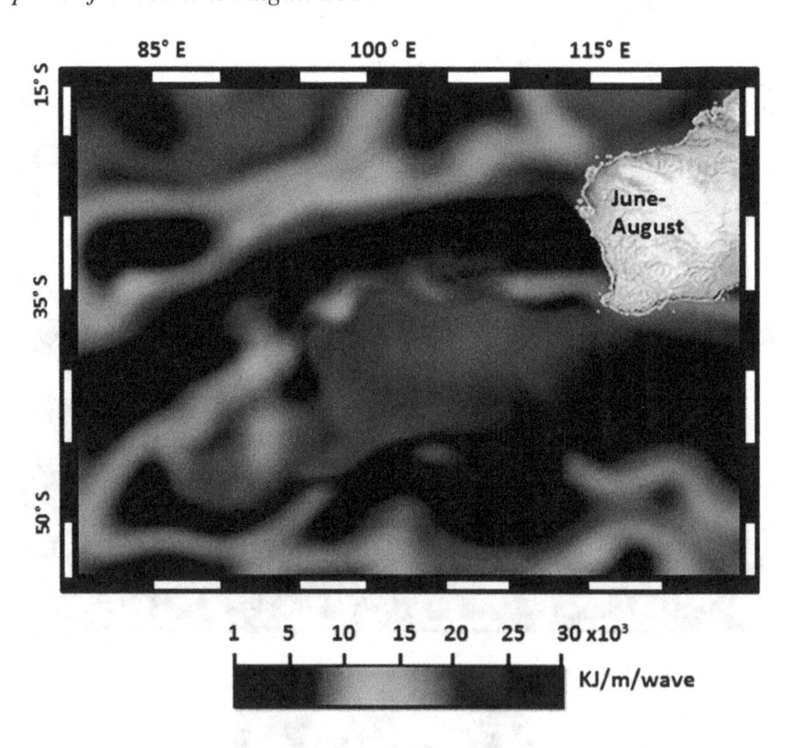

Figure 16. Wave power from September to October 2014

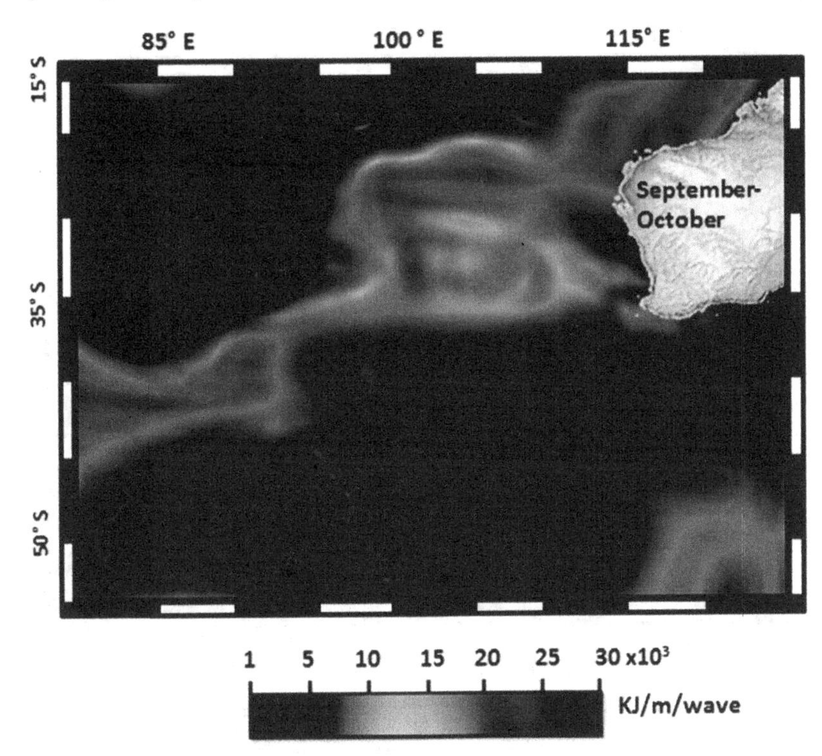

Figure 17. Wave power from November to December 2014

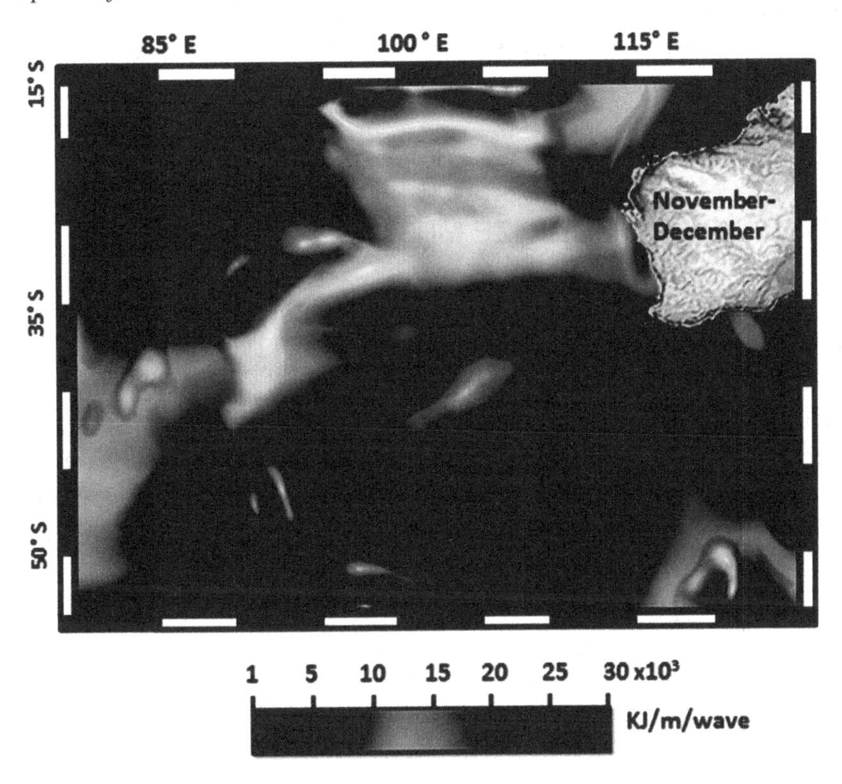

Figure 18. Wave power from January to March 2015

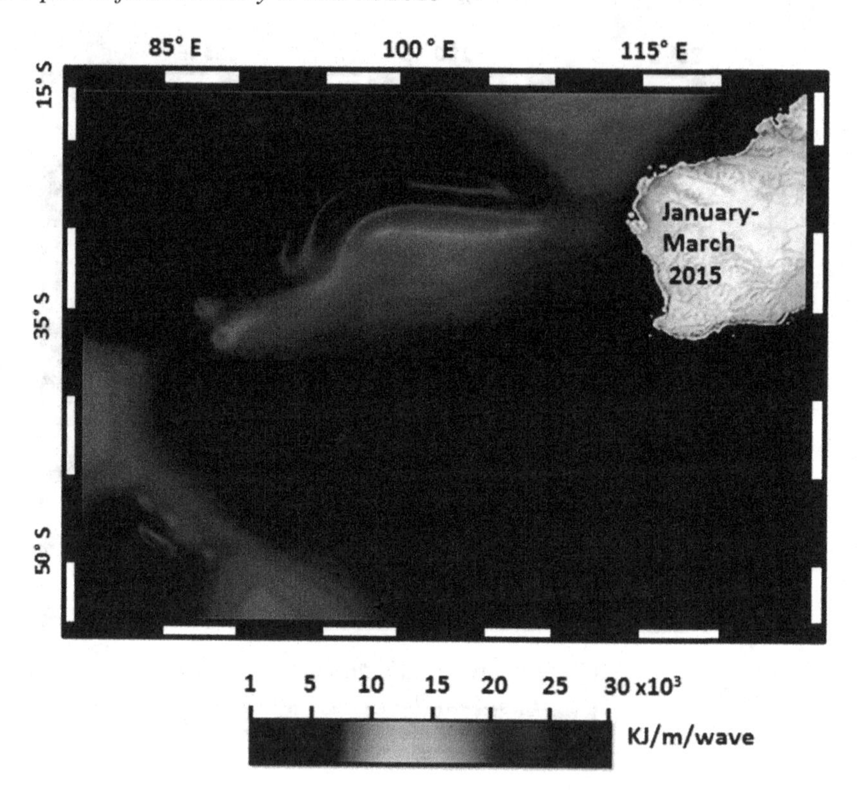

protection of the beaches along the southern coast. However, wave dissipation along the southern coast is considerably less than on the west coast and beaches there are subjected to a moderate to heavy south-westerly swell. Refraction around headlands is an important form of wave energy dissipation along the southern and eastern coasts of the continent.

HOW CAN WAVE POWER MOBILIZE MH370 FLAPERON?

The significant question can raise did the wave power impact the submerged of flaperon and other debris? Pareto optimal curve confirms that the submerged depth of flaperon increases as the wave power increases (Figure 19). In this circumstance, the flaperon could submerge to water depth less than 300 m due to the impact of wave power of 22000 KJ/m/wave. In this view, the flaperon could be submerged further to a water depth of 1000 m because of the wave power of 30000 KJ/m/wave. This proves by the submerged points are positioned entirely on the Pareto front curve than floating-point. In other words, the foremost points of floating investigation on the Pareto front and are the imprecise points. They are not correlated to the Pareto front curve.

The mathematical description between wave power (P)and the flaperon submerged depth (d_s) is given by:

$$d_s \alpha \text{ c } e^{bP} \qquad (11.24)$$

Figure 19. The Pareto curve for wave power impacts on the flaperon

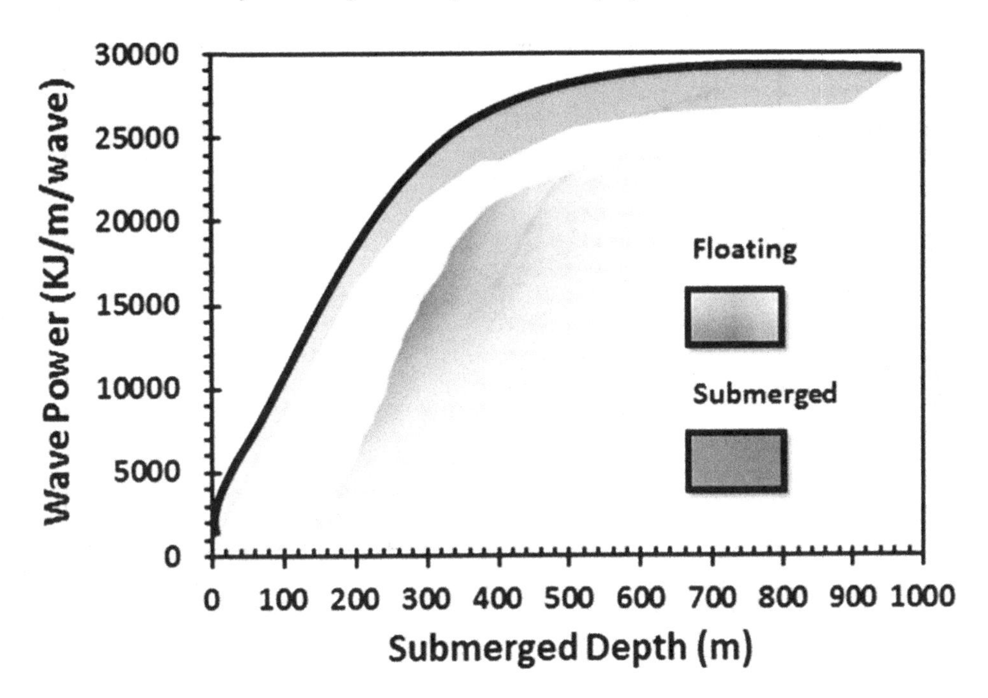

In equation 11.24, *c* and *b* are constant values, which can be determined from the exponential mathematical regression model developed by the author. In the case of the MH370, *c* and *b* equal 54.2 and 0.0001, respectively. In this regard, equation 11.24 can be written as:

$$d_s = 54.2\ e^{0.0001*P} \tag{11.25}$$

It worth mentioning that the wave power of less than 30000 KJ/m/wave, can submerge the flaperon within less than 1000 m depth within three months post the flaperon fell down on March 8 2014. The submerged duration is computed based on the powerful correlation with the wave power, which is given by:

$$T_s = 0.0372\ P^{0.4417} \tag{11.26}$$

This proves by the Pareto front of Figure 20 as the submerged duration of fewer than three months is positioned on the Pareto front.

WHY DID NOT FLAPERON REACH RÉUNION?

Winds rage in this region as pressures and temperatures change rapidly here, driving the winds frequently over 50-60kph, and give rise to storms. The Roaring Forties, at between roughly 40° and 50° latitude in the southern hemisphere, sits in the transition zone between the more tranquil, balmy subtropics, and frigid polar vortex zipping around the South Pole.

Figure 20. Pareto front for the flaperon submerged duration

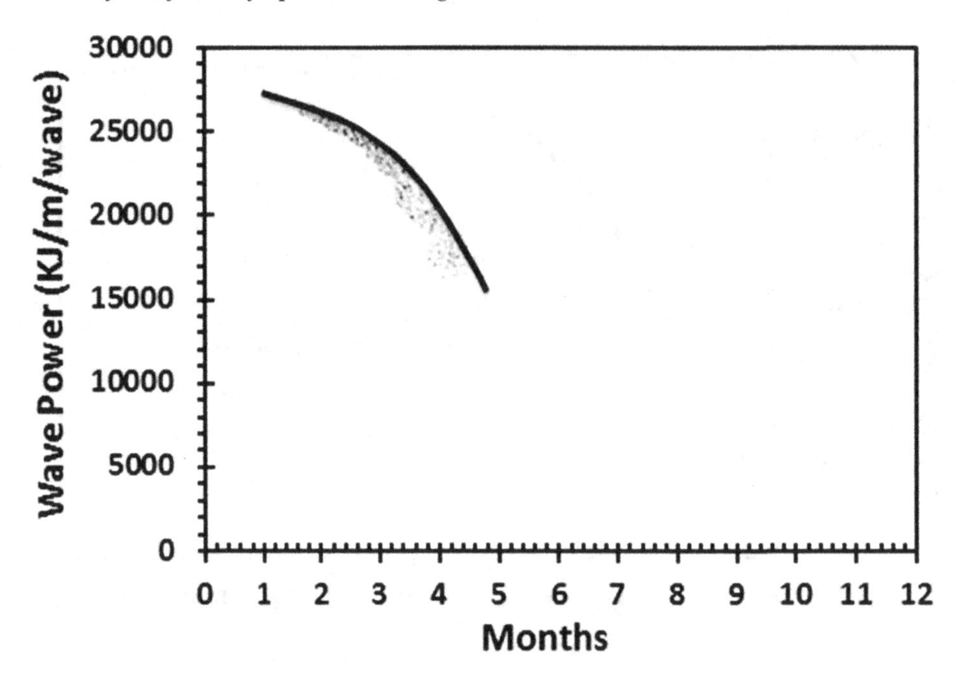

"That's the challenge – find the debris, then recover the debris. And it's going to be exceedingly difficult dispatching a submersible off a ship that is pitching and rolling in 60-foot swells.

In the sort of conditions out there, the sea is a cauldron of foam with white caps and crashing waves. It would be pointless and dangerous to be out there." And during winter comes, the southern Indian Ocean can be described as the worst place in the world.

Unstable weather, dangerous conditions, high waves, and wild seas all make for a terrifying scene, even for large ships. An Antarctic cold front hitting warm tropical air 2,500km off Australia's west coast is expected to severely affect the search over the coming weeks and months (David, 2016). Under these circumstances, the buoyant force comes from the pressure exerted on the flaperon by the ocean. Because the pressure increases as the depth increase, the pressure on the bottom of the flaperon is always larger than the force on the top - hence the net upward force. The buoyant force is present whether the flaperon floats or sinks. In these circumstances, the flaperon and MH370 debris had experienced a force due to gravity, a force due to buoyancy, forces due to wave power. In this understanding, the flaperon rotated and sunk due to high wave power. In other words, wave power affects MH370 flaperon mobilisation and then sunk. In this view, the flaperon could not float across the Southern Indian Ocean and positioned on Réunion Island.

In other words, the discovered flaperon both sides were combined with the presence of crustaceans, gender Lepas, which suggest that the flaperon did not float there from the plane's presumed impact point, but spent approximately four months tethered below the surface." This confirms the finding of the Pareto optimal solution (Figure 20)

CONCLUSION

This chapter introduced a new technique of interferometry based on using altimeter satellite data. To this end, dual SIRAL instruments on-board of CryoSat-2 are implemented to retrieve the annual cycle of significant wave height across the Indian ocean. In this chapter, a one-year significant wave height cycle, the swell remains propagating from the southwest to the northeast from January to March 2015 with a maximum significant wave height of 5 m in the northeast offshore of Australian Shelf and 7 m significant wave height southwest of Australian Shelf. In this circumstance, the Pareto algorithm proves that the flaperon would submerge to water depth less than 300 m owing to the impact of wave power of 22000 KJ/m/wave. In conclusion, the flaperon could be submerged further to a water depth of 1000 m because of the wave power of 30000 KJ/m/wave.

The critical question is "How did the flaperon get to Réunion if it did not float there?"Unanswered question concludes that the missing flight MH370 cannot be found in the search area in the Southern Indian Ocean. This ambiguity in information supported the theory derived in chapter 5, which ignored the vanishing of MH370 in the Indian Ocean. Conversely, the next chapters continue to investigate the actual reality of the crashing of MH370 in the Southern Indian Ocean.

REFERENCES

David, R. (2016). MH370 found in Cambodian JUNGLE? Search launched as Google Maps shows 'Boeing CRASH SITE'. *DailyStar*. https://www.dailystar.co.uk/news/world-news/727012/mh370-news-malaysia-airlines-flight-google-maps-cambodia-theory-latest

Galin, N., Wingham, D. J., Cullen, R., Fornari, M., Smith, W. H., & Abdalla, S. (2012). Calibration of the CryoSat-2 interferometer and measurement of across-track ocean slope. *IEEE Transactions on Geoscience and Remote Sensing*, *51*(1), 57–72.

Goda, Y. (2010). *Random seas and design of maritime structures*. World Scientific.

Gomes, R. P. F., Henriques, J. C. C., Gato, L. M. C., & Falcão, A. F. (2010, January). IPS two-body wave energy converter: acceleration tube optimization. In *The Twentieth International Offshore and Polar Engineering Conference*. International Society of Offshore and Polar Engineers.

Harries, D., McHenry, M., Jennings, P., & Thomas, C. (2006). Hydro, tidal and wave energy in Australia. *The International Journal of Environmental Studies*, *63*(6), 803–814.

Hemer, M. A., Simmonds, I., & Keay, K. (2008). A classification of wave generation characteristics during large wave events on the Southern Australian margin. *Continental Shelf Research*, *28*(4-5), 634–652.

Jensen, J. R. (1999). Angle measurement with a phase monopulse radar altimeter. *IEEE Transactions on Antennas and Propagation*, *47*(4), 715–724.

Louis, J. P., & Radok, J. R. M. (1975). Propagation of tidal waves in the Joseph Bonaparte Gulf. *Journal of Geophysical Research*, *80*(12), 1689–1690.

Martin-Puig, C., Marquez, J., Ruffini, G., Raney, R. K., & Benveniste, J. (2008b). *SAR altimetry applications over water.* arXiv preprint arXiv:0802.0804

Martin-Puig, C., Ruffini, G., Marquez, J., Cotton, D., Srokosz, M., Challenor, P., ... Benveniste, J. (2008a, July). Theoretical model of SAR altimeter over water surfaces. In *IGARSS 2008-2008 IEEE International Geoscience and Remote Sensing Symposium* (Vol. 3, pp. III-242). IEEE.

Michailides, C., & Angelides, D. C. (2015). Optimization of a flexible floating structure for wave energy production and protection effectiveness. *Engineering Structures*, *85*, 249–263.

Raney, R. K. (1998). The delay/Doppler radar altimeter. *IEEE Transactions on Geoscience and Remote Sensing*, *36*(5), 1578–1588.

Rey, L., de Chateau-Thierry, P., Phalippou, L., Mavrocordatos, C., & Francis, R. (2001, July). SIRAL, a high spatial resolution radar altimeter for the Cryosat mission. In *IGARSS 2001. Scanning the Present and Resolving the Future. Proceedings. IEEE 2001 International Geoscience and Remote Sensing Symposium (Cat. No. 01CH37217)* (Vol. 7, pp. 3080-3082). IEEE.

Sanderson, P. G., & Eliot, I. (1999). Compartmentalisation of beachface sediments along the southwestern coast of Australia. *Marine Geology*, *162*(1), 145–164.

Steedman, R. (1993). *Collection of wave data and the role of waves on nearshore circulation.* Report prepared for the Water Authority of Western Australia.

Sui, X., Zhang, R., Wu, F., Li, Y., & Wan, X. (2017). Sea surface height measuring using InSAR altimeter. *Geodesy and Geodynamics*, *8*(4), 278–284.

Tournadre, J., Girard-Ardhuin, F., & Legrésy, B. (2012). Antarctic icebergs distributions, 2002–2010. *Journal of Geophysical Research. Oceans*, *117*(C5).

Wingham, D. J., Francis, C. R., Baker, S., Bouzinac, C., Brockley, D., Cullen, R., & Phalippou, L. (2006). CryoSat: A mission to determine the fluctuations in Earth's land and marine ice fields. *Advances in Space Research*, *37*(4), 841–871.

Chapter 12
Pareto Optimization for Rossby Wave Pattern Impacts on MH370 Debris

ABSTRACT

This chapter censoriously appraises the comprehensive theories that specify that more concepts are needed to bridge the gap found between the dynamic of the Southern Indian Ocean and the actual MH370 vanishing mechanism. Thus, this chapter is devoted to the Rossby waves, which could attribute to the fact that the MH370 flaperon got to Réunion Island. In this view, Rossby waves generate growth of energy in the west of the ocean gyres and create the strengthening currents on the western side of the ocean basins. Pareto optimization algorithm of the impact power of Rossby waves proves that the flaperon could not drift across the Southern Indian Ocean and be positioned on Réunion Island.

INTRODUCTION

Indian ocean is a complicated water body similar to other oceans. In the previous chapter, the Southern West of Australia Shelf, where the MH370 plunged into the water as claimed by experts, is governed by the equatorial and southern ocean wave climates. It was obvious that the Southern Ocean governed the wave power input in the search area and the area where the flaperon fell down. Consequently, the wave power is one component of the multi complicated dynamic system of the Southern Indian Ocean. The experts never consider the wave power impact on the MH370 debris and flaperon unsteadiness. They considered the trajectory movements due to surface current. This chapter raises critical questions regarding the Rossby wave impacts on the stability of the flaperon or debris in the search area.

This chapter censoriously appraises the comprehensive theories that specify that more concepts are needed to bridge the gap found between the dynamic of the Southern Indian Ocean and the actual MH370 vanishing mechanism. Thus, this chapter is devoted to the Rossby waves, which could attribute to the fact that the MH370 flaperon got into Réunion Island.

DOI: 10.4018/978-1-7998-1920-2.ch012

The main question of what is meant by Rossby waves? Waves in the ocean generate numerous unique shapes and dimensions. Slow-moving oceanic Rossby waves are essentially exclusive from ocean surface waves. Dissimilar waves that smash alongside the shore, Rossby waves are huge, the undulating dynamic of the ocean that stretches horizontally throughout the planet for thousands of kilometers in a westward direction (Figure 1). They are so massive and large that they can alternate Earth's local weather conditions. Along with rising sea levels, King Tides, and the consequences of El Niño, oceanic Rossby waves make a contribution to excessive tides and coastal flooding in some zones of the global (Polvani et al., 1991 and Lovelace et al., 1999).

Figure 1. Rossby wave westward flow

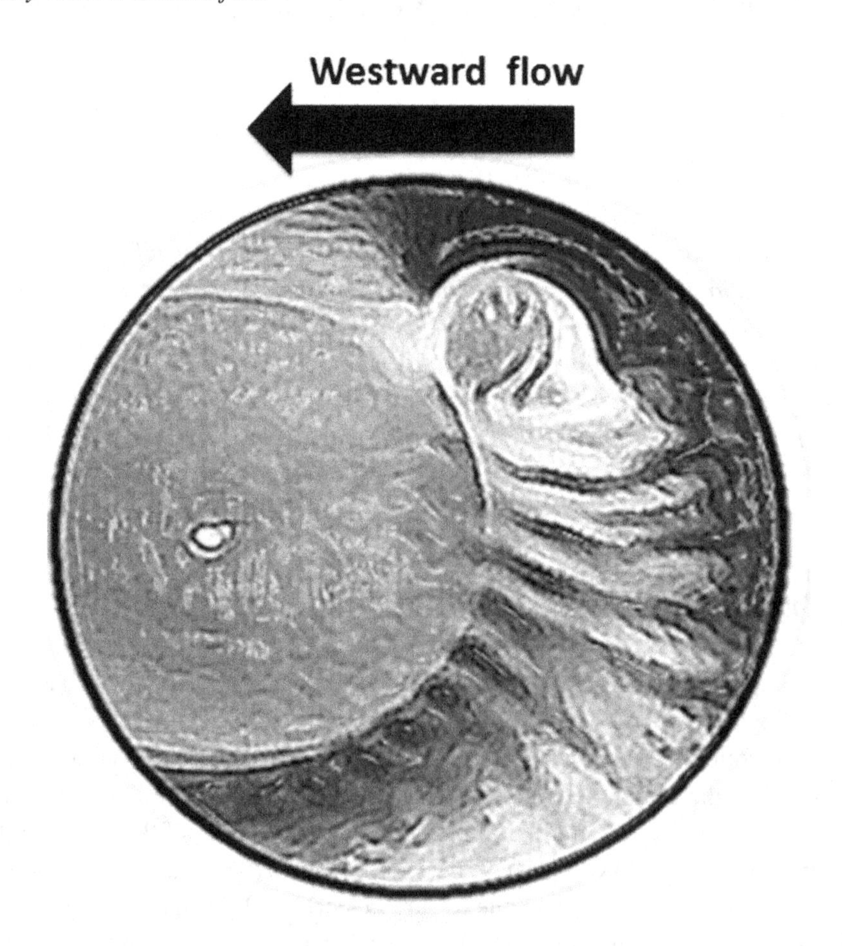

In this understanding, Rossby waves, additionally are recognized as planetary waves (Figure 2), inherently transpire in gyrating fluids. Within the Earth's ocean and atmosphere, these waves are generated as a consequence of the rotation of the planet (Lorenz,1972 and Woollings et al., 2008)

Rossby wave dynamic movement is complex. The horizontal wave velocity of a Rossby (the quantity of time it takes the wave to travel across an ocean basin) is a function of upon the latitude of the wave. In the Pacific, for instance, waves at lower latitudes (closer to the equator) (Figure 3) might also take

Figure 2. Example of planetary waves

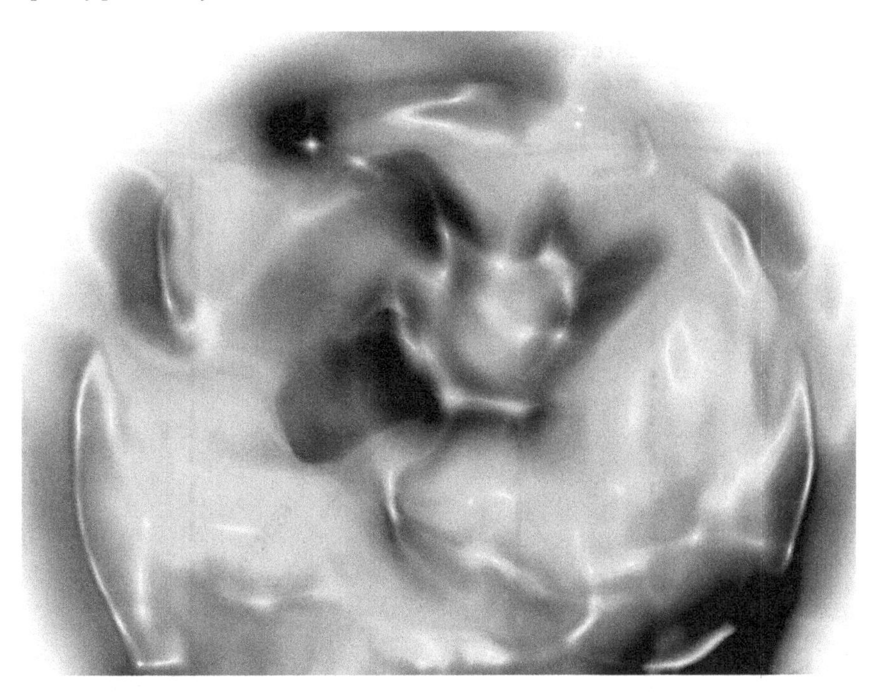

Figure 3. Equatorial Rossby wave

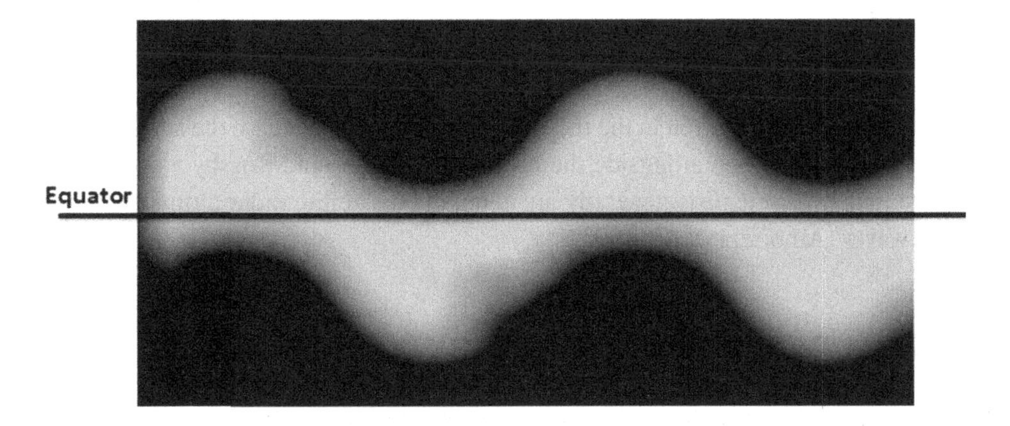

months to a year to travel across the ocean. Waves that structure farther away from the equator (at mid-latitudes) of the Pacific can also take nearer to 10 to 20 years to make the journey (Lovelace et al., 1999 and Woollings et al., 2008).

The vertical dynamic of Rossby waves is small alongside the ocean's floor and massive alongside the deeper thermocline (Figure 4) — the transition vicinity between the ocean's heat upper layer and chillier depths. This version in vertical movement of the water's surface can be quite dramatic: the regular vertical dynamic of the water's surface is commonly four inches or less, whilst the vertical motion of the thermocline of the equatorial wave is about 1,000 instances greater. In other words, for a four inch or much less surface displacement alongside the ocean surface, there may additionally be greater than 300

Figure 4. The vertical dynamic of Rossby wave

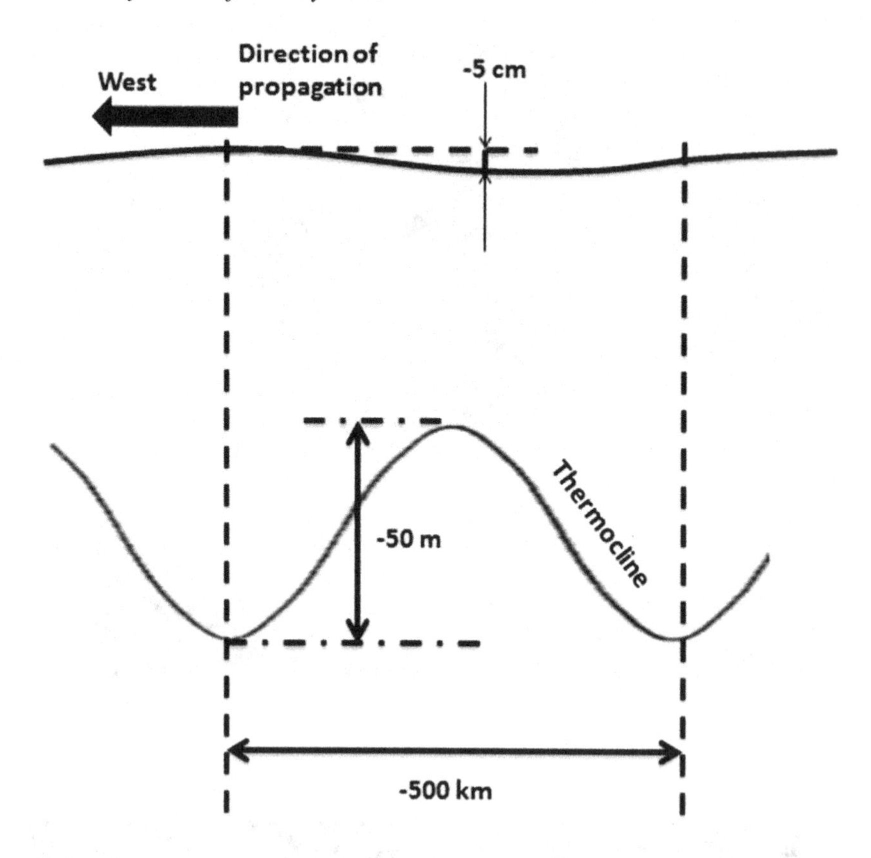

feet of corresponding vertical motion in the thermocline a long way beneath the surface (Lorenz, 1972)! Due to the small vertical motion alongside the ocean surface, oceanic Rossby waves are undetectable by way of the human eye. Scientists generally are counted on satellite radar altimetry to become aware of the massive waves (Ambrizzi, et al., 1995).

MATHEMATICAL DESCRIPTION OF ROSSBY WAVES CORIOLIS

The Coriolis parameter is the keystone to understand the Rossby waves. In this view, The Coriolis force f_{cor} is an imaginary force, which generates due to the Earth's rotation around its axis with constant angular velocity Ω, which equals 7.292 x 10^{-5} s^{-1} (Figure 5). In this sense, the scientific explanation of f can mathematically be written as:

$$f_{cor} = 2\Omega \sin(\phi) \tag{12.1}$$

Equation 12.1 demonstrates that the Coriolis force equals zero along the equator as its latitude ϕ is zero. In this regard, what is the role of Ω? The fixed frame of the Earth rotation causes the potential vorticity of the parcel of fluid in the ocean, which sustained.

It is well known that the latitudinal variation (β) in f_{cor} is simplified using:

Figure 5. Coriolis force concept

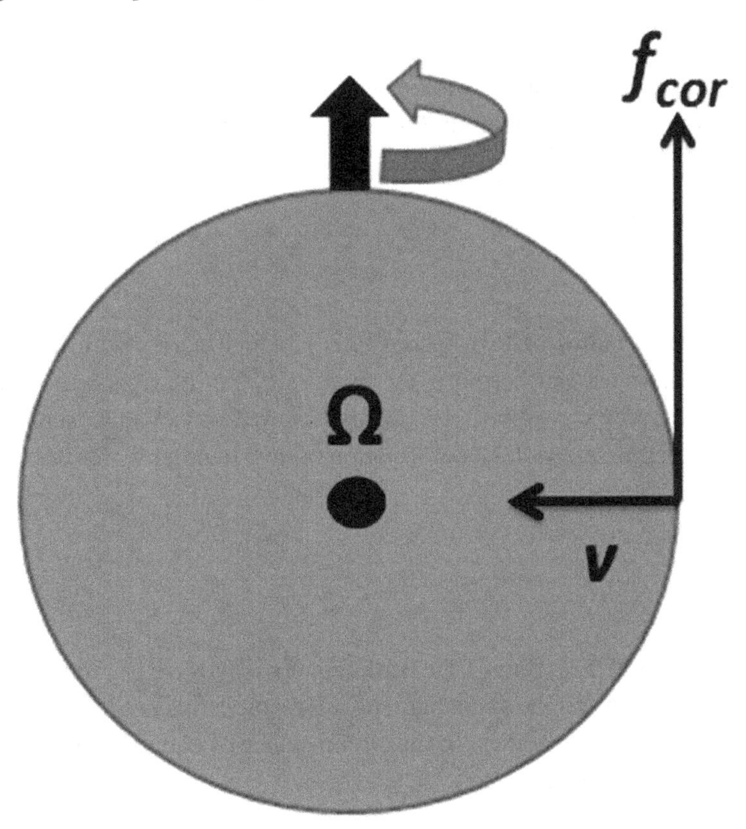

$$\beta = \frac{df_{cor}}{d\phi} = 2\Omega\cos(\phi) \tag{12.2}$$

Then the potential vorticity across the shallow water and a homogenous layer of thickness H is defined as:

$$\frac{\partial\left(\xi - \frac{f_{cor}\eta}{H}\right)}{\partial t} + \beta v = 0 \tag{12.3}$$

Equation 12.3 considers a continuity equation. In this equation, the vertical component of the potential vorticity can present by ξ through the free height of the shallow layer η. Consequently, v is the meridional (north-south) component of the fluid speed.

The potential vorticity is conserved because of a restoring force. In other words, a parcel of fluid latitudinally is a function of a restoring force. A propagation of Rossby wave is the resultant effect of the initial disturbance, the restoring force, and the inertia of the fluid parcel. These resultant effects are caused by atmospheric forcing or a change in ocean currents. Therefore, the propagation of Rossby waves can be described as:

$$\eta \sim e^{[i(kx+ly-\omega t)]} \tag{12.4}$$

here k and l are the zonal (east-west) and meridional wavenumbers, respectively. Consequently, the wave frequency (ω) suggests the dispersion relation for zero backgrounds mean flow, which is casted as:

$$\omega = -\frac{\beta k}{\left(k^2 + l^2 + r^{-2}\right)} \tag{12.5}$$

being the local deformation radius of Rossby waves is r, which varies with latitude, ϕ local density stratification, and mode number. Furthermore, the minus sign on the right-hand side of (Woollings et al., 2008) indicates that Rossby waves have westward phase velocity (Van Gysen and Coleman 1999).

In this view, each baroclinic mode has a deformation radius. In other words, mode baroclinic Rossby radius is:

$$r = V_n \left| f_{cor} \right|^{-1}, \qquad n = 1, 2, 3, \ldots.. \tag{12.6}$$

Equation 12.6 suggests that the radius of the baroclinic Rossby wave is a function of the propagation velocity V with the nth mode. In this view, the baroclinic Rossby radius is a length scale often associated with boundary currents, fronts, and eddies. In the circumstance of equations 12.5 and 12.6, the longest wavelength of Rossby waves is constrained to

$$k^2 + l^2 \ll r^2 \tag{12.7}$$

Equation 12.7 suggests that Rossby waves are non-dispersive. In the circumstances of l=0 and $k = \frac{1}{r}$, respectively, Rossby waves have a maximum or cut-off frequency ω_c, which is given by:

$$\omega_c = -0.5(\beta r) \tag{12.8}$$

In this view, Rossby waves have a revolving latitude, ahead of which Rossby wave clarifications grow into momentary. In other words, the Rossby wave propagation rate turns into a zero. Therefore, the dynamic ocean due to wind stress curl, buoyancy forcing, coastally trapped waves, and reversals in coastal currents are the forces, which generate Rossby waves. Moreover, the dependence of transport on f_{cor} means that on large "planetary" scales, variation in Coriolis causes convergence and divergence to the west and east of eddies such that their pattern propagates westward (Van Gysen et al., 1997).

CONVERGENCE AND DIVERGENCE DUE TO ROSSBY WAVES

Rossby waves generate growth of energy in the west of the ocean gyres and create the strengthening currents on the western side of the ocean basins (Van Gysen et al., 1992).

It must take into account the rate of the thermocline displacement, the convergence and the divergence processes to calculate the speed of the Rossby wave propagations.

Let assume that the propagation of the Rossby waves between two latitudes ψ_1 and ψ_2, respectively, in which f_{cor} changes by an amount $\beta\Delta\phi$ i.e. $f_{cor}=f_0+\beta\Delta\psi$. In other words, the β-plane approximation which is just a convenient way of representing the variation of f_{cor} with latitude. In this understanding β can be identified by:

$$\beta = \frac{\partial f_{cor}}{\partial \psi} = 2\Omega \cos\phi \frac{\partial \phi}{\partial \psi} \tag{12.9}$$

where $\frac{\partial \phi}{\partial \psi} = R_{Earth}^{-1}$ (12.9.1)

here R_{Earth} is the radius of the Earth, which is 6371 km. In this view, β the value can be given by:

$$\beta = \frac{2x7.292x10^{-5}}{6371\,x\,10^{3\cos\phi}} \tag{12.10}$$

$$\beta = 2.28x\,10^{-11\cos\phi} = 2.28 \text{ x } 10^{-11} cos\phi \text{ m}^{-1}\text{s}^{-1}$$

For instance, at 20°S $\beta = 2.15 \text{ x } 10^{-11}$ $\text{m}^{-1}\text{s}^{-1}$ while at 40°S $\beta = 1.75 \text{ x } 10^{-11}$ $\text{m}^{-1}\text{s}^{-1}$

The net volume convergence N_c between two latitudes ψ_1 and ψ_2 is calculated using:

$$N_c = g\bar{H}\frac{\Delta\rho}{\rho_0}\Delta H\frac{\beta\Delta\psi}{f^2} \tag{12.11}$$

here ρ_0 is seawater density, and g is the gravitational constant. Equation 2.11 suggests that the volume convergence must be balanced by a pycnocline that is being driven down at a vertical velocity of $\frac{\partial H}{\partial t}$.

Under the circumstance of balancing, the equation 12.11 becomes:

$$\Delta\psi_1\Delta\psi_2\frac{\partial H}{\partial t} = g\bar{H}\frac{\Delta\rho}{\rho_0}\Delta H\frac{\beta\Delta\psi}{f^2} \tag{12.12}$$

Equation 12.1 suggests that on the east side, *ΔH* changes sign and this means that $\frac{\partial H}{\partial t}$ is negative, consistent with a thinning upper layer. In this regard, the vertical displacement of the layer *H* is casted as:

$$\frac{\partial H}{\partial t}=\frac{\beta g\bar{H}}{f^2}\frac{\Delta\rho}{\rho_0}\frac{\partial H}{\partial x} \tag{12.13}$$

Being x is the longitudinal distance across the ocean. In this sense, the ratio of $\frac{\partial H}{\partial t}/\frac{\partial H}{\partial x}$ demonstrates the velocity v at which a line of constant H (a wave crest for instance) moves eastward. On the contrary, the planetary eddy pattern moves westward at a speed of $-v$, with

$$v=\beta\frac{\Delta\rho}{\rho_0}\frac{gH}{f_{cor}^2}\ \text{m/s} \tag{12.14}$$

Equation 12.14 demonstrates that the thermocline displacements have small sea surface displacements associated with them, which can be observed using a satellite altimeter.

In an interval time of Δt, the Rossby wave moves west with a distance off Δx, so the slope of these lines is estimated via:

$$\frac{\Delta x}{\Delta t}=\frac{\partial H}{\partial t}/\frac{\partial H}{\partial x}=-v \tag{12.15}$$

For instance, equation 12.15 can fit the observations in the Indian Ocean as:

$$v=\Delta x/\Delta t=\frac{30o\ \text{lon}*111\text{km}*\cos(25)}{60\ \text{cycles}*10\ \text{days}*86400\ \text{sec}}$$

$= 5.8 \times 10^{-2}\ \text{m s}^{-1}$

In the circumstance of the Indian Ocean, which has a thermocline depth of 1000m and the average surface water density ρ_0 of 1026.6, and a depth of 1027.8, so Rossby wave velocity is estimated through equation (12.14):

$v = 2.1 \times 10^{-11} * (27.8-26.6)/1027.8 * 9.81 * 1000\ \text{m} / (6.16 \times 10^{-5})^2$

$= 6.3 \times 10^{-2}\ \text{m s}^{-1}$

The application of equation 12.14 suggests that the Rossby wave speed gets larger when it is approaching the equator, always positive i.e. westward propagation in our sign convention. Moreover, this is an approximate equation for very long wavelength, long period (many months) Rossby waves, for the idealized 1½ layer ocean. In this understanding, the Rossby wave velocities for along period, perhaps impact the drift of the MH370 debris and flaperon. This might be observed precisely in the following sections.

MODELLING ROSSBY WAVE PATTERNS FROM SATELLITE ALTIMETER

The significant question of how can Rossby wave simulate using satellite altimeter?

Estimation Sea Surface Height Using Collinear Analysis

The initial altimetric investigations castoff Geosat altimeter satellite data to the express of the westward propagation of Sea surface Height signals in expressions of annual Rossby Waves (White et al.,1990). Even though these consequences had been particularly ambiguous, owing to the aliasing belongings of the Geosat data. However, westward propagation of signals has been confirmed by means of numerous studies based on the usage of T/P altimeter and in situ data, as well as with the aid of numerical models (Vivier e al., 1999; Strub et al., 2000; Kelly et al., 2002). These signals originate subsequent to the coast due to the annual alternate in the alongshore winds. Thus, along with the Australia west coast, the annual and interannual variability in the coastal currents and water mass characteristics are transferred westward, into the deep ocean (Marchesiello et al., 2003).

Let assume that I_i is the vector of altimeter measurements in the ith and I_{ij} is the measurements at the jth grid point. In this understanding, N is the number of grid points along the pass, where the mean sea surface height h_j is determined with a certain error e_{ij}. At the jth grid point, $I_{ij} \in h_j \in e_{ij}$. Then the mean sea surface height is simplified using(] van Gysen et al., 1992):

$$h_j = I_{ij} - e_{ij} - C_j x_i \tag{12.16}$$

being C_j is the raw of the vector of model coefficients for the time-dependent radial orbital error. Moreover, C_j is a function of travel time t_j to the jth grid point from some reference point along the pass. In this manner, the travelling time can be expressed by:

$$C_j = \left(1, \sin(2x\pi \frac{t_j}{T}) \right) \tag{12.17}$$

$$C_j = \cos(2x\pi \frac{t_j}{T}) \tag{12.18}$$

where T is the orbital period of the satellite. The linear model uses to compile all the N repeat passes is formulated as:

$$I = \left[E_N \otimes I_n, I_n \otimes C \right] \begin{bmatrix} h \\ x \end{bmatrix} + e = X \begin{bmatrix} h \\ x \end{bmatrix} + e \tag{12.19}$$

here $\otimes$ is the Kronecker product and E_N is the N-vector of that element is a “1”. Equation 12.19 can be expressed based on the Gauss-Jordan elimination as (van Gysen et al., 1992):

$$X^g = \begin{bmatrix} N^{-1}E_N^T \otimes I_n \\ \left(I_N - N^{-1}E_N E_N^T \oplus \left(C^T C\right)^{-1} C^T\right) \end{bmatrix} \tag{12.20}$$

Equation 12.20 involves the transfer matrix T for the selected parameters, which are stated early in equations 12.16 to 12.19. The sea surface height variability H_j at grid point j can be calculated from

$$\begin{aligned} H_j = N^{-1}l^T &\left(I_{nN} - \begin{bmatrix} I_n & N^{-1}E_N^T \oplus C \\ 0 & \left(I_n - N^{-1}E_N E_N^T \oplus I_r\right) \end{bmatrix} \right) \\ \left(I_N \oplus E_{ij}\right) &\left(I_{nN} - \begin{bmatrix} I_n & N^{-1}E_N^T \oplus C \\ 0 & \left(I_n - N^{-1}E_N E_N^T \oplus I_r\right) \end{bmatrix} \right) \end{aligned} \tag{12.21}$$

Equation 12.21 is the inverse of the equation 12.20 in grid position *i,j* being E_{IJ} is the (n,n) elementary matrix with "1"(van Gysen et al.,1997 and Van Gysen and Coleman 1999).

Fast Fourier Transform For Rossby Wave Pattern

The Fast Fourier Transform (FFT) is used to estimate the frequency domain and the wavelength of the Rossby Wave. To this end, the FFT is a method, which is converted into a wave number and frequency space, which highlights the distinct spectral components. The FFT has the expansion of displaying single foundations of a wave, which may additionally translate to distinct baroclinic modes (Subrahmanyam et al.,2001). But due to the fact it maps a single wave frequency to every component, it requires that the propagation characteristics of the wave continue to be regular for the vicinity and time being studied (Cipollini et al., 2001). The FFT is used to calculate the wavelength, period, and amplitude of the first-mode baroclinic Rossby waves in the zonal gradient of every parameter. The wavelength and period are calculated by taking the inverse of the frequency components of the peak corresponding to the first-mode Rossby wave, then transformed to kilometres and days respectively. The amplitude is calculated with the aid of dividing the absolute cost of the Fourier Transform by means of one half of the product of the length of the x and y dimensions (Subrahmanyam et al., 2009).

Consequently, the transformation of H_j from the time domain to the frequency domain is based on the Fourier transform and its inverse, which are defined as:

$$S(\omega) = \int_{-\infty}^{\infty} s(t) e^{j2\pi f_r t} dt \tag{12.22}$$

where

$$s(t) = \int_{-\infty}^{\infty} s(f_r) e^{j2\pi f_r t} df_r \tag{12.22.1}$$

here, $s(t)$, $S(\omega)$, and f_r are the time signal, the frequency signal and the frequency, respectively, and $j = \sqrt{-1}$. We, the physicists and engineers, sometimes prefer to write the transform in terms of angular frequency $\omega = 2\pi f_r$. To restore the symmetry of the transforms, the convention [17-21]: Stearns et al., 1976; Stoica et al., 1997; Sevgi et al., 2002; Sevgi, 2003;Harries et al., 2006)

$$S(\omega) = \frac{1}{\sqrt{2\pi}} \int_{-\infty}^{\infty} s(t) e^{j\omega t} dt \tag{12.23}$$

where

$$s(t) = \frac{1}{\sqrt{2\pi}} \int_{-\infty}^{\infty} S(\omega) e^{j\omega t} d\omega \tag{12.23.1}$$

The mathematical description of the Rossby wave amplitude “a” and wavelength L in the angular frequency domain is formulated as

$$\eta = a.\cos(kx - \omega t) \tag{12.24}$$

where η is FFT parameter, k is the wavenumber, i.e. $k = \frac{2\pi}{L}$ and angular frequency is ω for a Rossby wave travelling over time t across distance x. For instance, the Rossby wave spectra have derived from the FFT suggests westward propagation with a maximum frequency of 0.2 (1/day) and wavenumber of -0.3 (1/degree) (Figure 6).

ALTIMETER DATA USED TO MODEL ROSSBY WAVE

The Jason-2 altimeter satellite data are used to reveal the sea surface height anomaly (SSHA) data. It has a spatial and temporal resolution of 9.9156 days. These data are retrieved with an accuracy of 0.25 to 0.4 m. The data are covered 1 year from March 2014 to March 2015 along the Southern Indian Ocean. The data, consequently, are obtained from “www.aviso.oceanobs.com”. These data used to retrieve the Rossby wave pattern in the western shelf of Australia in the locations where the flaperon fell down and the search zone of MH370 vanishing.

Figure 6. Derived Rossby wave spectra using FFT

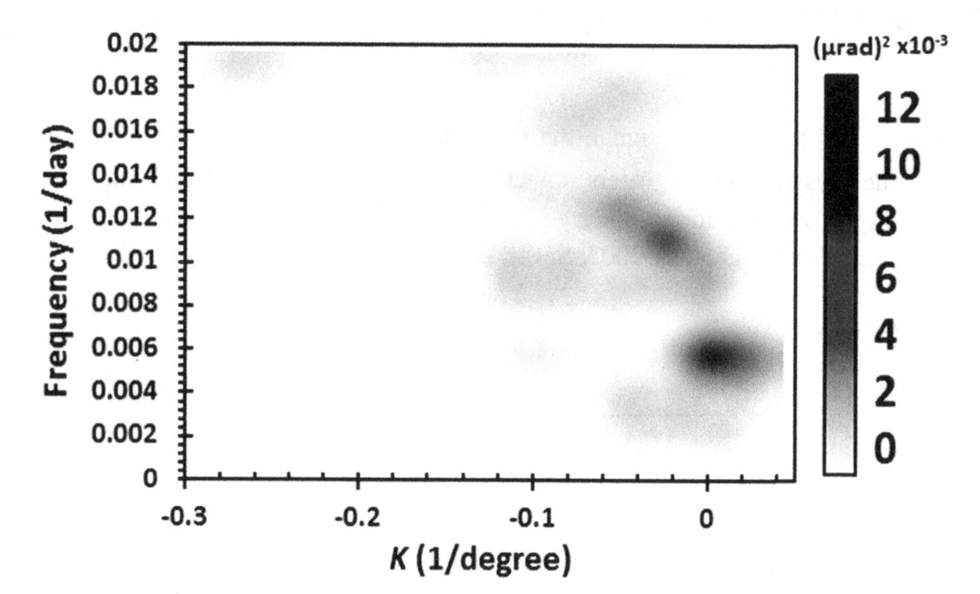

SIMULATION OF ROSSBY WAVE PATTERN BY MULTIOBJECTIVE ALGORITHM

From the point of view of mathematics, a multi-objective problem involves optimizing. In other words, optimization is minimizing or maximizing numerous objectives synchronously, with a quantity of dissimilarity or equality constraints. The problem can mathematically be written as:

Find $v=(v_i)$ $\forall i=1,2,\ldots,N_{param}$ such as
$f_i(v)$ is a minimum, $\forall i=1,2,\ldots,N_{obj}$
Subject to:

$$v = v_j - \frac{\beta}{R_1^{-2} + R_2^{-2}} + \frac{v_i - v_j}{1 + \frac{H_2}{H_1}} \tag{12.25}$$

$$R_n = \frac{\sqrt{gH_n}}{f_{cor}} \tag{12.26}$$

where v is the Rossby wave phase propagation in i and i directions as a function of the internal Rossby radius R_n for the nth layer. Moreover, H_n is the layer thickness and β is the meridional of the derivative of f_{cor}.

The potential velocity ϕ is a resultant of the Rossby wave propagation from the surface to layer thickness H_n is assumed to accelerate the plummeting of MH370 flaperon into Z sinking depth .

$\phi j_{(}v) \geq 0 \ \forall j=1,2,3,\ldots,M$ (12.27)

$Z_k(v) \geq 0 \ \forall K=1,2,3,\ldots,K$ (12.28)

Equations (12.27) and (12.28) suggest that the Rossby wave propagation increases the potential velocity. In this view, v is a vector containing the N_{param} design, parameters $f_i((v))_{i=1,\ldots\ldots,N_{obj}}$ the objective functions and N_{obj} the number of objectives. In this investigation, inequality constraints are only deliberated and are suggested as bounded domains. In other words, upper and lower limits are executed on all parameters:

$$-Z \in \left[v_{i,\min}; v_{i,\max}\right] \qquad i=1,\ldots\ldots,N_{param}. \tag{12.29}$$

The evolutionary algorithm is instigated to solve the equations from 12.27 to 12.29. To this end, an initial population of individuals (solutions) of retrieving velocity potential for Jason-2 data is generated from the matter domain which then endures evolution by suggesting the reproduction, crossover and mutation of individuals till an appropriate resolution exists.

Hence, similar to most alternative evolutionary algorithms, genetic algorithm demands that solely the parameters of the matter be quantified. Subsequently, the algorithmic program is smeared to attain a solution, which is usually a problem-independent. Genetic algorithms typically represent all solutions within the sort of fastened length character strings, analogous to the DNA that's found in living organisms. The rationale for the fastened length character strings is to permit easier manipulation, storage, modelling and implementation of the genetic rule (Marghany, 2018).

In this view, binary numbers conjointly afford straightforward conversion to and from the precise solution. Conversely, since there is evidently infinitely several real numbers between 1 and 2, fixed-length strings cause a further weakness for the computer programmer. To unravel this, the real number range should be discretized into a finite variety of constituent real segments, resembling every binary number utilized in the character string. Suppose that the character strings have a length of n=10. Then the potential values of the character string of the sinking depth of flaperon would be from 0000000000 to 1111111111(Figure 7)(Marghany,2018).

POPULATION OF SOLUTIONS

A collection of potential solutions is unbroken throughout the life cycle of the genetic algorithmic program. This assortment is mostly called the population since it's analogous to a population of living organisms. The population, typically, may be either of fastened or variable size; however, fastened size populations are used a lot often in order that the precise quantity of computer resources will be predetermined. The population of solutions is held on in main memory or on external storage, counting on the sort of genetic algorithm program and computer resources accessible.

At the starting of the algorithmic program, a population of solutions is generated indiscriminately. Within the case of the square root problem, a hard and fast variety of 10 character binary strings are

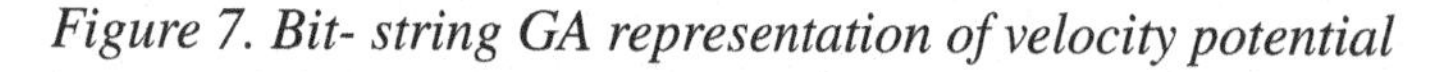

Figure 7. Bit- string GA representation of velocity potential

φ1
0 0 0 0 0 0
Gene
φ2
1 1 1 1 1 1
Chromosome
φ3
1 0 1 0 1 1
φ4
1 1 0 1 1 0
Population

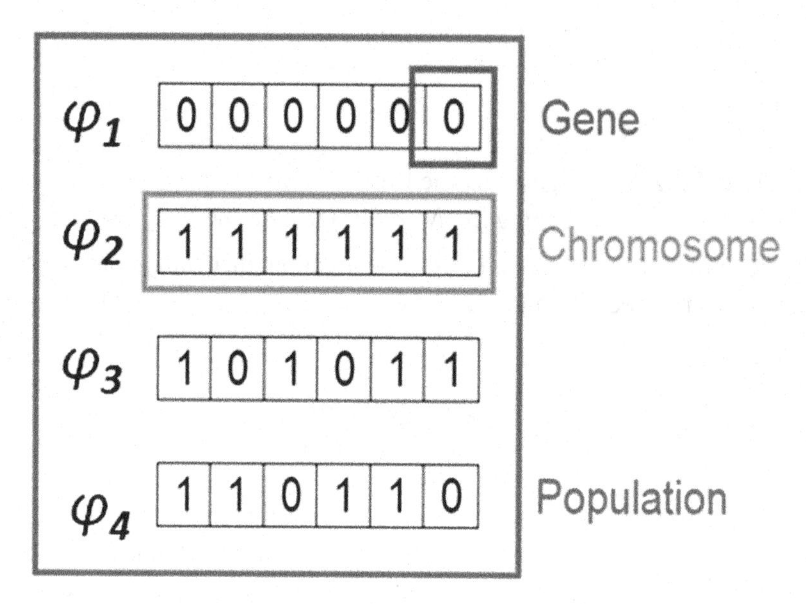

generated indiscriminately. This population is then changed through the mechanisms of evolution to result eventually in individuals that are nearer to the solution than these initial random ones (Figure 8).

Figure 10.8 represents the initial populations of 1000 for velocity potential individual generations from Jason-2 data. The binary number between 0 and 1 is randomly generated and sorted as a string

Figure 8. Initial populations

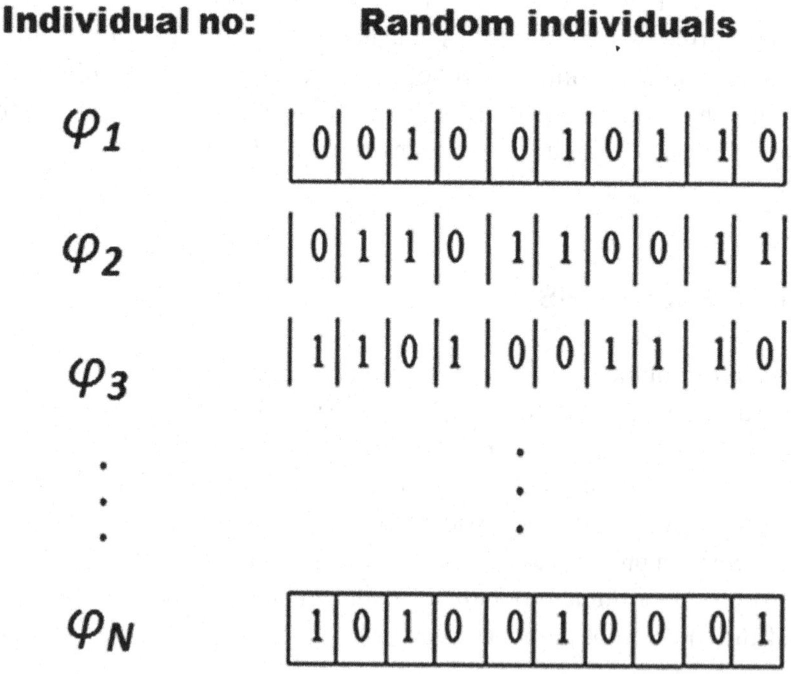

Individual no:
Random individuals
φ1
0 0 1 0 0 1 0 1 1 0
φ2
0 1 1 0 1 1 0 0 1 1
φ3
1 1 0 1 0 0 1 1 1 0
φN
1 0 1 0 0 1 0 0 0 1

in the computer memory. It can also be noticed that they are not sorted in order, but randomly which represent only the row of the Jason-2 data.

One of the explanations for the exploitation of binary numbers is to forbid incorrectly formatted solutions automatically. In binary, it is easier to visualize some characteristics of water density being present (by a 1) or absent (by a 0). This can be also applicable to non-numeric problem domains. There are solely two potential binary values of velocity potential (1 and 0). This suggests that every one of the potential binary values is often created by these two values. Consequently, the binary individuals of sinking water depth of flaperon 0000000000 and 1111111111 contain all the genetic material attainable i.e., they span the solution space (Figure 8).

FITNESS

Therefore, the range of the population is maintained by a widespread fitness sharing function. The satisfying N determination is used to determine the severe Pareto solution. In doing so, the blended crossover is used to generate children on fragment identified via two parents and a specific parameter. In this optimization, new plan variables of Rossby wave potential vorticity has a weight common as

$$Ch_1 = \varpi * P1 + (1 - \varpi) * P2 \quad (12.30)$$

$$Ch2 = (1 - \varpi) * P1 + \varpi * P2 \quad (12.31)$$

where $\varpi = (1+2\ell)\_ran1 - \ell$, Ch1 and Ch2 are child1,2, P1 and P2 are parent 1,2 which represent programmed scheme variables of the members of the new population and a reproduced pair of the old generation. Therefore, ran *is* a random number which is uniform in [0,1]. When the mutation takes place, equations 12.30 and 12.31 can be given as follows:

$$Ch1 = \varpi * P1 + (1 - \varpi) * P2 + \alpha(ran2 - 0.5) \quad (12.32)$$

$$Ch2 = (1 - \varpi) * P1 + \varpi * P2 + \alpha(ran2 - 0.5) \quad (12.33)$$

where ran2 is *a* random number which is uniform in [0,1], and α is set to 5% of the given range of each variable (Figure 9).

Subsequently, since the potential vorticity is a feature of Rossby wave propagation its diagram parameters have to be addressed predictably. Else, the computation deviates and infinite population cannot be weighed. Consequently, if set to 0.0, then mutation takes place at a likelihood of 10% .

In line with Sivanandam and Deepa [(2008), a genetic algorithm is commonly a characteristic of the reproducing step, which entails the crossover and mutation techniques in altimeter data. In the crossover step, the chromosomes interchange genes. A local fitness value results in every gene as given by:

$$f(P_i^j) = \left|\phi - P_i^j\right| \quad (12.34)$$

Figure 9. A genetic algorithm for initial velocity potential generation

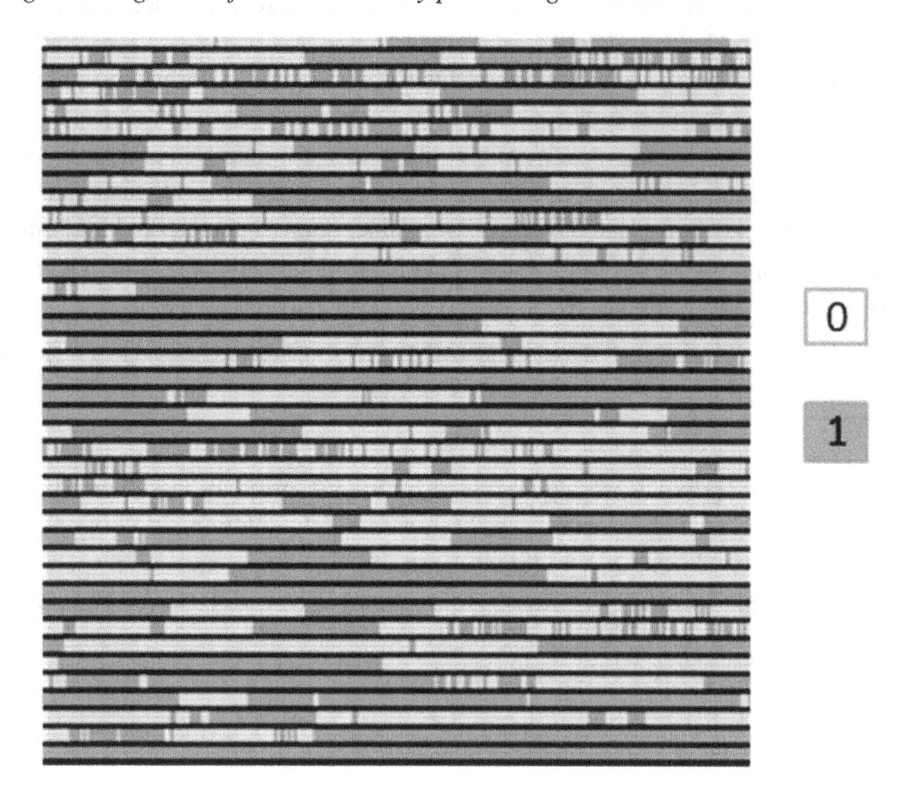

where ϕ is the velocity potential, $f(P_i^j)$ is local fitness value for every gene and P_i^j is a probability variation along *i* and *j*, respectively. In this view, the fitness values are changed between 0 and 1 (Figure 10) with iteration increments. The highest fitness value of 0.8 does not provide a clear information about ϕ (Figure 10a). As the fitness is gradually decreased with iteration increments, the ϕ is being to be calculated (Figure 10b). Lowest fitness value suggests that the clear ϕ pattern (12.10c).

Figure 10 shows the fitness of 0.69 and 0.9. In fact, the standardised fitness attempts to restrict the fitnesses to the range of positive real numbers only. The adjusted fitness changes the fitness value so that it lies strictly within the 0-1 range. Further, Figure 4 indicates whether the algorithm is convergent or not. If there are visible convergence and no solution has yet been found, then the algorithm can be extended over more generations. If convergence is not reached, then the parameters of the run can be tweaked to better suit the problem domain.

CROSS-OVER AND MUTATION

The crossover and mutation are described in short in the following sections which are difficult to a subsequent step of a genetic algorithm. In this understanding, the crossover operator constructs to converge around options with excessive fitness. Thus, the closer the crossover probability is to 1, the quicker is the convergence (Michalewicz, 1994; Deb et al., 2000; Mohanta and Sethi 2011).

Figure 10a. Fitness variations with iterations 0.9

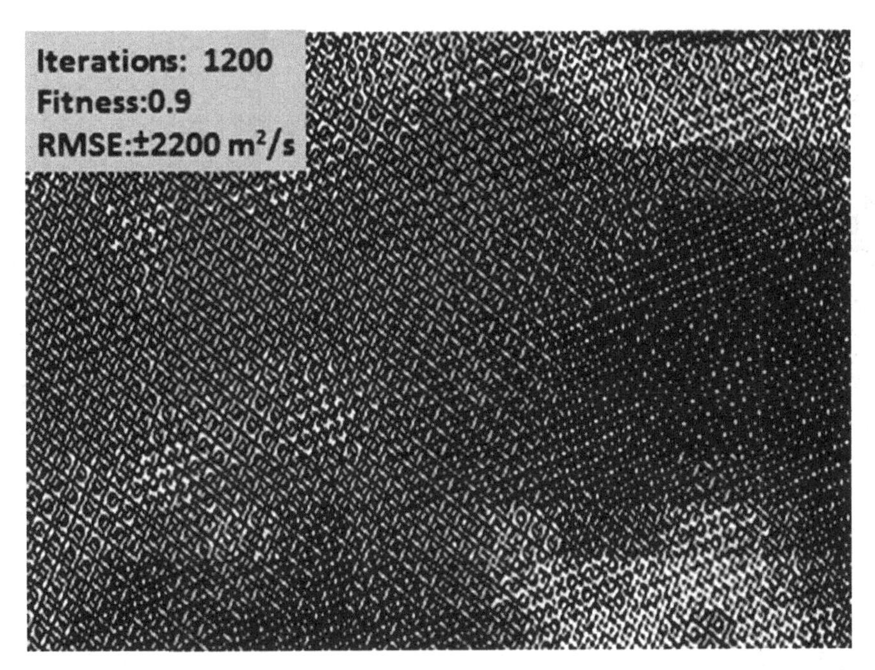

Then the crossfire between two individuals consists of keeping all individual populations of the first parent, which have a local fitness greater than the average local fitness $f(P^j_{av})$ and substitutes the remaining genes by the corresponding ones from the second parent. Hence, the average local fitness is defined by:

$$f(P^j_{av}) = \frac{1}{K}\sum_{i=1}^{K} f(P^j_i) \qquad (12.35)$$

Hence, the mutation operator denotes the phenomena of the great likelihood of the evolution process. Under this circumstance, the fitness value is reduced to 0.49, is a clear feature of the west Australian Shelf and ϕ is well identified (Figure 11). In addition, this step improves the fitness procedure by showing the lowest RMSE value of $\pm$ 0.03 m^2/s as compared to Figure 10.

Truly, some useful genetic records involving the chosen population may be required to replace the duration of reproducing step. As a result, the mutation operator introduces new genetic facts to the ordinary gene. Generally, the genetic algorithm will take binary match individuals and mate them (a process referred to as crossover). The offspring of the mated pair will acquire some of the traits of the parents. The methods of selection, crossover, and mutation are called genetic operators (Mohanta and Sethi 2011 and Marghany 2018).

Figure 12 shows the synoptic information retrieved velocity potential from Jason-2 satellite data, which is appropriately simulated using the multi-objective genetic algorithm. It is clear that the Rossby wave patterns are dominated in the Southern Indian Ocean, from March 2014 to March 2015. In this context, the velocity potential suggests rotational vorticity towards the westward with 100 x 10^5 m^2 s^{-1} from March to May 2014, which covers the search area and the suspected area of flaperon fell down.

Figure 10b. 0.78

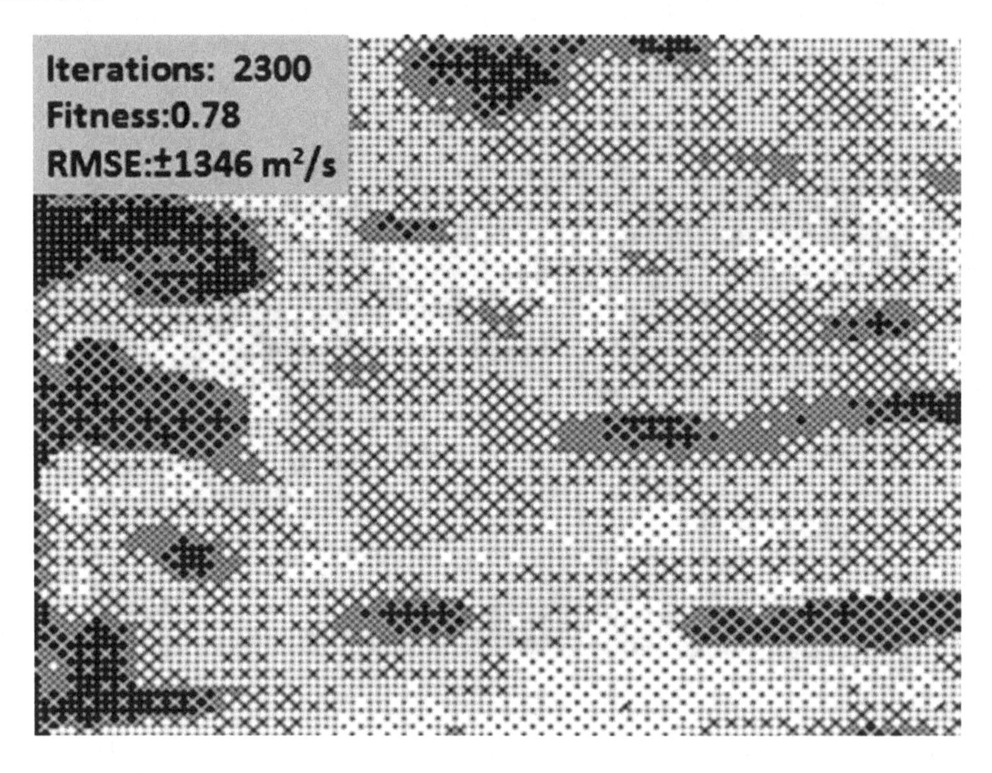

Figure 10c. 0.69

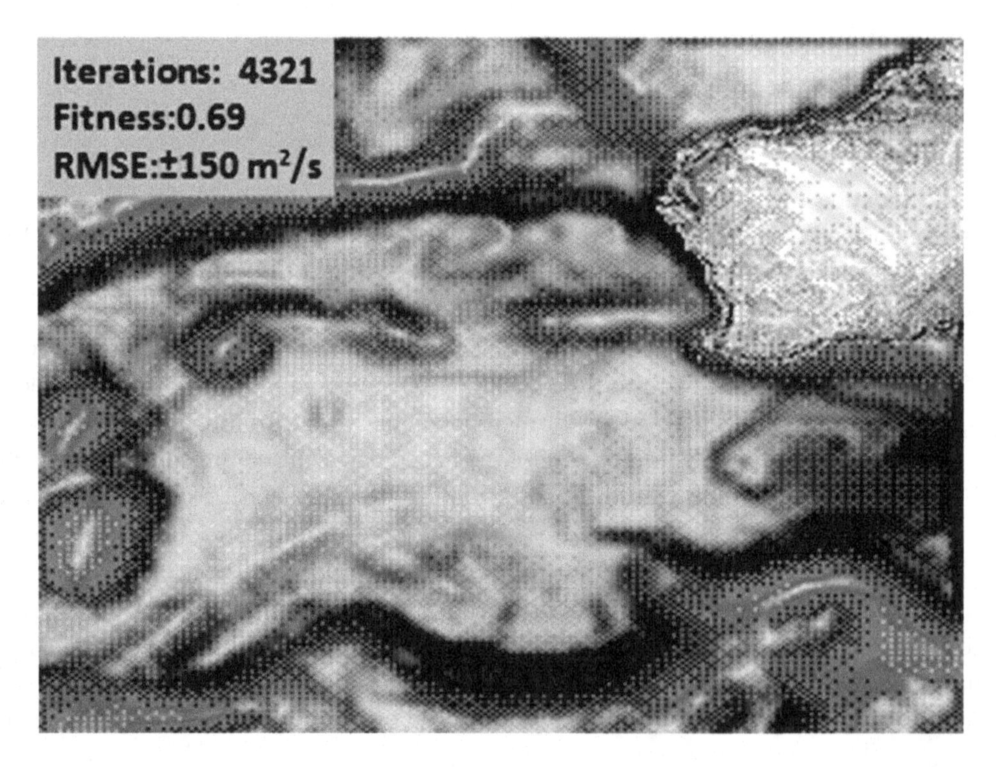

Figure 11. Cross-over improves the fitness value

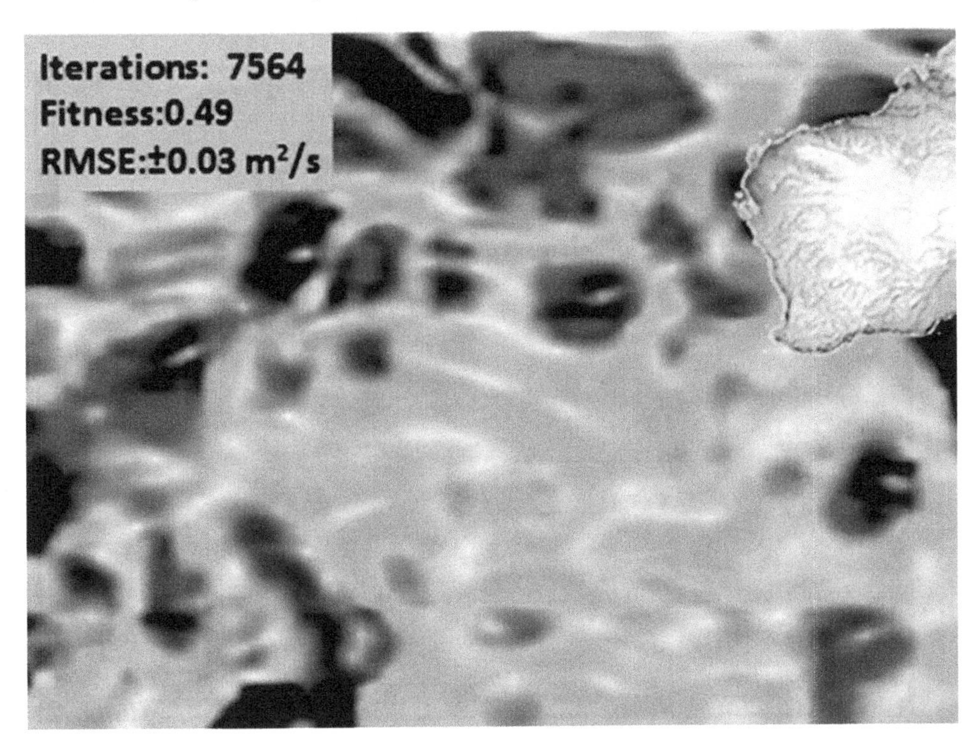

The velocity potential continues with -80 x 10^5 m^2 s^{-1} during June to August 2014. The potential flow rotates anticlockwise towards the west (Figure 3). The distance of 500 km, which covers the location where the flaperon fell down and the end location of MH370 is dominated by a strong potential flow which increases towards the west.

The velocity potential suggests an irregular irrotational flow from September to October 2014 (Figure 14). It is clear that the West Australian Shelf dominates by the lowest velocity potential of -40 x 10^5 m^2 s^{-1}. On the contrary, the maximum velocity potential archives from November to December 2014 where the highest retrieved ϕ of 120 x 10^5 m^2 s^{-1} (Figure 15).

In January to March 2015, the latitude of 25° S to 40° S is dominated by westward strong velocity potential of 120 x 10^5 m^2 s^{-1} (Figure 16). This pattern flow continues the widespread of potential vorticity along the West Australian Shelf.

The changing in the velocity potential pattern is created due to the propagation of the Rossby wave across the southern Indian Ocean. Indeed,

the Hovmöller diagram of the SSH (Figure 17) shows the Rossby waves propagating westward. This requires accurate steric height estimation from satellites because various kinds of studies of the oceans can be performed with this data. Further, SSH and the pressure are related, therefore the height field is also related to the velocity field. Similar to the atmosphere, the balance of pressure gradient force, Coriolis force and the frictional force to determine the kind of circulation that would dominate the ocean depths. Indeed, the time evolution of ocean circulation depends on the external factors, viz., wind stress, heating and cooling, evaporation and precipitation. Hence, the ocean temperature, salinity and other properties depend on these three external factors and also how the moving seawater parcels would bal-

Figure 12. Velocity potential from March to May 2014

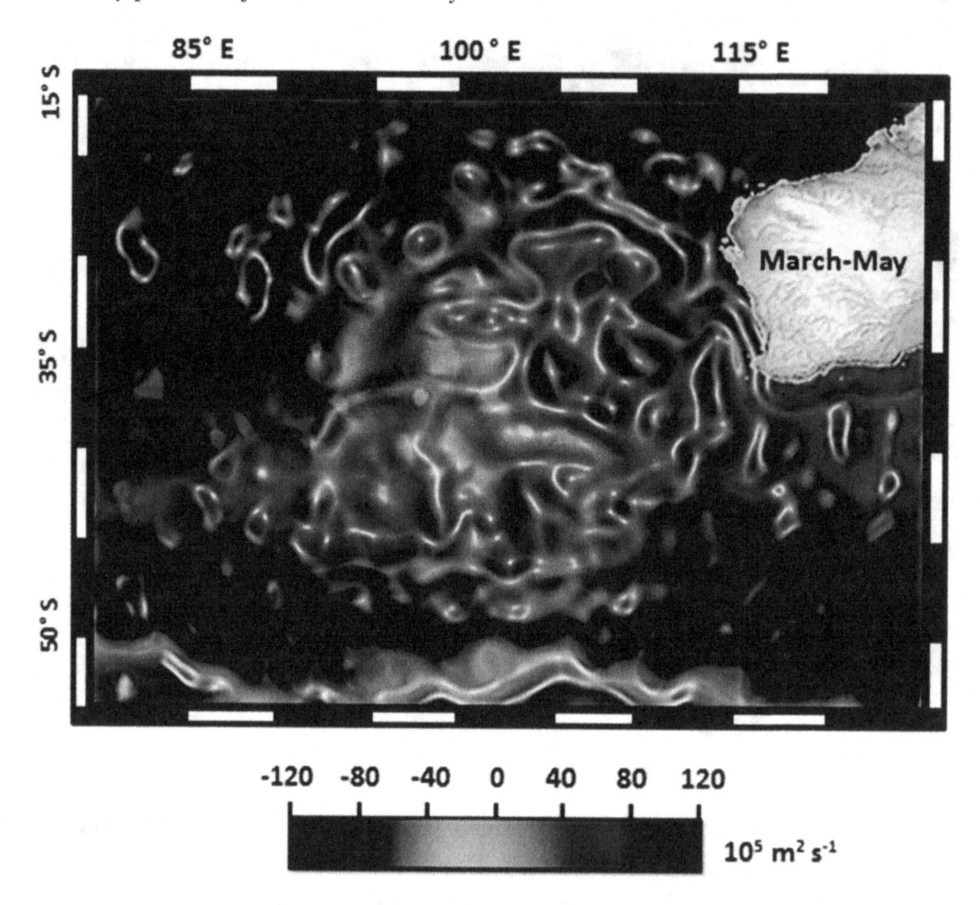

ance under the combined action of these forces. It is clear that the solution of the β-plane linear Rossby wave equation represents long periodic oscillations with periods of several months or longer (Figure 17).

It is worth mentioning that the impact of the rotation vorticity is dominated feature of Rossby wave propagation under the circumstance of $f_{cor} \rightarrow 0$, $R \rightarrow \infty$. In this context, the vorticity is occurred because of the large length scale comparable with R (Figures 12 to 17).

PARETO DOMINANCE

The exceptional query can increase up did the Rossby wave strength have an impact on the submerged of flaperon and different debris? The answer to this question can deliver by the Pareto optimal solution. In a multi-objective problem, the set of constraints, i.e. the individuals in the evolutionary algorithm terminology) can be compared according to Pareto's rule. In this understanding, the individual sinking depth Ψ_s dominates the individual velocity potential ϕ, if for at least one of the objectives, $\Psi_s \in H$ and $H \in \phi$. In this view, The submerged flaperon Ψ_s takes a place under the constrain circumstances of:

Figure 13. Velocity potential from June to August 2014

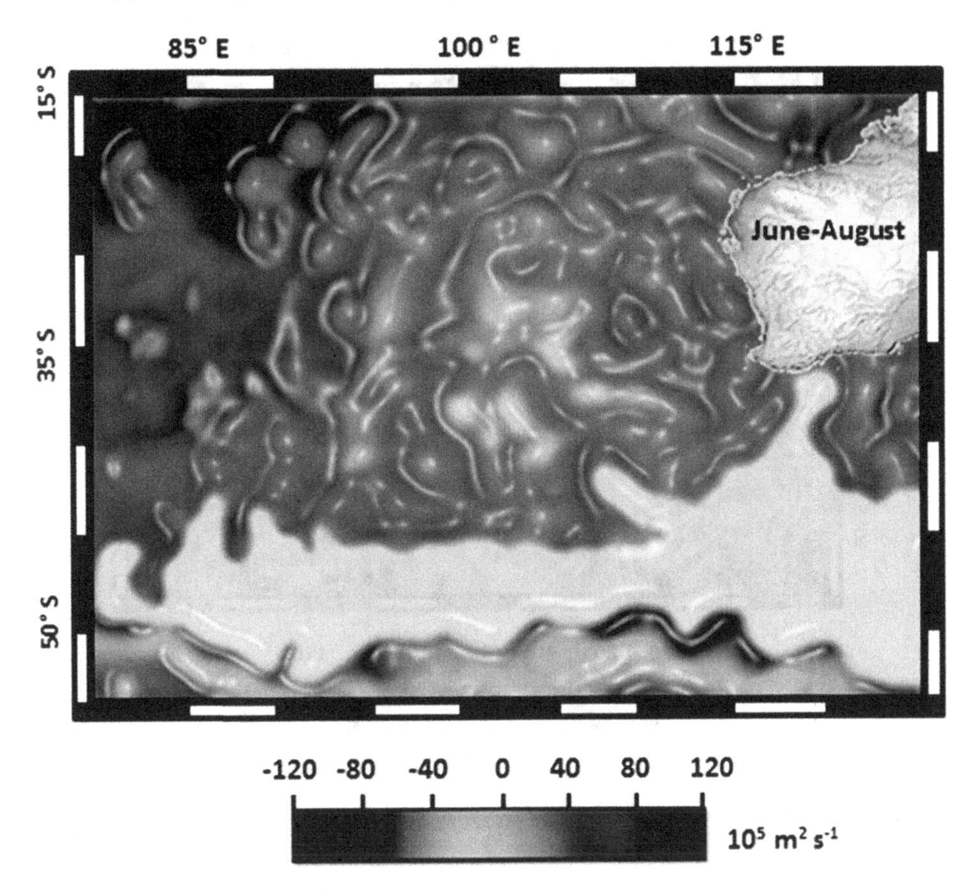

$$\max \Psi_s = \max \sqrt{2\phi \frac{H}{U}} \tag{12.36}$$

Subject to $\Psi_s \in \phi \in H$ (12.37)

$$\begin{matrix} U(i) < 0 < -1 \\ \psi_s \leq H \leq 0 \end{matrix} \quad i=1,2,3,..........k \tag{12.38}$$

Equation 12.36 can achieve a true multiobjective when the Ψ_s is dominated by other multiobjective. In other words, the other multiobjective of velocity potential, the velocity of the Rossby wave and thickness layer H are considered independent of Ψ_s. The Rossby wave propagates along with the water depth of approximately offshore of the West of Australia shelf with maximum vorticity velocity of 120 x 10^5 m^2 s^{-1} (Figure 18). Under this circumstance, the flaperon must sink down in water depth less than 1000 m (Figure 19).

Figure 14. Velocity potential from September to October 2014

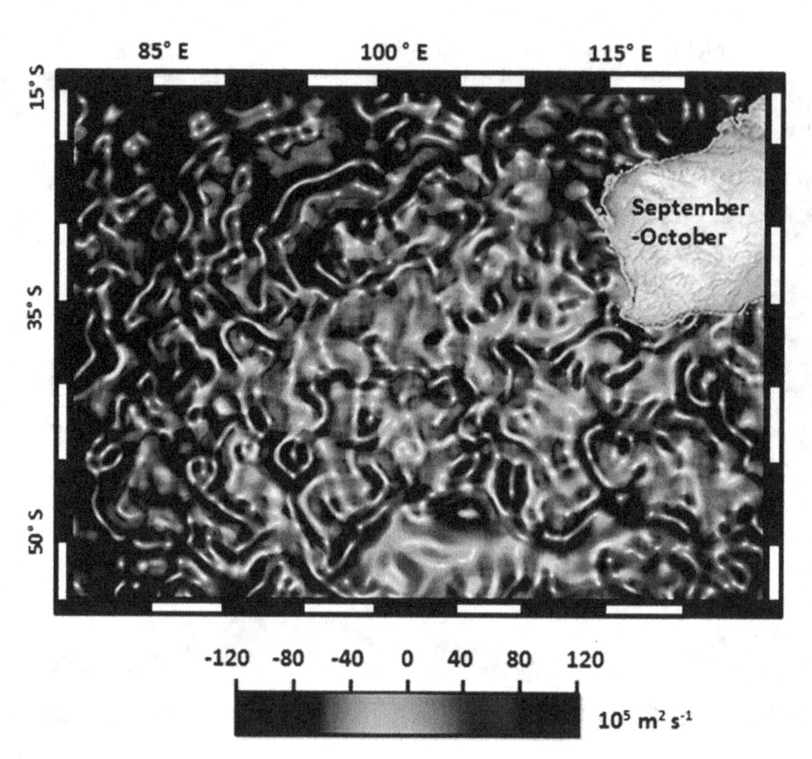

Figure 15. Velocity potential from November to December 2014

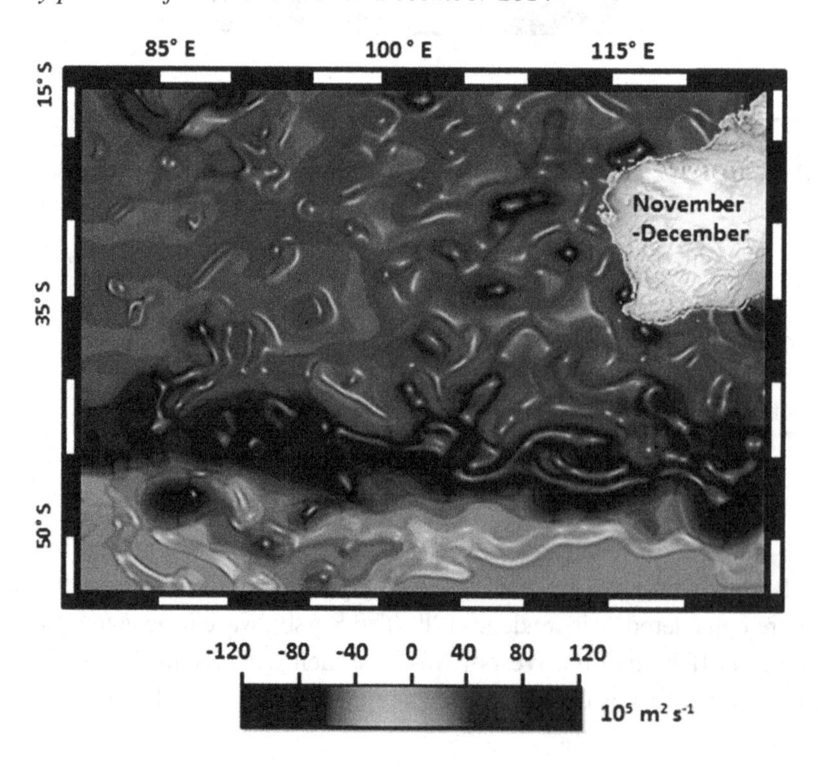

Figure 16. Velocity potential from January to March 2015

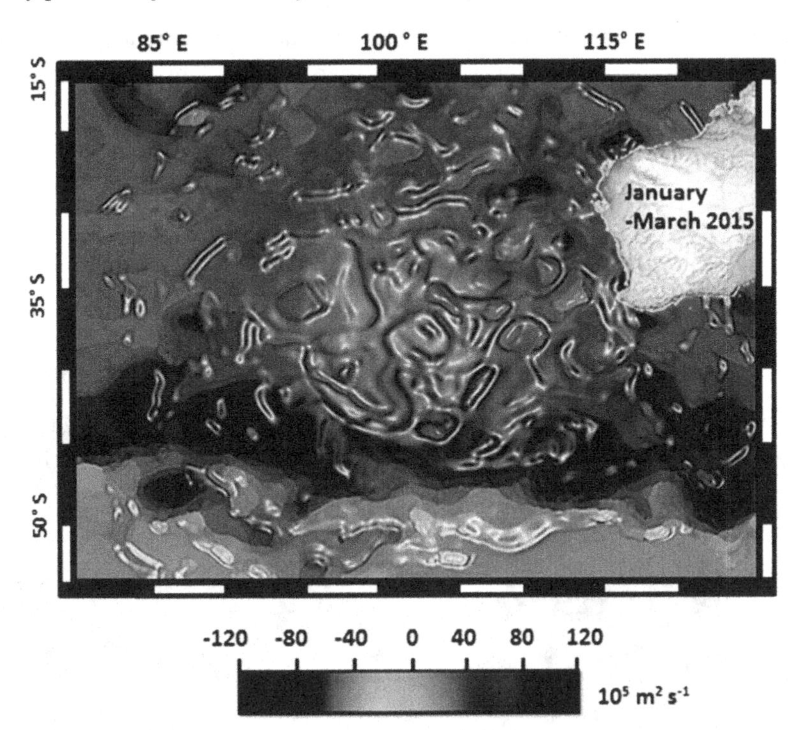

Figure 17. Rossby wave propagation simulated from Jason-2 satellite data

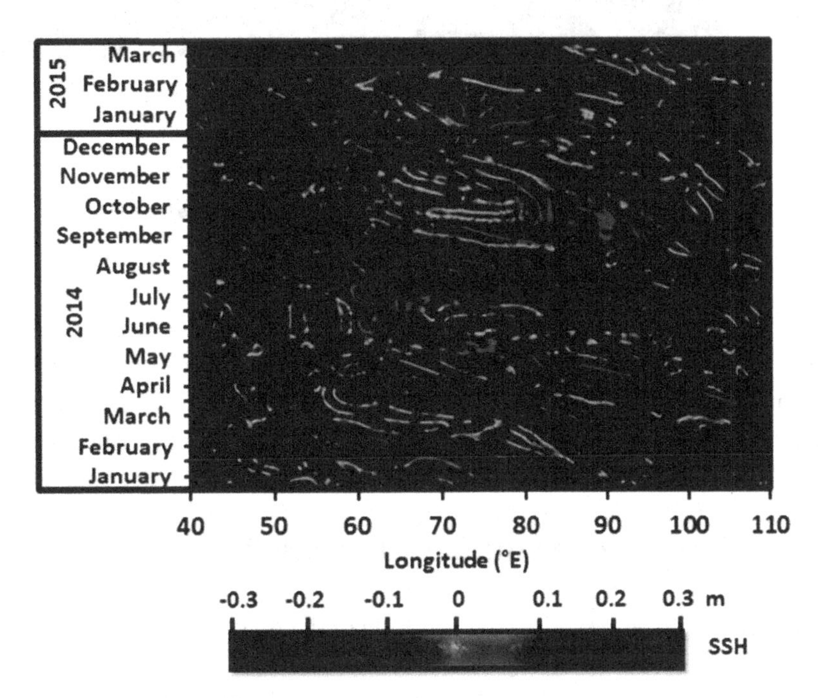

Pareto simplest curve confirms that the submerged intensity of flaperon will increase as the velocity potential, which generated by Rossby wave in the Southern Indian Ocean (Figure 18). In this circumstance, the flaperon should submerge to a water depth of more than 1000 m. This result agrees with the results of chapter 11. In this view, the velocity potential of approximately 120 x 10^5 m^2 s^{-1} in addition to wave

Figure 18.

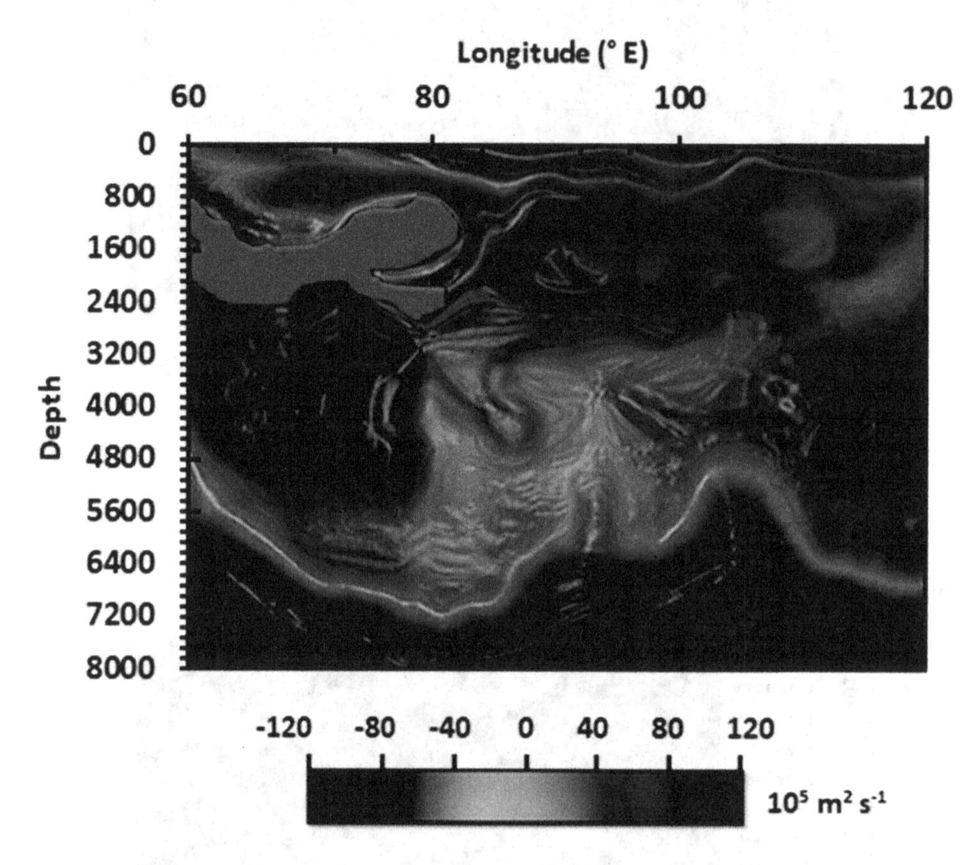

power of-of 22000 KJ/m/wave must lead to the sinking of the flaperon between the longitude of 99 to 110 °E and the latitude of 22°S and 28°S. This proves by using the usage of the submerged factors are placed absolutely at the Pareto the front curve two than floating factor (Figure 19). According to Figure 11.20, the flaperon can drift on the surface water no longer than 3 months. This confirms also by the impact of the velocity potential pattern from the surface to 8000 m across the southern Indian Ocean.

How Did Rossby Wave Mobilize the MH370 Debris?

The propagation of the Rossby wave causes the variation of the thermocline. In this view, the anomalies on the thermocline migrate 1° per 28.01 days in the zonal direction. In other words, the changes in thermocline can occur within three months. At some *large*-scale Rossby waves are likely to dominate whereas at small scale advection, and turbulence dominates. This can break down the thermocline layer(You and Tomczak 1993)., which leads to strong large-scale mixing (Talley 1996). In this circumstance, the

Figure 19.

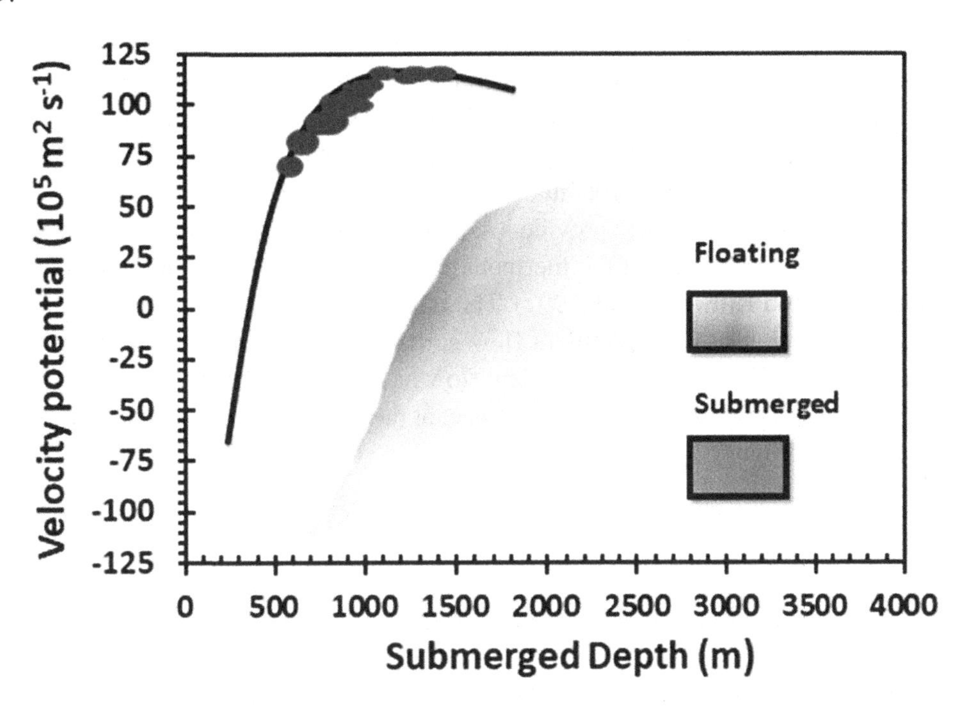

flaperon would sink deeply due to wave-turbulence cross-over. In this understanding, the fluid is stirred at some well-defined scale, producing an energy input ". In this regard, it assumes that no energy is lost to smaller scales i.e., energy cascades to large scales at that same rate. At some scale, the β term in the vorticity equation will start to make its presence felt. By analogy with the procedure for finding the viscous dissipation scale in turbulence, we can find the scale at which linear Rossby waves dominate by equating the inverse of the turbulent eddy turnover time to the Rossby wave frequency. Furthermore, the turbulent eddy transfer rate, proportional to $k^{2/3}$ in a $k^{-5/3}$ to the Rossby wave frequency $\beta k^x/k^2$ (You 1997). In this context, The inverse cascade plus Rossby waves thus lead to a generation of zonal flow (Figure 19). What occurs physically? The region inside the dumbbell shapes in Figure 19 is dominated by Rossby waves, where the natural frequency of the oscillation is *higher* than the turbulent frequency.

If the flow is stirred at a wavenumber higher than this the energy will cascade to larger scales, but because of the frequency mismatch, the turbulent flow will be unable to efficiently excite modes within the dumbbell. Nevertheless, there is still a natural tendency of the energy to seek the gravest mode, and it will do this by cascading toward the $k_x = 0$ axes; that is, toward the zonal flow (de Szoeke et al., 1999). Thus, the combination of Rossby waves and turbulence will lead to the formation of zonal flow and, potentially, zonal jets. In other words, at some large-scale Rossby waves are likely to dominate whereas at small scale advection, and turbulence dominates. The combination of small-scale advection and turbulence stir the flaperon downward along the mixing layer, which has a thickness of more than 1000 m (Figure 19).

The Antarctic Intermediate Water inflow creates patches of high potential vorticity at intermediate depths in the southern Indian Ocean, below, which the field becomes dominated by planetary vorticity, indicating a weaker meridional circulation and weaker potential vorticity sources(Dewar, 1986). Wind-driven gyre depths have lower potential vorticity gradients primarily due to same-source waters.

The Indian Ocean adds unique features to the study of potential vorticity with its complicated bathymetry, a prominent link to the Southern Ocean, and reversing (Figure 19) monsoonal currents. In this view, thermocline water masses have unique potential vorticity signals. The changes in the potential vorticity can cause high eddy variability along the MH370 search zone. The active wind-driven currents in the Indian Ocean overlie a strong abyssal circulation, allowing the possibility of comparison between the relative effects of each on potential vorticity (Holland et al.,1984). The potential vorticity is mapped on several neutral surfaces to provide a steady state picture of the potential vorticity as well as to develop a better understanding of the thermohaline circulation's effect on the potential vorticity signature (Dewar,1998 and Killworth et al.,1997). The ratio between vorticity velocity ψ and maximum vorticity velocity ψ_{max} suggests eddy turbulent flow across the suspected zone of flaperon fell down and the search zone of MH370. This eddy turbulent flow has approached 1 in the surface and 0.4 in the bottom. This could be sufficient to submerge any object at the water depth more than 2500 m across the Southern Indian Ocean (Figure 20).

Figure 20. ψ/ψ_{max} associated with the eddy turbulent flow across the MH370 search zone

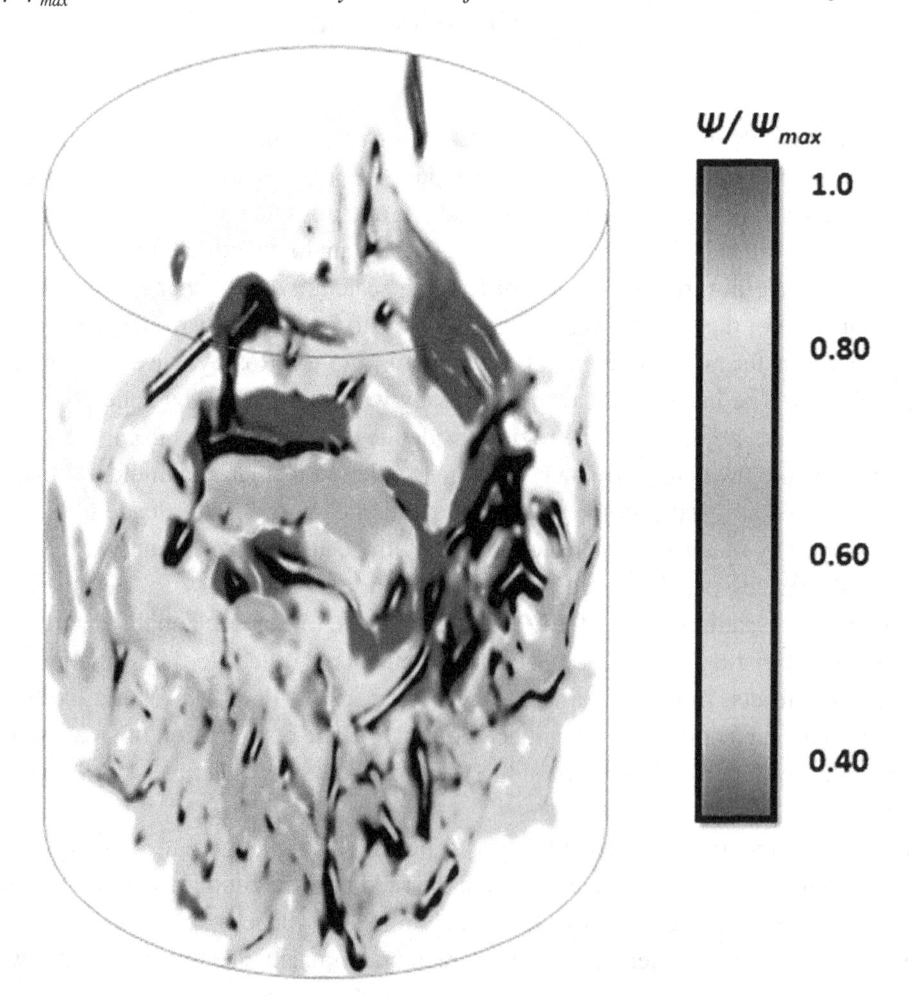

The mechanism of potential vorticity impact on the flaperon sinking is confirmed by the Pareto front where its ratio velocity of ψ/ψ_{max} which ranges between 0.6 to 1 pulls the flaperon down in a sinking depth of 3000 m (Figure 21).

Figure 21. Pareto front for flaperon sinking due to potential vorticity

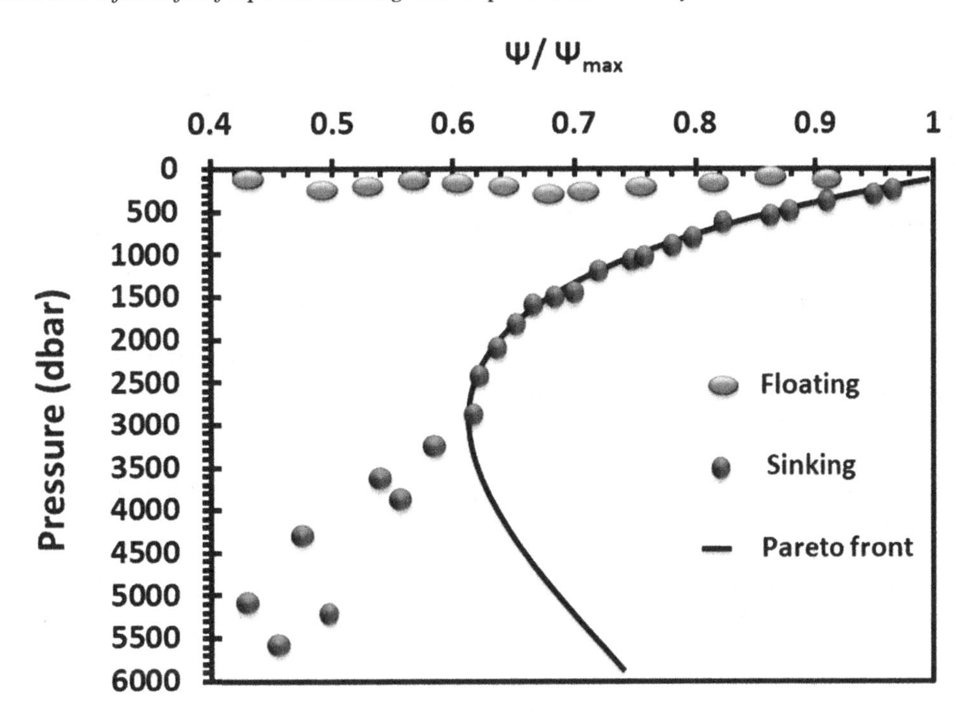

However, the floating of flaperon with different ranges ψ/ψ_{max} is not dominated on the Pareto front. Also, the sinking domain with ψ/ψ_{max} less than 0.6 is not dominated on the Pareto front, which confirms that $\psi/\psi_{max} \leq 0.6$ can be responsible for the sinking of flaperon in approximately a water depth of 3000 m.

CONCLUSION

Using the Pareto optimization algorithm with the impact power of Rossby waves proves that the flaperon could not drift crossways the Southern Indian Ocean and position itself on the Réunion Island. This proof is holding the critical question of how the MH370 debris and flaperon were found on the western side of Southern Indian Ocean i.e. Réunion Island, Tanzania, Madagascar, and South Africa. The remaining chapters perhaps deliver a logical answer of MH370 mysterious vanishing.

REFERENCES

Ambrizzi, T., Hoskins, B. J., & Hsu, H. H. (1995). Rossby wave propagation and teleconnection patterns in the austral winter. *Journal of the Atmospheric Sciences*, *52*(21), 3661–3672. doi:10.1175/1520-0469(1995)052<3661:RWPATP>2.0.CO;2

Cipollini, P., Cromwell, D., Challenor, P. G., & Raffaglio, S. (2001). Rossby waves detected in global ocean colour data. *Geophysical Research Letters*, *28*(2), 323–326. doi:10.1029/1999GL011231

de Szoeke, R. A., & Chelton, D. B. (1999). The modification of long planetary waves by homogeneous potential vorticity layers. *Journal of Physical Oceanography*, *29*(3), 500–511. doi:10.1175/1520-0485(1999)029<0500:TMOLPW>2.0.CO;2

Deb, K., Agrawal, S., Pratap, A., & Meyarivan, T. (2000, September). A fast elitist non-dominated sorting genetic algorithm for multi-objective optimization: NSGA-II. In *International conference on parallel problem solving from nature* (pp. 849-858). Springer. 10.1007/3-540-45356-3_83

Dewar, W. K. (1986). On the potential vorticity structure of weakly ventialted isopycnals: A theory of subtropical mode water maintenance. *Journal of Physical Oceanography*, *16*(7), 1204–1216. doi:10.1175/1520-0485(1986)016<1204:OTPVSO>2.0.CO;2

Dewar, W. K. (1998). On "too fast" baroclinic planetary waves in the general circulation. *Journal of Physical Oceanography*, *28*(9), 1739–1758. doi:10.1175/1520-0485(1998)028<1739:OTFBPW>2.0.CO;2

Harries, D., McHenry, M., Jennings, P., & Thomas, C. (2006). Hydro, tidal and wave energy in Australia. *The International Journal of Environmental Studies*, *63*(6), 803–814. doi:10.1080/00207230601046943

Holland, W. R., Keffer, T., & Rhines, P. B. (1984). Dynamics of the oceanic general circulation: The potential vorticity field. *Nature*, *308*(5961), 698–705. doi:10.1038/308698a0

Kelly, K. A., & Thompson, L. A. (2002). Scatterometer winds explain damped Rossby waves. *Geophysical Research Letters*, *29*(20), 52–1. doi:10.1029/2002GL015595

Killworth, P. D., Chelton, D. B., & de Szoeke, R. A. (1997). The speed of observed and theoretical long extratropical planetary waves. *Journal of Physical Oceanography*, *27*(9), 1946–1966. doi:10.1175/1520-0485(1997)027<1946:TSOOAT>2.0.CO;2

Lorenz, E. N. (1972). Barotropic instability of Rossby wave motion. *Journal of the Atmospheric Sciences*, *29*(2), 258–265. doi:10.1175/1520-0469(1972)029<0258:BIORWM>2.0.CO;2

Lovelace, R. V. E., Li, H., Colgate, S. A., & Nelson, A. F. (1999). Rossby wave instability of Keplerian accretion disks. *The Astrophysical Journal*, *513*(2), 805–810. doi:10.1086/306900

Marchesiello, P., McWilliams, J. C., & Shchepetkin, A. (2003). Equilibrium structure and dynamics of the California Current System. *Journal of Physical Oceanography*, *33*(4), 753–783. doi:10.1175/1520-0485(2003)33<753:ESADOT>2.0.CO;2

Marghany, M. (2018). *Advanced remote sensing technology for Tsunami modelling and forecasting*. CRC Press. doi:10.1201/9781351175548

Michalewicz, Z. (2013). *Genetic algorithms+ data structures= evolution programs*. Springer Science & Business Media.

Mohanta, R. K., & Sethi, B. (2011). A Review of Genetic Algorithm application for Image Segmentation. *International Journal of Computer Technology & Applications.*, *3*(2), 720–723.

Polvani, L. M., & Plumb, R. A. (1992). Rossby wave breaking, microbreaking, filamentation, and secondary vortex formation: The dynamics of a perturbed vortex. *Journal of the Atmospheric Sciences*, *49*(6), 462–476. doi:10.1175/1520-0469(1992)049<0462:RWBMFA>2.0.CO;2

Sevgi, L. (2003). *Complex electromagnetic problems and numerical simulation approaches*. John Wiley & Sons.

Sevgi, L., Akleman, F., & Felsen, L. B. (2002). Groundwave propagation modeling: Problem-matched analytical formulations and direct numerical techniques. *IEEE Antennas & Propagation Magazine*, *44*(1), 55–75. doi:10.1109/74.997903

Sivanandam, S. N., & Deepa, S. N. (2008). Genetic algorithm optimization problems. In *Introduction to Genetic Algorithms* (pp. 165–209). Springer. doi:10.1007/978-3-540-73190-0_7

Stearns, S. D., & Ahmed, N. (1976). Digital signal analysis. *IEEE Transactions on Systems, Man, and Cybernetics*, *SMC-6*(10), 724–724. doi:10.1109/TSMC.1976.4309433

Stoica, P., & Moses, R. L. (1997). *Introduction to spectral analysis* (Vol. 1). Prentice hall.

Strub, P. T., & James, C. (2000). Altimeter-derived variability of surface velocities in the California Current System: 2. Seasonal circulation and eddy statistics. *Deep-sea Research. Part II, Topical Studies in Oceanography*, *47*(5-6), 831–870. doi:10.1016/S0967-0645(99)00129-0

Subrahmanyam, B., Heffner, D. M., Cromwell, D., & Shriver, J. F. (2009). Detection of Rossby waves in multi-parameters in multi-mission satellite observations and HYCOM simulations in the Indian Ocean. *Remote Sensing of Environment*, *113*(6), 1293–1303. doi:10.1016/j.rse.2009.02.017

Subrahmanyam, B., Robinson, I. S., Blundell, J. R., & Challenor, P. G. (2001). Indian Ocean Rossby waves observed in TOPEX/POSEIDON altimeter data and in model simulations. *International Journal of Remote Sensing*, *22*(1), 141–167. doi:10.1080/014311601750038893

Talley, L. D. (1996). Antarctic intermediate water in the South Atlantic. In *The South Atlantic* (pp. 219–238). Springer. doi:10.1007/978-3-642-80353-6_11

Van Gysen, H., & Coleman, R. (1999). On the analysis of repeated geodetic experiments. *Journal of Geodesy*, *73*(5), 237–245. doi:10.1007001900050240

Van Gysen, H., Coleman, R., & Hirsch, B. (1997). Local crossover analysis of exactly repeating satellite altimeter data. *Journal of Geodesy*, *72*(1), 31–43. doi:10.1007001900050145

Van Gysen, H., Coleman, R., Morrow, R., Hirsch, B., & Rizos, C. (1992). Analysis of collinear passes of satellite altimeter data. *Journal of Geophysical Research. Oceans*, *97*(C2), 2265–2277. doi:10.1029/91JC02451

Vivier, F., Kelly, K. A., & Thompson, L. (1999). Contributions of wind forcing, waves, and surface heating to sea surface height observations in the Pacific Ocean. *Journal of Geophysical Research. Oceans*, *104*(C9), 20767–20788. doi:10.1029/1999JC900096

White, W. B., Tai, C. K., & Dimento, J. (1990). Annual Rossby wave characteristics in the California Current region from the Geosat exact repeat mission. *Journal of Physical Oceanography*, *20*(9), 1297–1311. doi:10.1175/1520-0485(1990)020<1297:ARWCIT>2.0.CO;2

Woollings, T., Hoskins, B., Blackburn, M., & Berrisford, P. (2008). A new Rossby wave–breaking interpretation of the North Atlantic Oscillation. *Journal of the Atmospheric Sciences*, *65*(2), 609–626. doi:10.1175/2007JAS2347.1

You, Y. (1997). Seasonal variations of thermocline circulation and ventilation in the Indian Ocean. *Journal of Geophysical Research. Oceans*, *102*(C5), 10391–10422. doi:10.1029/96JC03600

You, Y., & Tomczak, M. (1993). Thermocline circulation and ventilation in the Indian Ocean derived from water mass analysis. *Deep-sea Research. Part I, Oceanographic Research Papers*, *40*(1), 13–56. doi:10.1016/0967-0637(93)90052-5

Chapter 13
Multiobjective Algorithm-Based Pareto Optimization for Modelling Trajectory Movement of MH370 Debris

ABSTRACT

It is well-known that the altimeter satellite data can model the global world ocean circulation. In this view, the ocean dynamic circulation altimeter data is required to understand the drift movement of MH370 across the Indian ocean. The integration between the Volterra-Lax-Wendroff algorithm and Pareto optimal algorithm is used to investigate the dynamic movement of MH370 debris over annual current circulation across the Indian Ocean. This chapter shows that the maximum value of the hit-rate (HR) is 160%, which is occurring with an extreme rapidity of eddy current of 0.65 m/s. In conclusion, it is a great impossibility for the existence of the debris along Mozambique, Reunion Island, Madagascar coastal waters, and Mossel Bay, South Africa, as proven by the Pareto optimization.

INTRODUCTION

Up to date, the trajectory movement models have delivered to track the MH370 debris cannot identify the main crashing location. This leads to the challenge question where is the fuselage of MH370? Since March 8, 2014, no one can answer where MH370 and how did it crash? Moreover, researchers cannot make use of debris trajectory movement to understand how did MH370 vanish. In this regard, what is the impact of the Southern Indian Ocean surface circulation of the trajectory movements of the MH370 flaperon and debris?

The scientific answer to this question can solve the mystery of MH370disappearing. The aim of this chapter to model the impact of the ocean annual ocean dynamic circulation of the debris trajectory movements across the southern Indian ocean. To this end, the optimal Pareto optimization algorithm is examined based on altimeter satellite data.

DOI: 10.4018/978-1-7998-1920-2.ch013

Furthermore, this chapter tries to bridge the gap between the impact of the physical oceanography parameters, object detections in remote sensing data, and the search area, which is determined by Inmarsat satellite signals. In these regards, the previous chapters have demonstrated that the flaperon cannot be drifted towards Réunion Island. In other words, either the flight did not plunge into the Southern Indian Ocean or the debris and flaperon sunk in the deepest Indian Ocean floor.

This chapter curtly evaluates the widespread theories of ocean circulation current movements to bridge the gap flanked by the dynamic of the Southern Indian Ocean and proper MH370 vanishing mechanism. In this regard, a new approach is used to simulate the sea surface current flow from the sea level anomaly using the Volterra-Lax-Wendroff algorithm.

Consequently, this division is dedicated to the Southern Indian surface currents, which would accredit to the authentic fact that the MH370 flaperon bought into Réunion Island.

ALTIMETER THEORY FOR MEASURING SEA SURFACE CURRENT

The altimeter satellite data have been identified as an effective approach for modelling coastal hydrodynamic components such as ocean surface currents and wave patterns (Marghany et al., 2008). In this perspective, the altimeter satellite data are debatably the furthermost beneficial of totally the satellites for determining ocean surface currents. Even though radar altimeter data accessibility are nevertheless fairly brief in contrast to sets of tidal gauge information. This approach appears reasonably talented for the sea level investigation problems due to the fact it affords sea stage dimension with massive coverage. With an accuracy of about 1 mm/year of quantity, worldwide sea level variations can be acquired (Marghany, 2009). Accordingly, the algorithms that have been implemented to acquire the ocean surface current are grounded on radar pulse backscatter from the sea surface(Glenn et al. 1991). Consistent with Robinson (2004), a satellite altimeter considers as a nadir-viewing radar, which emanates a pulse within traveling period, and the backscattering signal from the sea surface. In this existence, the travel period (t) can be adapted to a distance when the delay pulses can be calculated (Figure 1). Martins et al. (2002) mentioned that the peak of the sea surface can be determined employing combining this distance with the location of the satellite that determined with the aid of precision orbit dedication and correcting for the earth and ocean tides and atmospheric loading.

The altimeter range (R)from the satellite to sea surface level is estimated from the round –trip travel time by equation 13.1:

$$R = 0.5ct - \Sigma\Delta R_j \tag{13.1}$$

Equation 13.1 suggests that c is the speed of light in the air, which approximately 299,700 km/s and ΔR_j is an alteration of the numerous components of the atmospheric refraction and for biases between the mean electromagnetic scattering surface (EM) and mean sea level at the air-sea interface (Hwang and Chen 2000). Then the height of the sea surface h, which is relative to the reference ellipsoid can be estimated by:

$$h = H - R \tag{13.2}$$

Figure 1. The mechanism of the altimeter to measure sea surface height

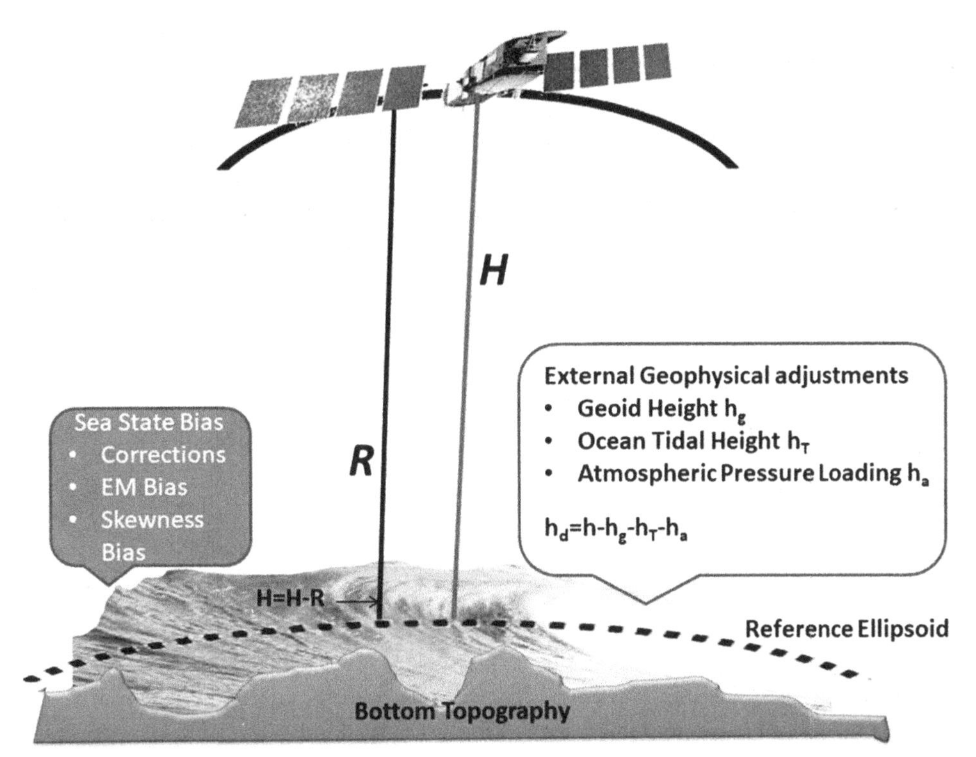

being H is the altimeter orbital height. In this view, equation 13.2 demonstrates that h is the superposition of a number of geophysical impacts i.e.; geoid undulations about the ellipsoidal approximation (h_g), tidal height variation (h_T) and the ocean surface response to atmospheric pressure loading (h_a) (Figure 1). In this understanding, the dynamic sea surface height (h_d) is known as the corrected height h, which is mathematically expressed as:

$$h_a = h - h_g - h_T - h_a \tag{13.3}$$

Equation 13.3 validates that to explode this range R into a useful measure of sea level requires a variety of 'tricks'. First of all, the orbit of the satellite should be comprehended with almost exceptional precision – we need to be aware of the place it flies (some ~1000 km over the surface) with an error of the order of 1 cm! Then we want to account for the truth that the microwave pulse is barely slowed down through the ionosphere, the atmospheric gases, and water vapour.

Even post these intricate corrections, the derived sea level is not solely due to currents: there is a supplementary contribution through the tides, plus the consequences of local variations in atmospheric pressure. These extra aspects are removed using precise tidal models and atmospheric pressure models. In the end, satellite oceanographers gain precise measurements of the sea level profile alongside the ground tracks of the satellite.

GEOID ERROR

Geoid and Ellipsoidal Concepts

The orthometric (geoid) is the height of any point on the Earth's surface is the space H_o from the point to the geoid. The ellipsoidal height of a point of the Earth Surface is the distance H_e from the point to the ellipsoid (Figure 2). An ellipsoid is a third-dimensional geometric discerns that resembles a sphere, nonetheless, whose equatorial axis (Figure 3) is barely extended than its polar axis (b). The World Geodetic System equatorial axis was established in 1984, which is approximately 22 kilometers longer than the polar axis.

Figure 2. Definition of geoid and ellipsoid

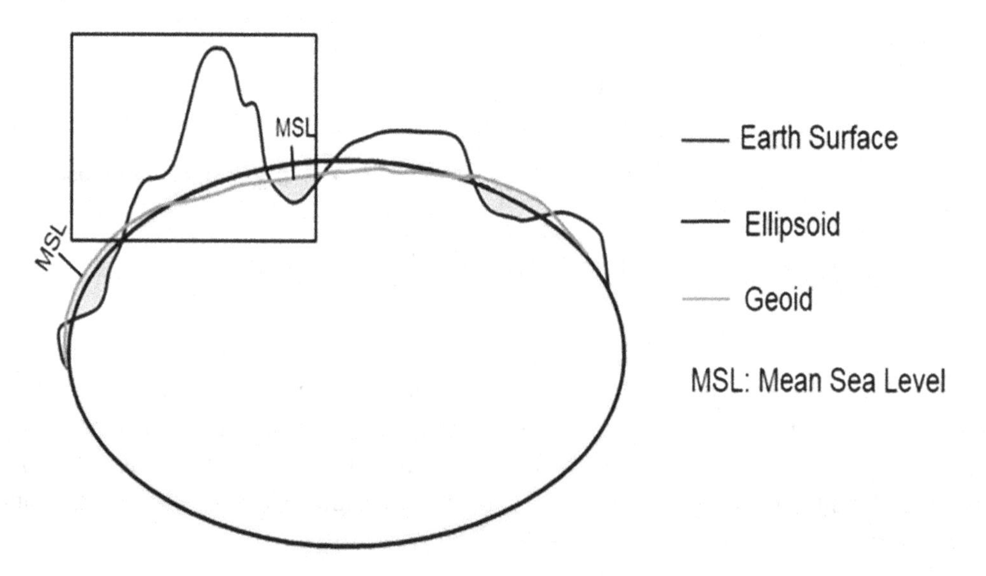

Ellipsoids are classically exploited as replacements for geoids to abridge the arithmetic alarmed in a coordinate tool grid with a dummy of the Earth's shape. Ellipsoids are worthy, nevertheless, not precise, presumptions of geoids (Figure 3). The diagram in Figure 4, recommends differences in elevation between a geoid model signified to for instance as GEOID96 and the WGS84 ellipsoid, respectively. The surface of GEOID96 growths up to 75 m surpassing the WGS84 ellipsoid over New Guinea as shown in red colour. In the Indian Ocean with the map is colored purple, which indicates that the floor of GEOID96 falls about 104 m underneath the ellipsoid surface.

Geoid Error Impacts

From the vision of equation 13.3, the real ocean flow, conversely, can't be represented the use of measuring ocean cutting-edge from altimeter data. Consequently, the along-track-only quantities of sea surface heights estimate fully the cross-track component of the floor geostrophic speed in which the sea-level

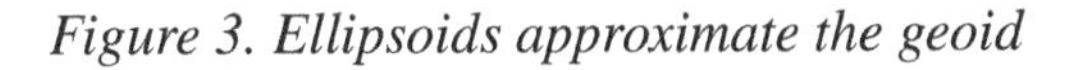

Figure 3. Ellipsoids approximate the geoid

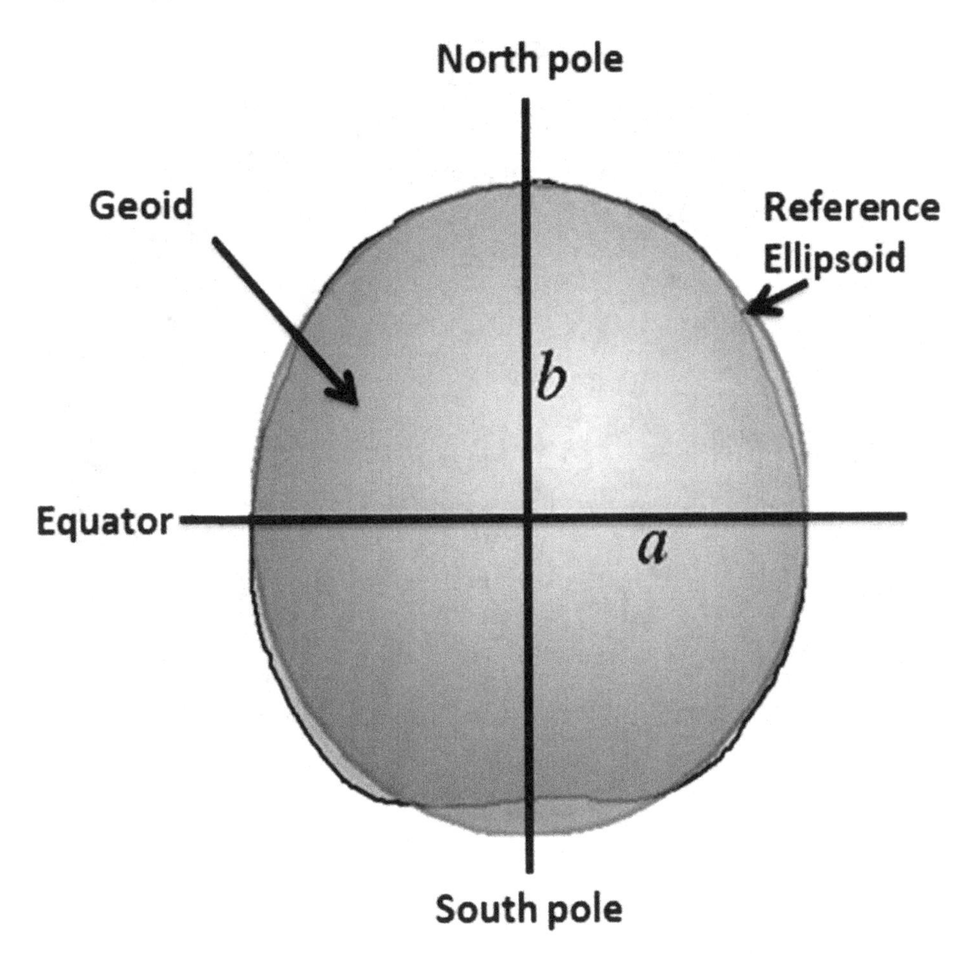

variants drifting the flow is balanced predominantly through the use of the Coriolis force (Hwang and Chen 2000). In this circumstance, the sea surface and the horizontal stress gradient is proportional to the sea-surface slope measured relative to the equipotential surface (that is, the geoid). In this circumstance, Balaha and Lunda (1992) concluded that a major assignment with altimeter measurements is that sea surface peak have to be relative to the geoid, which is designated as the form of the ocean surface in the absence of all exterior forcing and indoors motion. The sea surface height, however, is dominated via the (time-invariant-at least on oceanographic time scales) Earth geoid . On the word of Hu et al. (2001), the surface currents ought to be calculated if the geoid used to be precisely known. Consequently, Glenn et al., (1991) delivered artificial geoid methods to compute the correct geoid. In this method, the floor dynamic top is measured via in situ statement in which correlated by means of the use of a regression model with one is measured from radar altimeter. Conversely, the challenge of this method is that the unknown barotropic currents can end result in errors in the dynamic peak subject in which are blanketed in the geoid. Thus, Blaha and Lunde (1992) have used the logical reverse of the artificial geoid approach in which to use the altimeter height measurements to estimate the depth of thermocline; for instance for every 10 cm upward thrust in altimeter height.

Figure 4. Differences between an ellipsoid and a geoid

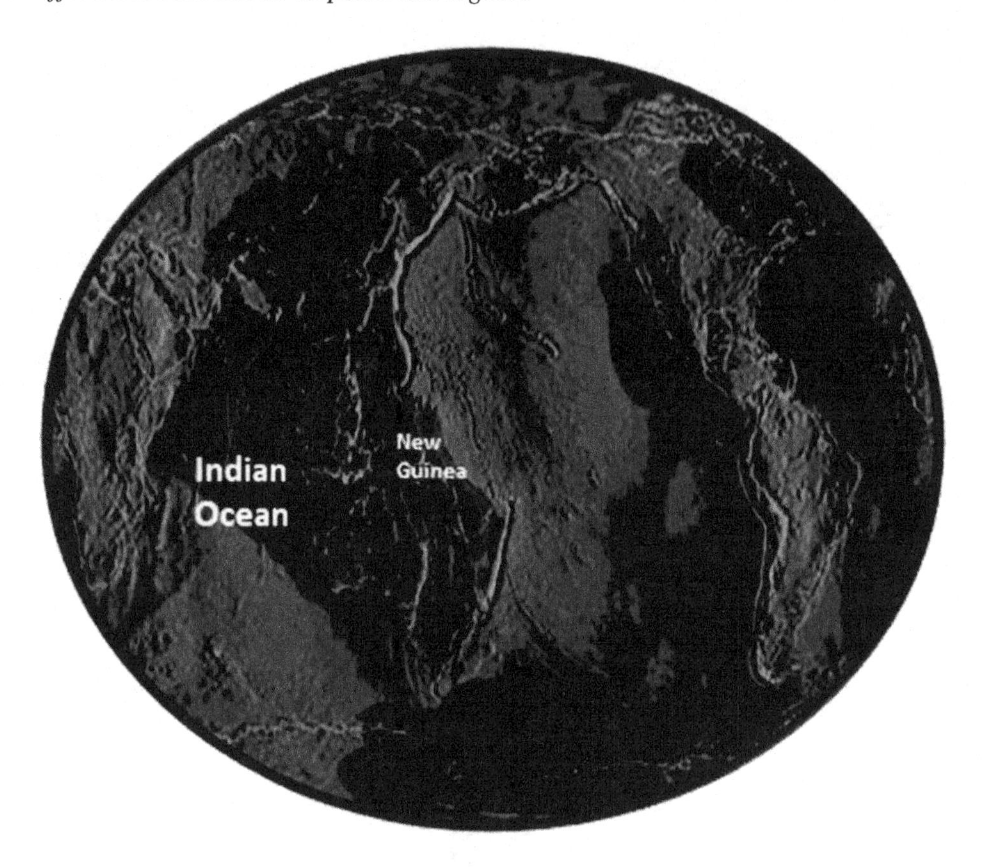

One of the methods that is more and more used for the ignition of the ocean fashions for prediction of ocean cutting-edge is the assimilation of satellite altimeter facts into complicated ocean models. Researchers have agreed that data assimilation into a numerical mannequin is a beneficial device to make use of the altimeter data further. As said by Ghil and Malanotte-Rizzoli, (1991), the model displays the data and extracts solely the elements that are defined through the physics retained in the model. Researchers have divided assimilation strategies into two categories: a sequential approach and a variation method. A sequential approach is dominated through the achievable problem of the low cross-track decision (Ikeda, 1995). The low cross-track resolution has been solved through the usage of the Nyquist wavelength, which is twice the cross-track interval (White et al., 1990). In this perspective, White et al. (1990) established that Nyquist wavelength improves decrease horizontal resolution, which is suitable for measuring ocean contemporary from altimeter Geosat data with space and time decorrelation scales of ~ 400 km and 40-135°. This method, nevertheless, is no longer capable to reconstruct mesoscale eddies due to the fact the cross-track interval is massive relative to eddy size.

In contrast, the parties' techniques perhaps be suitable for data assimilation. The less well-recognized constituents of the model are frequently selected as regulator variables. As said by Ikeda (1995), the space of control variables is often constituted of preliminary and boundary conditions. The variation rule and optimal control have exploited to originate a connect equation system. In this perspective, Moore (1991) exercised the variation technique, in which a feature model with arranging vertical profiles is required to produce the first guess of control variables for initializing feature assemblies. This approach, conversely, has a serious problem is how to minimize the cost function. Perhaps the simulation the cost function gradient is an effective approach to solve the boundary equations. Ikeda (1995), consequently, recommended that solving boundary equations can attain by placing partial derivatives of a Lagrange function to zero.

Moreover, Ghil and Malanotte-Rizzoli (1991) realized a dynamical interruption means, for example, the Klaman filter to obtain precise information of ocean currents from radar altimeter satellite data. In this phase, nonetheless, data assimilation of the satellite altimeter is still the principal research tool. This is since data assimilation methods differ from comparatively simple bumping procedures to further sophisticated algorithm corresponding to Volterra filter and attribute models i.e., finite difference model (Marghany 2009).

Currently, few research have utilized exceptional altimeter sensors for sea surface investigations in the South China Sea (Saraceno et al., 2008). Hu et al. (2001) investigated the sea surface peak with a duration of 3-6 months the usage of six years of TOPEX/POSEIDON altimeter data. They mentioned that the sea surface height variants are related strongly with cutting-edge and eddy features. Further, Imawaki et al. (2003) delivered a new method for obtaining high-resolution imply surface velocity. To this end, they blended the makes use of of TOPEX/POSEIDON and ERS-1/2 altimeter information and drifter information received from 1992 through 2001 to observe the undulation of mean-surface dynamic topography down to the ocean current. Indeed, the ocean cutting-edge is continually allied with geoid via the quantity of sea surface dynamic topography. Marghany (2009) advised the amendment of the hydrodynamic equation, options of the imply sea-surface to invert accurate pattern of sea surface current. Therefore, the altimeter data that are both extra accurate and extra extensively allotted in time are required. Furthermore, Ikeda (1995) mentioned that geoid mistakes can be considerably smaller than the corresponding values of the sea surface topography if know-how of the long-wavelength geoid is improved.

The foremost contributions of this work are to layout a scheme to limit the affects of geoid and Coriolis parameters in the continuity equation. In doing so, this locates out about encompasses the preceding principle of geostrophic current through imposing the Volterra model and Lax-Wendrof scheme. Indeed, a decrease altitude is dominated through susceptible Coriolis parameter. In this context, a susceptible geostrophic current may be took place (Marghany 2009). The major speculation in this idea established by Marghany (2009) is the truth of modelling the altimeter data as a sequence of the nonlinear filter. In this view, there is a sizeable distinction between the ocean current velocities that can be received by means of the inverse filter of linear and nonlinear Volterra kernels. Besides, the utilization of the Lax-Wendroff scheme can suppress the numerical solution of current speed is obtained through the inverse filter of Volterra kernels.

MODELLING SEA SURFACE CURRENT USING JASON-2 ALTIMETER

This section presents an altered formulation for a geostrophic current. The method is mainly grounded on the exploitation of the Volterra series expansion into the geostrophic current calculation. The foundation of this method is to completely modification the time series of JASON-2 satellite altimeter into a real ocean surface flow. Then, the Volterra kernel inversion exploited to obtain the sea surface current velocity. To this end, the finite element model of the Lax-Wendorff scheme exercised to regulate the spatial deviation of surface current flow across the Indian ocean.

In this view, the satellite data feed into the Lax-Wendorff scheme is the sea level anomaly from March 2014 to March 2015. They were acquired from the live access server of "Archiving, Validation, and Interpretation of Satellite Data in Oceanography, which is known as AVISO. The address of the website is "http://opendap.avio.oceanbs.com" charts were created in intermissions of each three months as of March 2014 with a resolution of 1/3° in both latitude and longitude and referenced relation to the 2014 - 2015 meantime duration.

Volterra-Lax-Wendroff Algorithm

The Volterra model is grounded on the Volterra series and also termed as "memory power series ", which is a practical expansion of a linear (first-order) model to bilinear (second-order), trilinear (third-order), and higher-order models (Fu et al., 2000). Liska and Wendroff (1998) presented the multidimensional frequency functions as *H(f,g)*, *H(f,g,h)*, and *H(f)* to designate the bilinear system, the trilinear system, and linear system respectively (Rugh, 1981).

Let consider that the output *Y(t)* owing to input *X (t)* is described by a total of Volterra functions as obeys:

$$Y(t) = y_0+y_1(t) + y_2+y_3(t) +\dots\dots\dots \quad (13.4)$$

Equation 13.4 reveals that three yields: $y_1\ (t)$ is a linear yield, $y_2\ (t)$ is bilinear yield and $y_3\ (t)$ is tri-linear yield, respectively (Figure 5) and y_0 is constant. Consequently, the mean value $\bar{Y} = E[Y(t)]$ is given by:

Figure 5. Volterra nonlinear input/output block diagram model

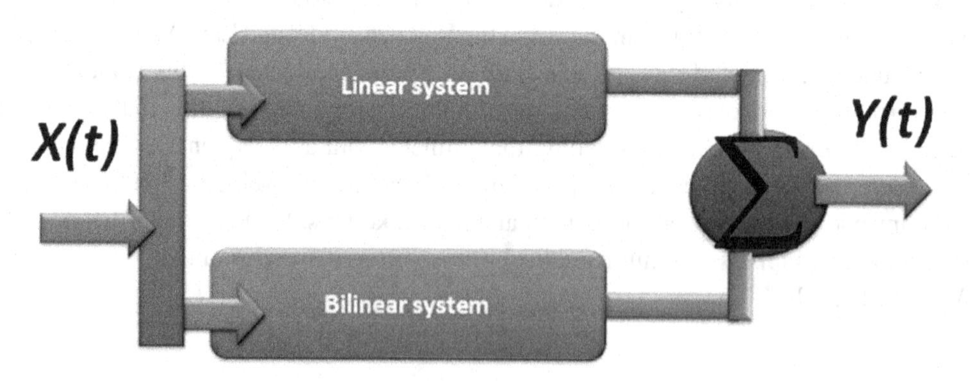

$$\bar{Y}(t)-\bar{Y} =[y_1(t)-\bar{y}_1]+[y_2(t)-\bar{y}_2]+[\bar{y}_3(t)+\bar{y}_3]+\dots\dots\dots\dots (13.5)$$ here,

$$\bar{Y}(t) = y_0 +\bar{y}_1 +\bar{y}_2 +\bar{y}_3 +\dots\dots\dots\dots \quad (13.5.1)$$

Consequent Marghany (2009), both sides of the equation (13.5) are implemented by using Fourier transforms to transfer them in the frequency domain that satisfies:

$$Y(f) = Y_1(f) + Y_2(f) + Y_3(f) + \ldots\ldots\ldots \quad (13.6)$$

where,

$$Y(f) = F[Y_1(t) - \bar{Y}] = \int_{-\infty}^{\infty} [Y(t) - \bar{Y}] e^{-j2\pi ft} dt \quad (13.6.1)$$

with

$$Y_1(f) = F[Y_1(t) - \bar{Y}_1] \quad (13.7)$$

$$Y_2(f) = F[Y_2(t) - \bar{Y}_2] \quad (13.8)$$

Grounded on Equations 13.6 to 13.8, Marghany (2009) invented a new algorithm to determine the geostropic current rate as a function of a nonlinear series of Volterra model. A comprehensive, nonparametric context to designate the input-output u and v geostrophic current constituents as a function of nonlinear system time-invariant. In the distinct formula, both input, geostrophic current $\vec{u}(t)$, and output, $Y(t)$ as a function of the Volterra series can be formulated as (Marghany,2009):

$$\begin{aligned} Y(t) = h_0 &+ \sum_{\tau_1=1}^{\infty} h_1(\tau_1)\vec{u}(t-\tau_1) + \sum_{\tau_1=1}^{\infty}\sum_{\tau_2=1}^{\infty} h_2(\tau_1,\tau_2)\vec{u}(t-\tau_1)\vec{u}(t-\tau_2) \\ &+ \sum_{\tau_1=1}^{\infty}\sum_{\tau_2=1}^{\infty}\sum_{\tau_3=1}^{\infty} h_3(\tau_1,\tau_2,\tau_3)\vec{u}(t-\tau_1)\vec{u}(t-\tau_2)\vec{u}(t-\tau_3) + \ldots\ldots\ldots\ldots \\ &+ \sum_{\tau_1=1}^{\infty}\sum_{\tau_2=1}^{\infty}\ldots\ldots\ldots\ldots\sum_{\tau_k=1}^{\infty} h_k(\tau_1,\tau_2,\ldots\ldots\ldots,\tau_k)\vec{u}(t-\tau_1)\vec{u}(t-\tau_2)\ldots\ldots\ldots\ldots\vec{u}(t-\tau_k) \end{aligned} \quad (13.9)$$

Equation 13.9 describes the discrete-time lags as t, τ_1, τ_2, …, and τ_k. The Volterra kernel of the kth-order is presented by $h_k(\tau_1, \tau_2, \ldots, \tau_k)$. The kernel of the first-order Volterra functional is h_1, which accomplishes a linear function on the contribution and h_2, h_3,…,h_k confine the nonlinear exchanges between input and output sea level dissimilarities, which have assimilated from JASON-2 satellite data. Subsequently, the directive of the non-linearity is the maximum operative imperative of the multiple synopses in the practical series (Rugh, 1981).

In general, equation 13.9 can be implemented as real sea level in the circumstance of obtaining a discrete values in the time domain $[h_k(\tau_1, \tau_2, \ldots, \tau_k)]$ and in the frequency domain $[H_2(f_1, f_2,\ldots, f_k)]$. In this regard, both first (h_1) and second (h_2) kernels are obtained by the Fourier transform as follows:

$$H_1(f) = \int h_1(\tau)e^{-j2\pi f\tau}d\tau \quad (13.10)$$

$$H_2(f_1, f_2) = \iint h_2(\tau_1, \tau_2)e^{-j2\pi(f_1\tau_1 + f_2\tau_2)}d\tau_1 d\tau_2 \quad (13.11)$$

Both equations 13.10 and 13.11contains the frequency-domain components of $H_1(f)$ and $H_2(f_1,f_2)$, respectively. In other words they are also known as the first, and second-order frequency–domain kernels, correspondingly. An adequate circumstance $H_1(f)$ to occur is that the linear system is steady. $H_2(f_1,f_2)$, nonetheless, is transpired once the bilinear system is steady. Furthermore, $H_2(f_1,f_2)$ is proportioned about its binary frequency variables since $h_2(\tau_1,\tau_2)$ is exclusive and fulfills $h_2(\tau_1,\tau_2)= h_2(\tau_1,\tau_2)$. In the words of (Liska and Wendroff 1998), $h_2(\tau_1,\tau_2)$ is proportioned for its binary components. Equations 13.10 and 13.11, consequently, can yield the prescriptions for the initial dual Fourier transform functions in equation (13.6). In this unerstanding, the modification of equation 13.6 in which grounded on equations 13.10 and 13.11 is formulated as:

$$Y(f) = H_1(f)X(f) + \int H_2(\alpha, f-\alpha)x(f-\alpha)d\alpha - \overline{y}_2\partial_1(f) \quad (13.12)$$

Along with Rugh (1981), the variable $\partial_1(f)$ is an approximation to the usual hypothetical delta function and befalls in the binary calculation of Fourier transforms with sub-records of the finite delta function and $\overline{y}_2$ is formulated as:

$$\overline{y}_2 = \iint h_2(f_1,f_2)a_{xx}(f_1, f_2)df_1 df_2 \quad (13.13)$$

here a_{xx} presents the autocorrelation function of the $x(t)$ as-is given by:

$$x(t) = e^{j2\pi f_1 t} + e^{j2\pi f_2 t}. \quad (13.13.1)$$

Geostrophic Current Simulation Grounded on Volterra Model

The instant geostrophic surface speed U_g can be acquired using sequential mean speed $\overline{U}(x,t)$ and the variance of sea-surface geostrophic velocity $\overline{U}(x,t) = (u',v')$ by the formula is introduced by Marghany (2009):

$$U_g(x,t) = \overline{U}(x) + \overline{U}(u',v') \quad (13.14)$$

Equation 13.15 reveals the components of the geostropic sea surface current from the radar altimeter data (u',v') in x and y directions. Moreover, the anomaly of sea surface dynamic topography ξ, which essentially corresponding to the sea-surface height anomaly ζ_s can be used to estimate U_g as follows;

$$U_g(x,t) = N(\xi,\zeta_s) + \sum_{i=1}^{\infty} \int_R h_i(\bar{\tau},\xi) \prod_{j=1}^{i} \bar{U}_a(u'-\tau_j)(v'-\tau_i)d\tau \tag{13.15}$$

Equation 13.15 shows that the sea surface current can be also estimated from the geoid height $N(\xi,\zeta_s)$ as termed by Imawaki et al. (2003). Succeeding Marghany (2009), the mathematical expressions for first-order (D_{1x} and D_{1y}) and second-order Volterra kernels (D_{2xx} and D_{2yy}) of instantaneous surface geostrophic velocity U_g are as follows:

$$D_{1x}(f_x,f_y) = \left[\xi_x \bar{f} + \frac{\partial N(\vec{\xi})}{\partial \vec{\xi}}.(\bar{U}_{t_0} - \beta(\frac{\partial \zeta_s}{\partial y} - \frac{\partial \xi}{\partial y}) \right] \tag{13.16}$$

$$D_{1y}(f_x,f_y) = \left[\xi_y \bar{f} + \frac{\partial N(\vec{\xi})}{\partial \vec{\xi}}.(\bar{U}_{t_0} - \beta(\frac{\partial \zeta_s}{\partial x} - \frac{\partial \xi}{\partial x}) \right] \tag{13.17}$$

$$\begin{aligned} D_{2xx}(f_{1x},f_{1y},f_x-f_{1x},f_y-f_{1y}) = \Bigg[& \xi_x \frac{\partial D_{1x}(f_{1x},f_{1y})}{\partial \zeta}(f-f_1) - \xi_x \bar{f} \\ & + \frac{\partial N(\vec{\xi})}{\partial \vec{\xi}}.(\bar{U}_{t_0} - \beta\left(\frac{\partial \zeta_s}{\partial y} - \frac{\partial \xi}{\partial y}\right)\Bigg] \times \beta \frac{\partial \xi}{\partial y} \bar{f} \end{aligned} \tag{13.18}$$

$$\begin{aligned} D_{2yy}(f_{1x},f_{1y},f_x-f_{1x},f_y-f_{1y}) = \Bigg[& \xi_y \frac{\partial D_{1y}(f_{1x},f_{1y})}{\partial \zeta}(f-f_1) - \\ & - \left[\xi_y \bar{f} + \frac{\partial N(\vec{\xi})}{\partial \vec{\xi}}.(\bar{U}_{t_0} - \beta\left(\frac{\partial \zeta_s}{\partial x} - \frac{\partial \xi}{\partial x}\right)\right]\Bigg] \times \beta \frac{\partial \xi}{\partial x} \bar{f} \end{aligned} \tag{13.19}$$

In equation 13.19 the components of the frequency domain in x and y dimensions are presented by f_x and f_y. Conversely, an average frequency domain is $\bar{f}$. Moreover, β presents the gravity acceleration ratio to the Coriolis due to the Earth's rotation. To obtain sea surface current from radar altimeter data, the nonlinear relationship between sea surface slope and 2-D FFT (two-dimensional Fourier transform) is taken into account. In this circumstance, let the relationship between $F(f_x f_y)$, the first-order (D_{1x},D_{1y}) and second-order Volterra Kernels (D_{2xx}, D_{2yy}) is formulated as:

$$F(f_x f_y) = U_x(f_x f_y).D_{1x}(f_x f_y) \tag{13.20}$$

$$F(f_x f_y) = U_y(f_x f_y).D_{1y}(f_x f_y) \tag{13.21}$$

$$F(f_{xx} f_{yy}) = U_x(f_{xx} f_{yy}).D_{2xx}(f_{xx} f_{yy}) \tag{13.22}$$

$$F(f_{xx} f_{yy}) = U_y(f_{xx} f_{yy}).D_{2yy}(f_{xx} f_{yy}) \tag{13.23}$$

As stated by Marghny (2009), the results of current velocity $\bar{U}(x,t) = (u', v')$ can be obtained by the inverse of the Volterra model as

$$\vec{U}(f_x, f_y) = \frac{FT\left[\prod_{j=1}^{i} x(t)\right]}{D(f_x, f_y)} \tag{13.24}$$

where $FT\left[\prod_{j=1}^{i} x(t)\right]$ is the linearity of the Fourier transform. Consistent with Marghany (2009), the inverse filter $P(f_x f_y)$ is exploited subsequently $D(f_x f_y)$ is not equal a zero for $(f_x f_y)$ which specifies that the average current speed would vary along x and y directions. The inverse filter $P(v_x, v_y)$ can be expressed as

$$P(f_x, f_y) = \begin{cases} [D_x(f_x, f_y)]^{-1} & if(f_x, f_y) \neq 0, \\ [D_y(f_x, f_y)]^{-1} & otherwise. \end{cases} \tag{13.25}$$

Therefore, the resultant current velocity can be calculated by using equation 13.25 into equation 13.24 as:

$$\vec{U}(f_x, f_y) = FT\left[\prod_{j=1}^{i} x(t)\right].P(f_x, f_y) \tag{13.26}$$

Following Marghany (2009), the current direction can be obtained by using this formula,

$$\theta = \tan^{-1}(\frac{U_y}{U_x}) \tag{13.27}$$

Equation 13.26 requires an accurate mathematical approach to fulfill the request of the precise sea surface current circulation in the large-scale basin,for instance, the Indian Ocean. In this understanding, the discrete finite difference approach such as Lax- Wendrof algorithm is operated to simulate the annual Indian ocean circulation pattern.

Lax-Wendrof Scheme

A staggered grid is used to address the annual current flow obtained from the second-order accurate dispersive Lax-Wendrof scheme. Following surface current flow in JASON-2 altimeter data which can be written in the predictor-corrector form in a staggered grid

Following Liska and Wendroff (1998), JASON-2 altimeter data are coded into staggered grid. In fact, The staggered grid contains of a major grid where JASON-2 data are categorized in *(i, j, k)* and a binary grid where points are categorized in (i + 12, j + 12, k + 12) (Figure 6). following Marghany(2009),

Figure 6. Presenting Lax-Wendroff scheme in Staggered grid

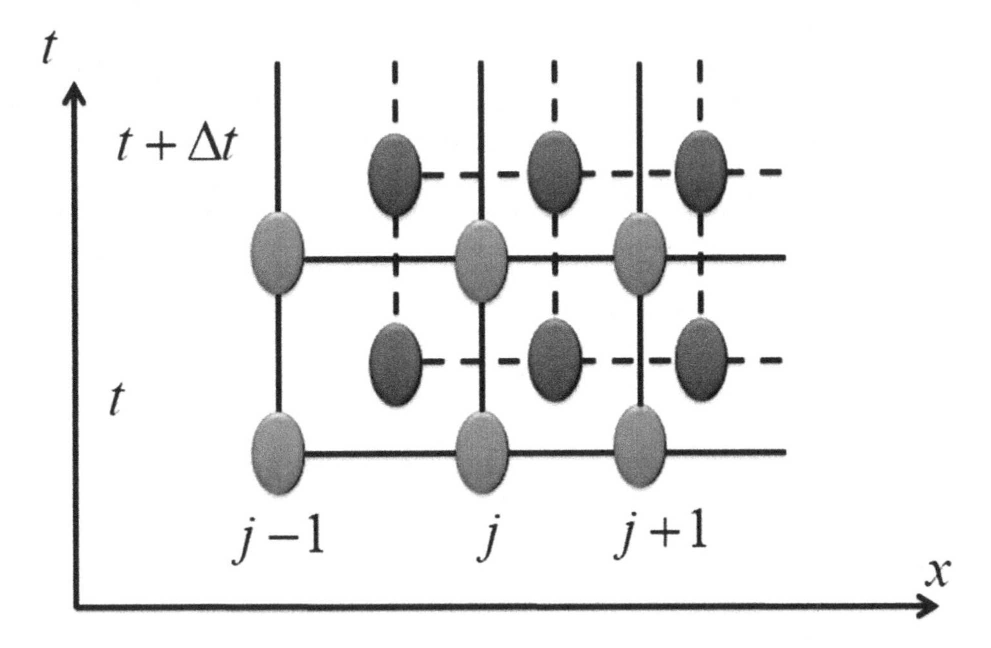

2-D, numerical sea surface current for the two-step forms of Lax-Wendroff scheme is formulated as:

$$\begin{aligned} U_{i,j}^{n+1} = U_{i,j}^{n} + \frac{\Delta t}{2\Delta x}\Big[\big(v(U_{i+0.5,j+0.5}^{n+0.5}\big) + \big(v(U_{i+0.5,j+0.5}^{n+0.5}\big) - \big(v(U_{i-0.5,j+0.5}^{n+0.5}\big) - \big(v(U_{i-0.5,j-0.5}^{n+0.5}\big)\Big] \\ + \frac{\Delta t}{\Delta y}\Big[\big(G\big(v(U_{i+0.5,j+0.5}^{n+0.5}\big) + \big(G(U_{i-0.5,j+0.5}^{n+0.5}\big)\Big] - \Big[\big(G(U_{i+0.5,j-0.5}^{n+0.5}\big) - G(U_{i-0.5,j-0.5}^{n+0.5}\big)\Big] \end{aligned} \tag{13.28}$$

Liska and Wendroff (1998) considered v and G as smooth functions. the analog of 2-D predictor, therefore, is implemented to compute the quantities on time level $n + ¼$, which is positioned at the center of all faces of the primary cell as follows:

$$U^{n+0.25}_{i,j+0.5,k+0.5} = 0.25\left[\left(U^{n}_{i,j,k} + U^{n}_{i+1,j,k} + U^{n}_{i,j+1,k} + U^{n}_{i,j,k+1}\right)\right] + \frac{\Delta t}{4\Delta y}\left[\left(G(U^{n+1/6}_{i,j,k+0.5})\right)\right] \tag{13.29}$$

Equation 13.29 denotes the finite difference scheme model which implemented to conventionally rectangular grids owing to most of their straightforwardness. A numerical solution is then improved by expending the curvilinear approximately orthogonal, and the coastline-succeeding grids. This algorithm is instigated to overwhelmed the issues elevated owing to the complicated boundary conditions of the Southern Indian Ocean besides its complicated topography.

The Solution of Boundary Data

The algorithm is formulated in equation 13.29 is the keystone of the boundary condition appropriately, which is a prerequisite to acquiring a solution using the staggered grid. Four-point boundary data would be itemized at the apiece boundary to obtain a distinctive solution. On the left-hand boundary, merely U is quantified at time interval $n+1/4$, then both are requested to attain v and G at this grid point. Consistent with Imawaki et al., (2003), an approximately exceptional boundary approach is then required to obtain U. Let consider $U(0,t)$ as precise the left-hand boundary data. As shown in Figure 6, roughly interpolation frequently lacking iteration owing to the initial setting of the G-characteristic that prominent the boundary of $(n+\frac{1}{4})(\frac{\Delta t}{4})$, which is unidentified. Accordingly, let consider an preliminary presumption of $\varphi = \frac{(\Delta x/4)}{4}$, $U(\phi)$ and $G\text{–}(\phi)$ are formulated as:

$$G- = \frac{-\varphi}{\Delta t/4} \tag{13.30}$$

$$U(\varphi) = \frac{\varphi U_4^{\,n} + (\frac{\Delta x}{4} - \varphi)U_1^{\,n}}{\frac{\Delta x}{4}} \tag{13.31}$$

Equation 13.31 is then exploited to simulate the definite space ϕ and the iteration maintains along with the roughly convergence condition. Generally, equation 13.31 offers the second boundary quantity as a fragment of the solution scheme and in particular, can contain sea surface current pattern variations.

The mean sea velocity derived using equation 13.31 suggests a turbulent eddy flow along the search area of MH370 and the suspected area where the flaperon fell down. The current flows have a maximum speed of 1.20 m s^{-1} along the western Australian Shelf (Figure 7). In this view, the second-order kernel can deliver a clear pattern of ocean current from altimeter satellite data. It can reduce the impact of the nonlinearity between altimeter return pulse and sea surface geophysical parameters. Consistent with Marghany (2009), the uncertainty of Merged Sea Level Anomaly (MSLA) obtained from altimeter data is due to geoid and mean dynamic topography that is allied with pulse time delay. In this regard, the real annual sea surface current pattern would compute using the nonlinearity of the second-order kernel.

The second-order kernel can determine a smooth current flow pattern, in which the sharpest edge of the eddy flows discriminate clearly.

Figure 7. Sea surface current simulated by the second-order kernel

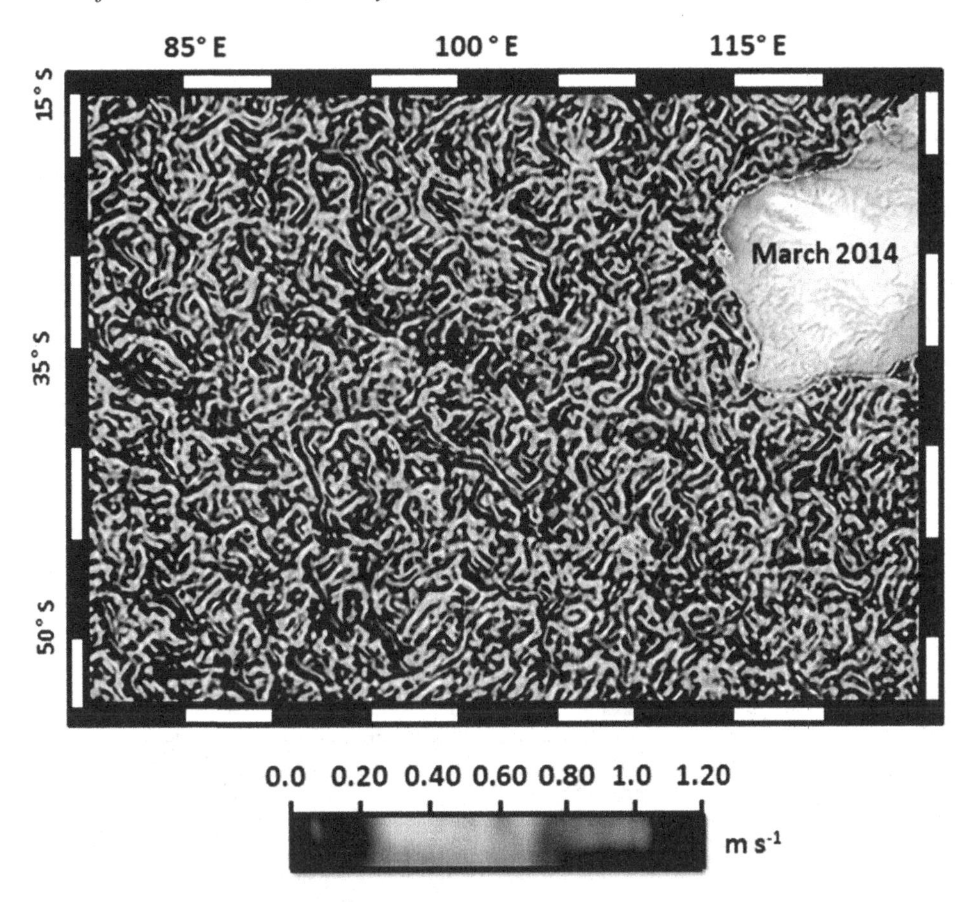

CURRENT PATTERN SIMULATION USING MULTI-OBJECTIVE EVOLUTIONARY ALGORITHM (MOE)

Let undertake that the huge parameter space $\mathbb{R}$ would-be reconnoitered by the genetic algorithm (GA) to normalize the operative solutions. In this sight, the extrapolative algorithm contains the nonlinear guess function which is grounded on archived time-series information on the Indian ocean current $U(\phi)$ and the Indian Ocean floor features to predict the final destination of MH370 debris (Anderson,2003). Let assume that $\{U(\phi)_i\}$ be the observation made with a generic function ϕ which can state as follows:

$$\bigcap_{m=1}^{\infty}\bigcup_{n=1}^{\infty}\left(U(\phi)_n-\left(2^{n+m}\right)^{-1},U(\phi)_n+\left(2^{n+m}\right)^{-1}\right) \tag{13.32}$$

The sequence observations that itemize the rational numbers are represented by $\{U(\phi)_n\}_{n=1}^{\infty}$. This signifies that generic function ϕ is satisfying:

$$\{U(\phi)_n\}_{n=1}^{\infty} = \sum_{n=1}^{\infty} \varphi(U(\phi)_i) \tag{13.33}$$

In the genome, for individual members of the population, the population is adjusted by arbitrary transfer of a 0 or 1 for every one of the 32 bits (Figure 8). In the initial stage, the rate of RMSE is ±900 m s $^{-1}$ which presents a fuzzy pattern of ocean surface flow as compared to Lax-Wendroff scheme (Figure 7). Subsequently, the first 20 and 12 bits are transcribed into an integer representing the *i,j* coordinates, respectively to evaluate the fitness (Figure 8). The locations of the trajectory movement of debris thenceforth are simulated post 34000 iterations, with a fitness value of 0.8 and RMSE of ±0.005 m s $^{-1}$ (Figure 9). This indicates the ability of the multiobjective algorithm to construct the trajectory movement of any object across the surface ocean current of the Southern Indian Ocean.

Let assume that *m* is hydrodynamic parameters of the Indian ocean and *n* presents in the flight debris. In thi regard, every hydrodynamic parameter has a utility function as:

Figure 8. The initial stage of a multiobjective algorithm for simulating trajectory movement of MH370 debris

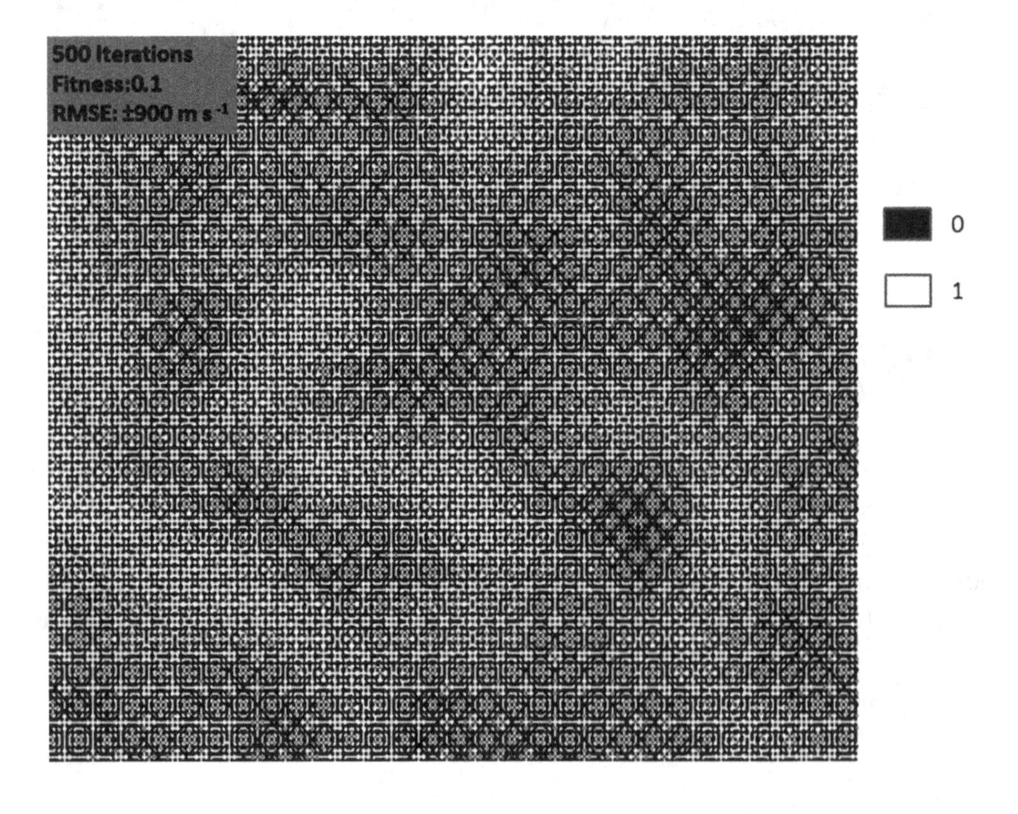

Figure 9. Approximate Southern Indian Ocean flow simulated based on a multiobjective algorithm

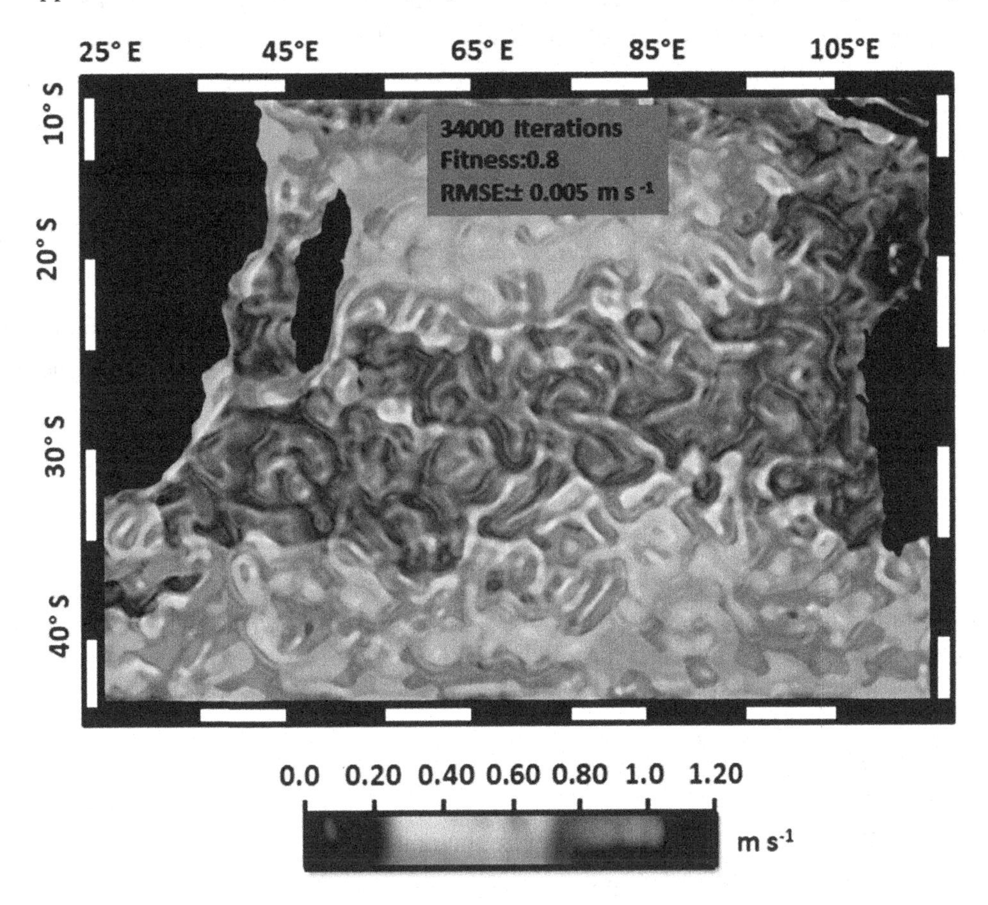

$$\psi = f(\vec{v}_i) \tag{13.34}$$

Equation 13.34 indicates that MH370 debris, vertical mixing, sea surface current, significant wave height, wind speed, sea-level variations are presented within vector $\vec{v}$. In other words, the numerous hydrodynamic parameters including sinking depth would present as $\vec{v}_i = (\vec{v}_1, \vec{v}_2, ..\vec{v}_n)$. In this regard, the feasibility constraint is considered as $\sum_{j=1}^{m} \vec{v}_i = b_j$ for $j = (1,2,3,....,n)$. Lastly, the Euler–Lagrange equations are exploited to find the Pareto optimal solutions of the trajectory movements of debris across the southern Indian Ocean.

$$L_i\left(\left(U(\phi)_j^k\right)_{kj}, (\lambda_k)_k (\mu_j)_j\right) = f(\vec{v}) + \sum_{k=2}^{m} \lambda_k (\psi_k - f^k(v^k) + \sum_{j=1}^{n} \mu_j (b_j - \sum_{k=1}^{m} \vec{v}_j) \tag{13.35}$$

whereby, L is Lagrangian for each debris v^k for $k=1,....,m$ and the vectors of multipliers are λ_k and $(\mu_j)_j$, respectively and $k \neq j$. The archived time series significant wave heights, sea surface current, sea-level variations, and wind speed data from March 2014 to March 2016 are gathered from the Jason-2/Ocean

Surface Topography Mission (OSTM), and QuikSCAT, respectively. These data are used to simulate the possible debris trajectory movement pattern (Marghany, 2015).

Let assume that T is debris total numbers as a function of vertical mixing due to wind effects that is formulated as

$$T = \int_{-z}^{0} \int_{-\infty}^{0} \bar{n} dz [1 - e^{(-zw_b 1.5 u_{*_w} k H_s)}] \quad (13.36)$$

where z presents the water depth, u_{*_w} shows the water friction velocity, H_s presents the significant wave height, k is constant value 0.4 presents the von Karman constant and $\bar{n}$ for the buoyant increase w_b is expressed by

$$\bar{n}(z) = \bar{n}_0 e^{(-zw_b (1.5 u_{*_w} k H_s)^{-1})} \quad (13.37)$$

Here T is proportional with water depth and time t, respectively. Subsequently, the total number of MH370 debris relies on vertical mixing, wind stress, and significant wave height fluctuations. In this understanding, debris design parameters have to be tackled certainly.

PARETO FRONT

Multi-objective optimizers are demanded to acquire the non-dominated set as close as possible to the accurate Pareto front. Additionally, optimizers are claimed to preserve the constant feast of the non-dominated set endways the exact Pareto front.

Definition 1. Let assume that $\mathbb{R}^n$ in the Euclidean space of Southern Indian Ocean which includes a compact set of feasible decisions $T(L)$. It has a closed interval unit of [0, 1], and l is the feasible set of criterion vectors $\mathbb{R}^m$. The MOE is used to determine a vector $\vec{l} \in \mathbb{R}$ that optimizes the vector function $\vec{f}(\vec{l})$.

Definition 2. Pareto dominance. Let A vector $\vec{l}$ dominates $\vec{l}'$ which is signified $\vec{l} \prec \vec{l}'$.

1. If $f_i \leq f_i(\vec{l}')$ for entirely i functions in $\vec{f}$,and
2. There is slightest one i such that $f_i(\vec{l}) < f_i(\vec{l}')$.

Definition 3. Pareto optimal. A vector $(\vec{l}^*)$ is Pareto optimal if it does not occurs a vector $\vec{l} \in \mathbb{R}$ such that $\vec{l} \prec \vec{l}^*$.

Definition 4. Pareto optimal set. The Pareto optimal set for a MOE is formulated as:

$$P^* = \{\vec{l}^* \in \mathbb{R}\} \quad (13.38)$$

Definition 5. Pareto front. Given a MOE and its Pareto optimal set are given by:

$$P^*F = \{f(\vec{l})\,|\,\vec{l} \in P\}. \tag{13.39}$$

Starting Definition 2 it perceived that a solution is enhanced than another if it is enhanced in one objective and in the other would equivalent. Conversely, in definition 3 the Pareto optimal is only a solution as there is no other existing solution. Nevertheless, definitions 4 and 5 suggest that the Pareto optimal set is completed that is known as the Pareto front. In fact, the Pareto front is the optimal solution of a MOE. Then, the dimension of the Pareto front s is given by:

$$s = \sum_{p=2}^{|P|} \sqrt{\sum_{m=1}^{M} (f_m(p) - f_m(p-1))^2} \tag{13.40}$$

Equation 13.40 denotes the number of non-dominated solutions P and the number of objectives M, respectively. In this regard,$f_m(p)$ presents the mth objective function of the pth solution. Then, the perfect space between dual consistently scatter solutions s_u is expressed as:

$$s_u = \frac{s}{N_{exf} - 1} \tag{13.41}$$

where N_{exf} is the anticipated number of Pareto solutions, which have minimal and maximal values of objective functions. In this regard, functions are automatically allocated as the primary member and the former member of the last non-dominated set P_f (deliberating the issue with dual objective functions). The jth member of the P is that one with the minimal value dlt:

$$dlt(j) = |s_j - (j-1)\bullet s_u|,\ j=2,3,\ldots,N_{exf} - 1 \tag{13.42}$$

where s_j is computed using (13.40) while j replaces p. This approach has to be extended for problems with more objective functions. If the Pareto front consists of more parts, all the detected discontinuous parts have to be treated separately.

The hit-rate (HR) was used to determine the best solution performs with Pareto front:

$$HR = \frac{|P|}{FFC} \cdot 100\% \tag{13.43}$$

Marghay (2015) and Serafino (2015) stated that equation 13.43 contains the non-dominated set the total number of members as $|P|$ while FFC represents the total number of evaluations of objective functions. Definitely, the greater values of HR specify the superior efficiency of the algorithm subsequently merely the area comprising Pareto front members is sought.

OCEAN CIRCULATION IMPACTS ON MH370 DEBRIS

The Malaysian Flight MH370 has been reported vanished in the Southern Indian Ocean close to Perth Australia on March 8, 2014. In this view, March presents the north-east monsoon season. Nevertheless, the south Indian Ocean does not experience monsoon winds, the seasonal circulation is not varied as in the north Indian Ocean. The mean water flows from March to May 2014 dominated by anticlockwise eddies, which concentrated in the area where the flaperon fell down. These eddies have maximum speed of approximately larger than 0.5 m s^{-1}. In this circumstance, the flaperon must be sunk due to the impact of eddies within a month of May 2014 offshore western shelf of Australia and Perth.

Consequently, the mean surface flow increases towards the western boundary with the maximum speed of 1.2 m s^{-1}, which is clearly shown between Madagascar and east of Africa i.e. Mozambique (Figure 10).

Evidently, when the current approach the northwest landfill in Madagascar, it breaches into two streams in which one stream spins southward to form East Madagascar Current. Finally, the second stream moves towards the African Coast, where it splits again into northern and southern streams. The southward stream travels through the Mozambique Channel. Nonetheless, it does not continue as a con-

Figure 10. Mean current flows from March to May 2014

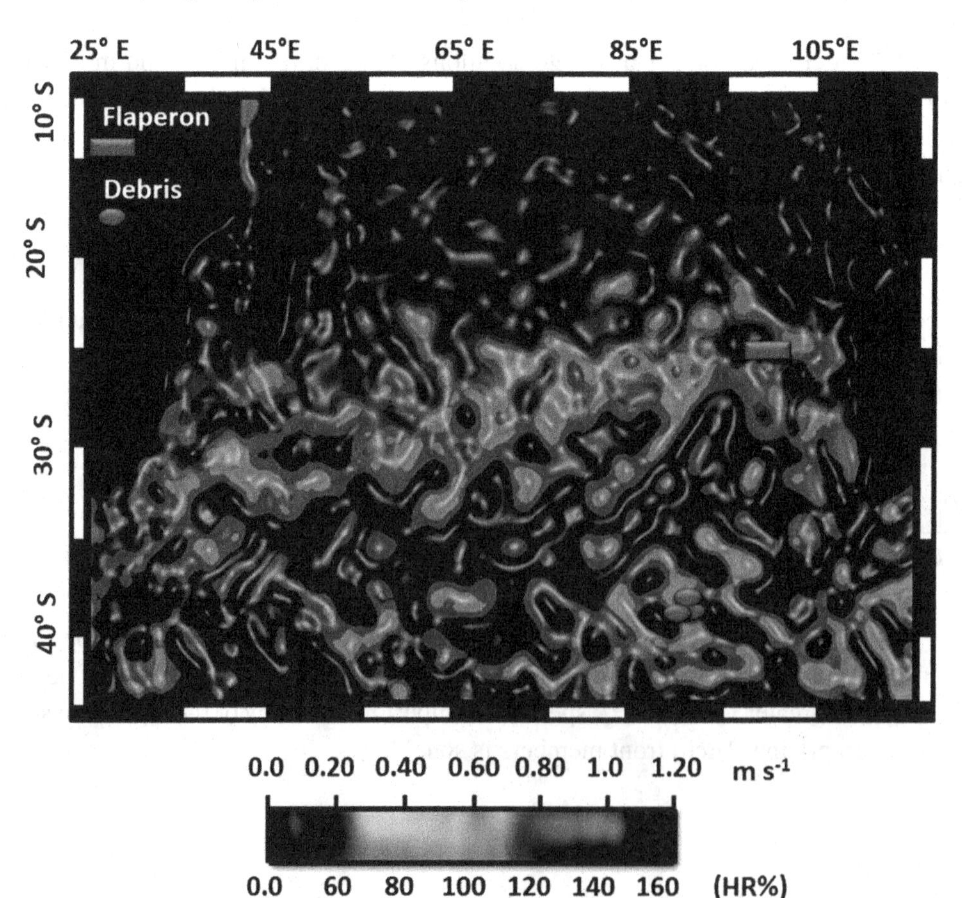

stant western boundary current and as an alternative breaks up into a series of eddies around the thin section of the Mozambique Channel (Figure 11 and Figure 12).

Figure 11. Mean current flows from June to August 2014

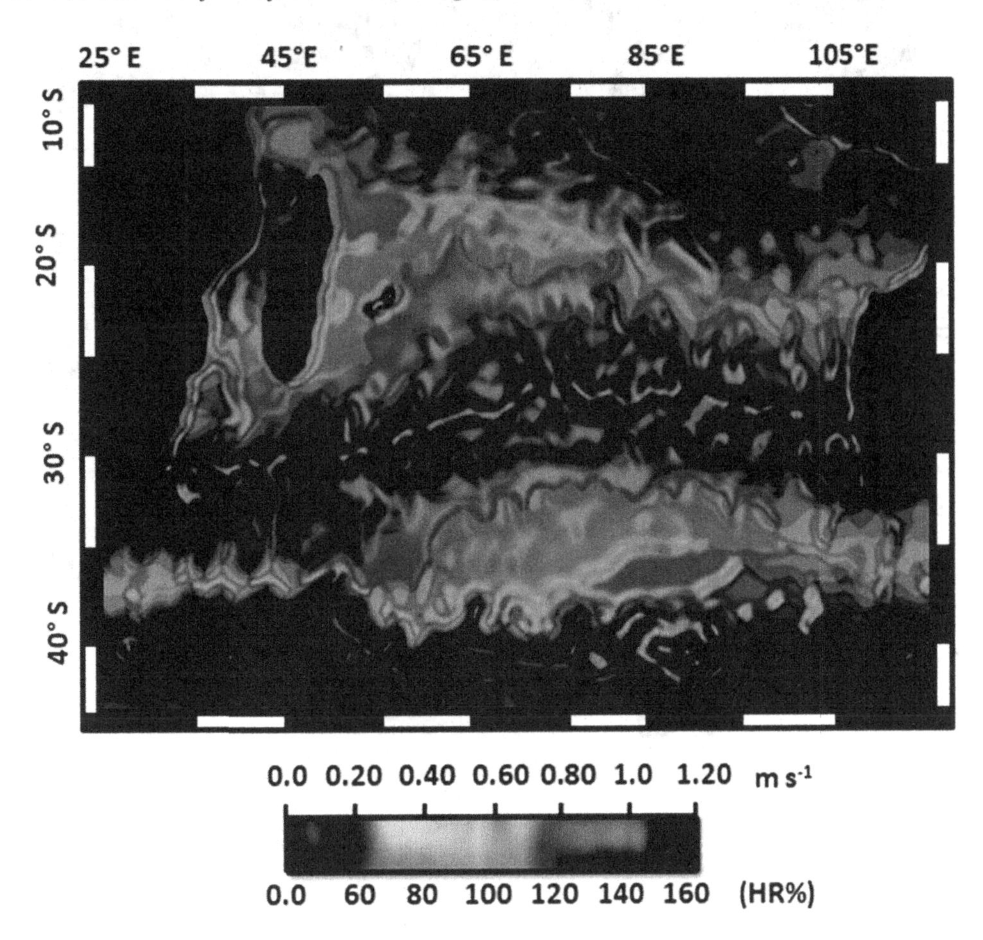

From November to December the eddies are increased across the southern Indian Ocean. The strength of the eddies is suggested through the Mozambique Channel as a strength the Agulhas Current (Figure 13). In other words, suggest that seasonal anomaly in the tropical circulation runs southern poleward through the Mozambique Channel and influences the mass transport of the Agulhas Current. This phenomenon continues from January to March 2015. In this view, Agulhas Current runs along the east coast of Africa as a narrow fast boundary current (Figure 14)

The main source of water flow across southern the equator is the northern Indian flow. In this regard, in the eastern part of the southern Indian Ocean, the coastal currents south of Java and the Leeuwin Current along the western coast of Australia also donate expressively to the inter-basin interactions. The South Java Current is an eastward stream with strong semi-annual and intraseasonal inconsistency near the coast of Java and Sumatra. The eastward flow of the South Java Current is robust during May (Figure 15) - June (Figure 16) and October (Figure 17) - November (Figure 18) periods and the flow reverses westward in other seasons. The seasonal reversal of the current is attributed to several reasons

Figure 12. Mean current flows from September to October 2014

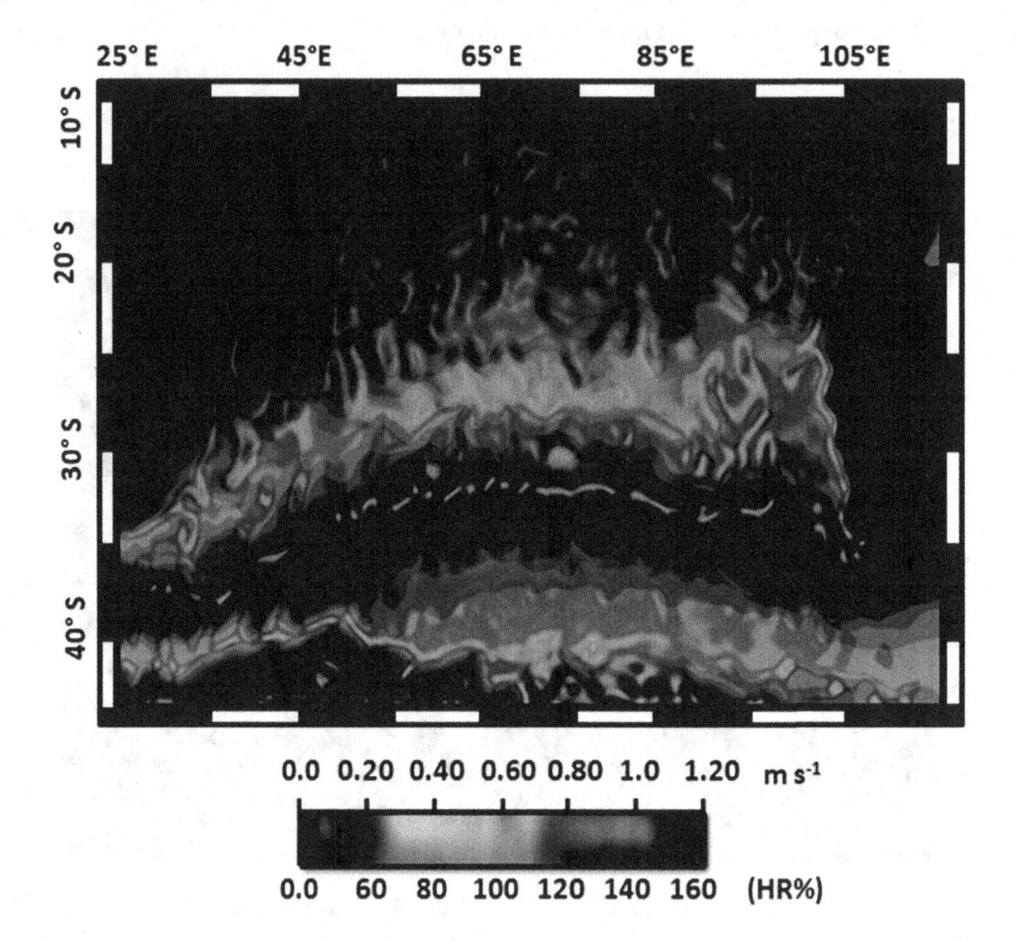

like changing monsoon winds, freshwater flux as well as to the equatorial wave activity (Marghany 2014 and Marghany 2017).

WHY THE FLAPERON CANNOT DRIFT TOWARDS RÉUNION ISLAND

Consistent with above prospective, the debris fragments would not float on Indian ocean surface for several months, but they would have sunk deeply thousand of meters. Indeed, the debris fragments would experience istabilities on the ocean surface due to the Antarctic Circumpolar Current (ACC) (Wyrtki,1971 and Reverdin et al., 1983). In this understanding, the MH370 debris would spin in a large scale counter-clockwise eddies and cannot drift westward to the African east i.e. Mozambique and Madagascar coastal waters. In this scenario, if the debris less buoyant, they would sink deeply, but if they more buoyant than water, they would float temporarily.

Consequently, the turbulent dynamic instability with 50 km/ day of the large southern Indian gyre with a width of 100 km would pull down the debris to depth more than 3,000 m through the Southern Indian Ocean (Marghany, 2017).

Figure 13. Mean current flows from November to December 2014

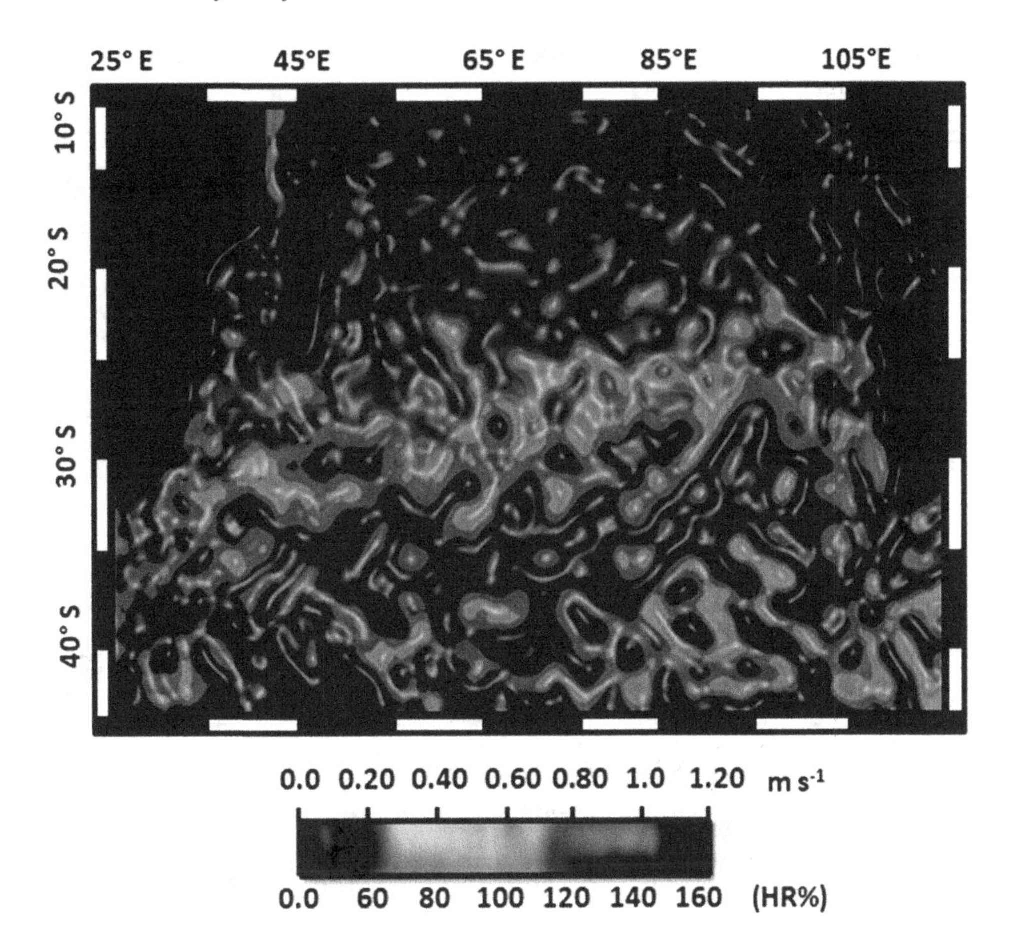

This can be approved with the maximum value of the hit-rate (*HR)* of 160% which is happened with a extreme rapidity of eddy current of 0.65 m/s (Figures 13.13-13.18). Albeit furthermost debris is completely buoyant and are detached by wind and ocean currents, noteworthy quantities develop neutrally buoyant and submerefed deeply and merged into sediments. Further, biofouling can play a important responsibility in governing a negative buoyancy.

Micro- and microorganisms are well known ocean surface pollution.

Biofilm formation is causing biofouling, that cultivates via four dissimilar phases; adsorption of dissolved organic molecules, attachment of bacterial cells, attachment of unicellular eukaryotes and attachment of larvae and spores. Microbial biofilms can successively cause the connection of definite invertebrates and algae, which growths the levels of biofouling.

In these circumstances, the debris that washed in Réunion Island is not the part of MH370. This is attributed to that the microbial biofilms are not keystone features cover debris. The MH370 debris, subsequently, would sink deeply at 1500 water depths within less than a few months as described in the previous chapter. In this concern, a several oceanographic processes deliver the debris deeply at 1500 m water depth or more. These processes comprise dense shelf water flowing, offshore convection and saline subduction. All these dynamic processes cause vertical and horizontal transfers of large volumes

Figure 14. Mean current flows from January to March 2015

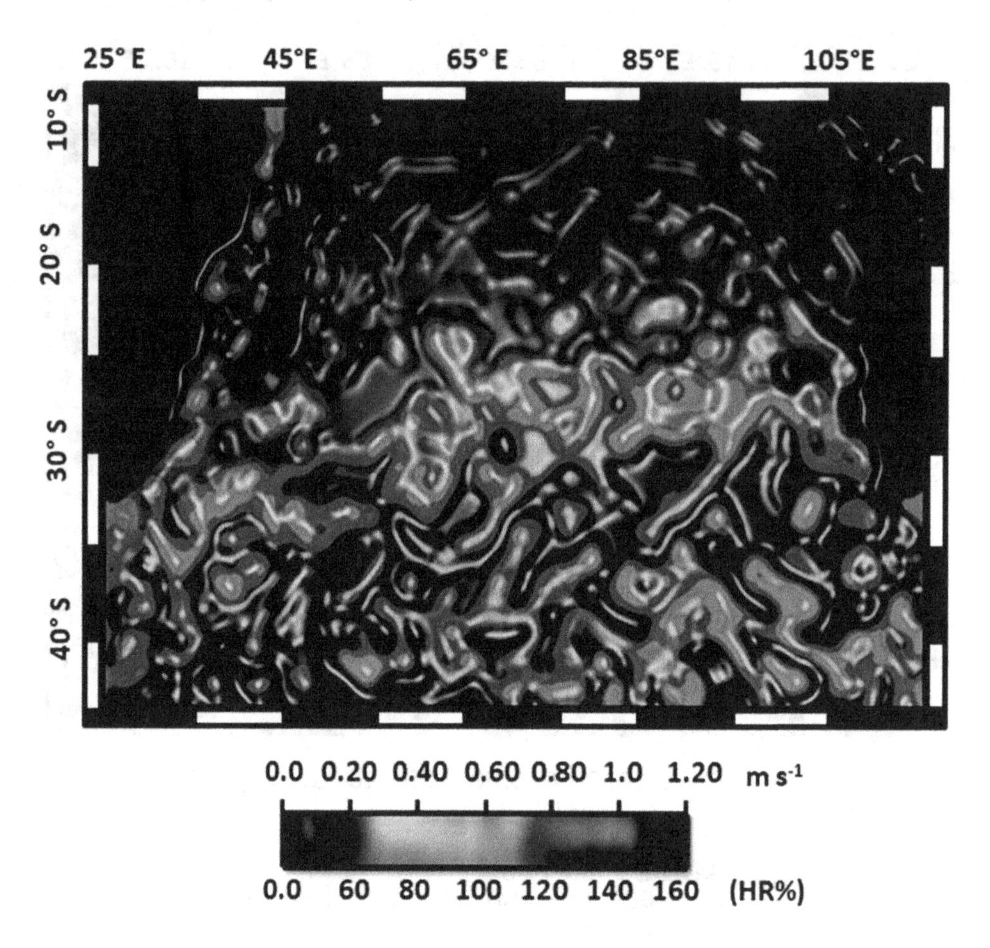

of particle-loaded waters, containing a full spectrum of grain sizes (from sand to clay), as well as litter and contaminants, from shallow ocean layers and coastal regions to deeper ones, with submarine canyons acting as preferential conduits. Submarine topographic features may also develop downwelling streams and growth the retaining of debris at precise positions for example Taylor columns over seamounts. Debris fragments are also influenced by advection and, more generally, circulation patterns. Nonetheless, this work does not agree with the study of Trinanes et al., (2016). The trajectory movements of debris were introduced by Trinanes et al., (2016) was grounded on the surface movement effects. They did not consider the influence of the vertical mixing process. Moreover, they did not accurately inspect the historical debris, which is positioned in dissimilar sites such as Mozambique, Reunion Island, Madagascar coastal waters, and Mossel Bay, South Africa. Accordingly, Trinanes et al., (2016) have itemized 6 debris, which has raised the critical question: where the remain of other debris?

Recent studies and Trinanes et al., (2016) exposed that the drift of sea surface current can correspondingly carry the MH370 debris in coastal water of Perth, Australia with May 2014 (Figure 10). Conversely, the most appealed debris is located on the western boundary of the southern Indian Ocean.

Lastly, the Pareto optimization demonstrated 60% of debris would sink deeply at 1500 m water depth close to Perth (Figure 19). This is confirmed with the highest cumulative percentage of 95%. In fact,

Figure 15. South Java strong current from March to May 2014

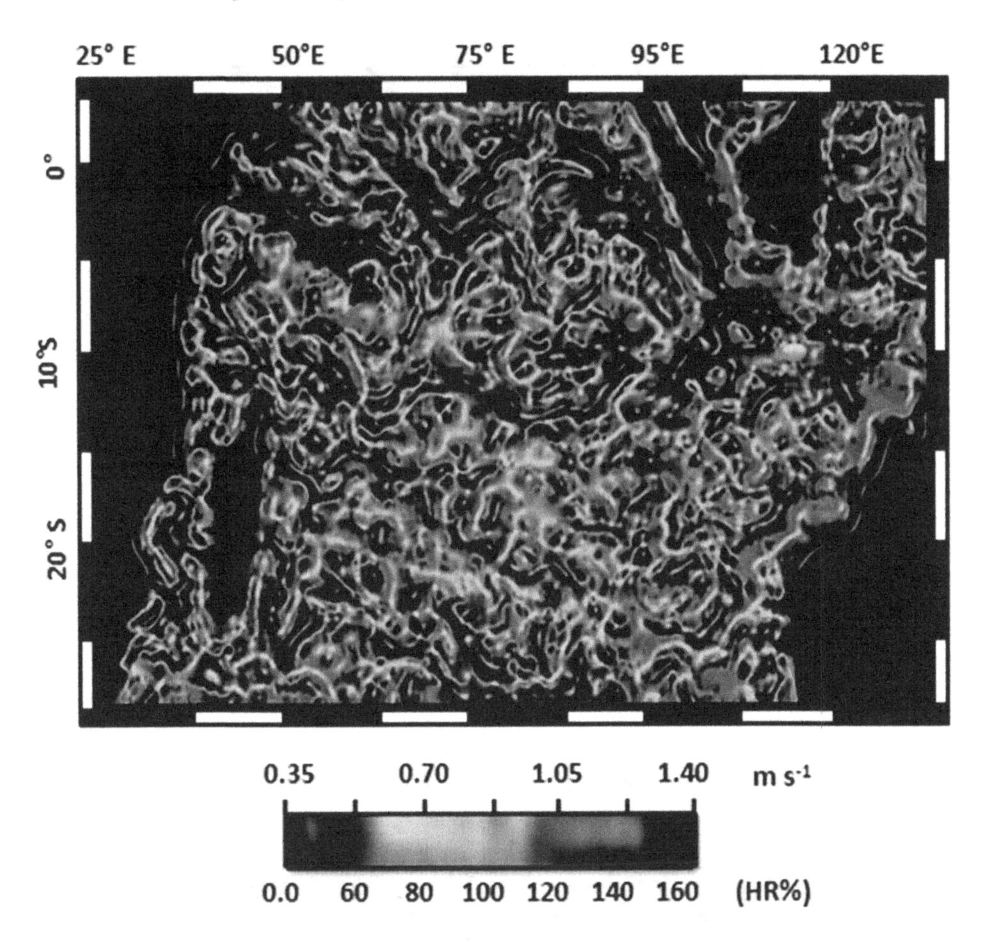

the turbulent flows across the southern Indian ocean is corestone of sinking the debris deeply across the 1500 m water depth or more.

CONCLUSION

This chapter concluded the difficulties of the existence of the debris along Mozambique, Reunion Island, Madagascar coastal waters, and Mossel Bay, South Africa based on the Pareto optimization. In this regard, there is no evidence verifiying the presence of debris either from remote sensing data or ground search diagonally through the southern Indian Ocean. In this sense, the MH370 never went missing in the Indian ocean close to Perth, Australia. In the various circumstances of the ditching angle, the MH370 must be broken down into several pieces through the water surface and column. The ambiguity of debris recognition by visual clarification of optical data which contains heavy cloud covers can predict that the MH370 never ever reach offshore of Perth, Australia.

The next chapter will answer the critical question: Did the MH370 ditch into the southern Indian Ocean? The next chapter will attempt to answer also why the southern Indian Ocean cannot be the last destination of MH370.

Figure 16. South Java strong current from June to Augst 2014

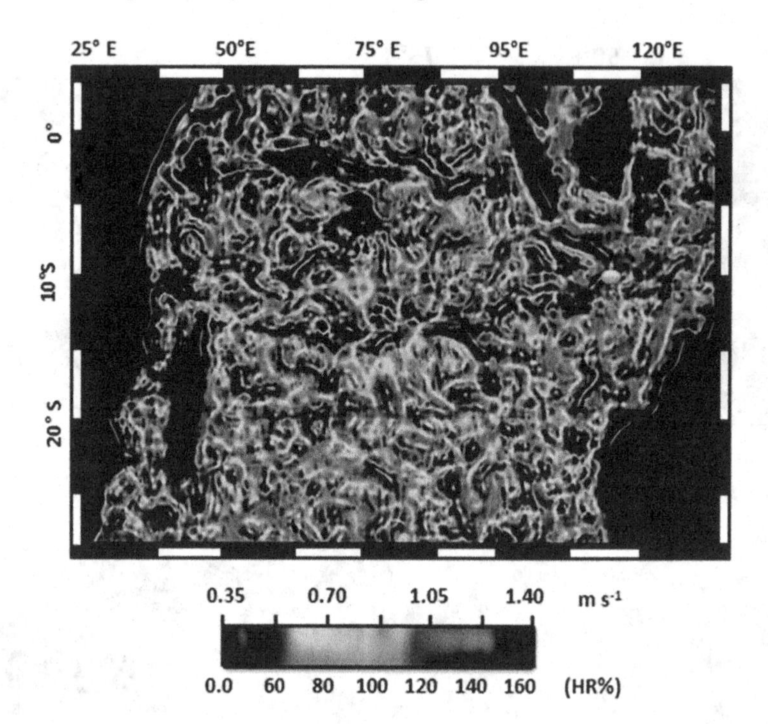

Figure 17. South Java strong current from September to October 2014

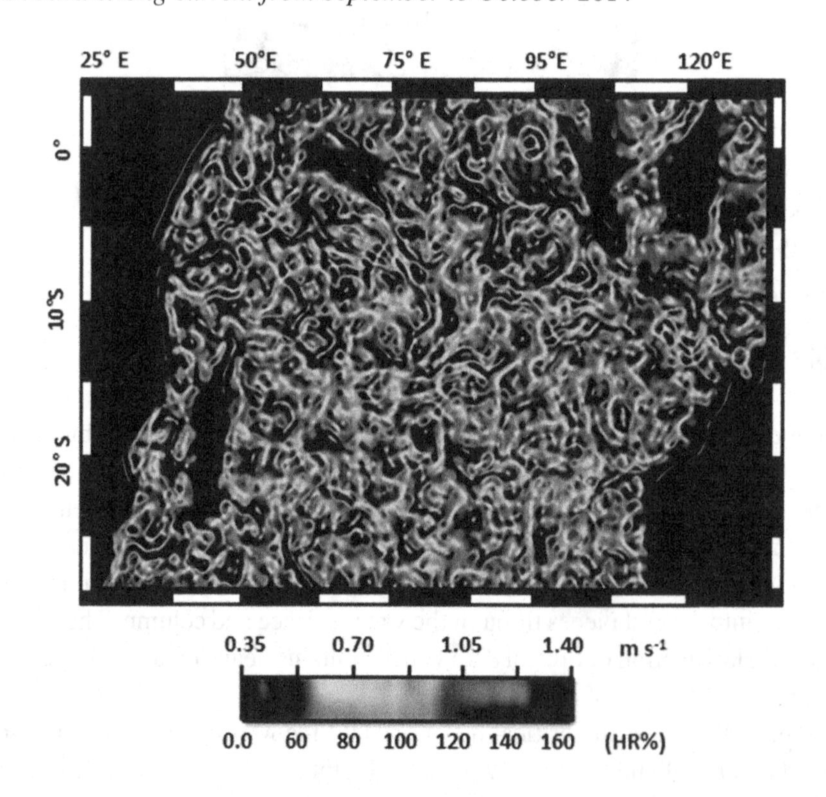

Figure 18. South Java strong current from November to December 2014

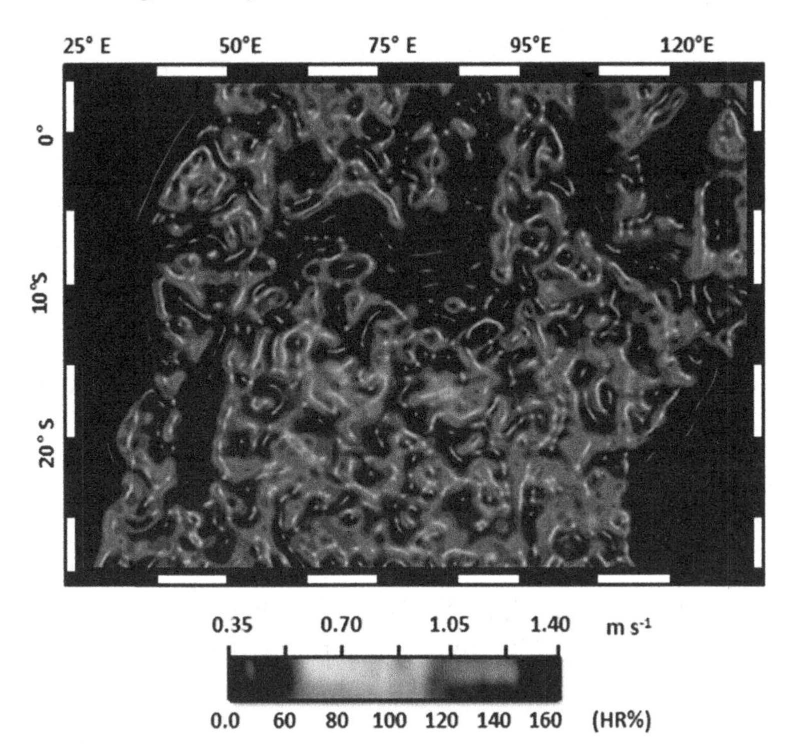

Figure 19. Pareto optimization for debris concentration in water depth

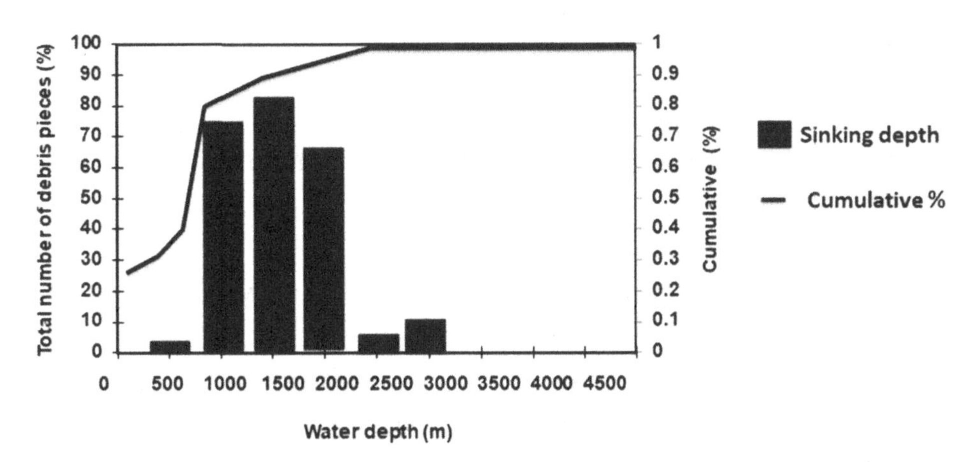

REFERENCES

Anderson, S. J., Edwards, P. J., Marrone, P., & Abramovich, Y. I. (2003, September). Investigations with SECAR-a bistatic HF surface wave radar. In *2003 Proceedings of the International Conference on Radar (IEEE Cat. No. 03EX695)* (pp. 717-722). IEEE. 10.1109/RADAR.2003.1278831

Blaha, J., & Lunde, B. (1992). Calibrating altimetry to geopotential anomaly and isotherm depths in the western North Atlantic. *Journal of Geophysical Research. Oceans*, *97*(C5), 7465–7477. doi:10.1029/92JC00249

Fu & Cazenave (Ed.). (2000). *Satellite altimetry and earth sciences: A handbook of techniques and applications*. Elsevier. doi:10.1016/S0065-2687(08)60442-2

Ghil, M., & Malanotte-Rizzoli, P. (1991). Data assimilation in meteorology and oceanography. [). Elsevier.]. *Advances in Geophysics*, *33*, 141–266.

Glenn, S. M., Porter, D. L., & Robinson, A. R. (1991). A synthetic geoid validation of Geosat mesoscale dynamic topography in the Gulf Stream region. *Journal of Geophysical Research. Oceans*, *96*(C4), 7145–7166. doi:10.1029/91JC00078

Hu, J., Kawamura, H., Hong, H., Kobashi, F., & Wang, D. (2001). 3≈ 6 months variation of sea surface height in the South China Sea and its adjacent ocean. *Journal of Oceanography*, *57*(1), 69–78. doi:10.1023/A:1011126804461

Hwang, C., & Chen, S. A. (2000). Fourier and wavelet analyses of TOPEX/Poseidon-derived sea level anomaly over the South China Sea: A contribution to the South China Sea Monsoon Experiment. *Journal of Geophysical Research. Oceans*, *105*(C12), 28785–28804. doi:10.1029/2000JC900109

Ikeda, M. (1995). Mesoscale Variability revealed with sea surface topography measured by altimeters. CRC Press, Inc.

Imawaki, S., Uchida, H., Ichikawa, K., & Ambe, D. (2003). Estimating the high-resolution mean sea-surface velocity field by combined use of altimeter and drifter data for geoid model improvement. In *Earth Gravity Field from Space—From Sensors to Earth Sciences* (pp. 195–204). Springer. doi:10.1007/978-94-017-1333-7_16

Liska, R., & Wendroff, B. (1998). Composite schemes for conservation laws. *SIAM Journal on Numerical Analysis*, *35*(6), 2250–2271. doi:10.1137/S0036142996310976

Marghany, M. (2009). Volterra–lax-wendroff algorithm for modelling sea surface flow pattern from jason-1 satellite altimeter data. In *Transactions on Computational Science VI* (pp. 1–18). Springer. doi:10.1007/978-3-642-10649-1_1

Marghany, M. (2014, October). Developing genetic algorithm for surveying of MH370 flight in Indian Ocean using altimetry satellite data. In *35th Asian conference of remote sensing, at Nay Pyi Taw, Mynamar* (pp. 27-31). Academic Press.

Marghany, M. (2015, October). Intelligent optimization system for uncertainty MH370 Debris Detec-tion. In *36th Asian conference of remote sensing, acrs2015. ccgeo.info/proceedings/TH4-5-6.pdf*

Marghany, M. (2017). Multi-Objective Evolutionary Algorithm for Mh370 Debris. *Ann Mar Biol Res*, *24*(1), 1020.

Marghany, M., Mazlan, H., & Cracknell, A. (2008). Volterra algorithm for modeling dea surface current circulation from satellite altimetry data. In O. Gervasi, B. Murgante, A. Lagana, A. Taniar, Y. Mun, & M. Gavrilova (Eds.), Computational Science and Its Applications – ICCSA 2008 (pp. 119–128). Springer.

Martins, C. S., Hamann, M., & Fiúza, A. F. (2002). Surface circulation in the eastern North Atlantic, from drifters and altimetry. *Journal of Geophysical Research. Oceans*, *107*(C12), 10–11.

Moore, A. M. (1991). Data assimilation in a quasi-geostrophic open-ocean model of the Gulf Stream region using the adjoint method. *Journal of Physical Oceanography*, *21*(3), 398–427. doi:10.1175/1520-0485(1991)021<0398:DAIAQG>2.0.CO;2

Reverdin, G., Fieux, M., Gonella, J., & Luyten, J. (1983). Free drifting buoy measurements in the Indian Ocean equatorial jet. In *Elsevier Oceanography Series* (Vol. 36, pp. 99–120). Elsevier.

Robinson, I. S. (2004). *Measuring the oceans from space: the principles and methods of satellite oceanography*. Springer Science & Business Media.

Rugh, W. J. (1981). *Nonlinear system theory*. Johns Hopkins University Press.

Saraceno, M., Strub, P. T., & Kosro, P. M. (2008). Estimates of sea surface height and near-surface alongshore coastal currents from combinations of altimeters and tide gauges. *Journal of Geophysical Research. Oceans*, *113*(C11).

Serafino, G. (2015, April). Multi-objective aircraft trajectory optimization for weather avoidance and emissions reduction. In *International Workshop on Modelling and Simulation for Autonomous Systems* (pp. 226-239). Springer. 10.1007/978-3-319-22383-4_18

Trinanes, J. A., Olascoaga, M. J., Goni, G. J., Maximenko, N. A., Griffin, D. A., & Hafner, J. (2016). Analysis of flight MH370 potential debris trajectories using ocean observations and numerical model results. *Journal of Operational Oceanography*, *9*(2), 126–138. doi:10.1080/1755876X.2016.1248149

White, W. B., Tai, C. K., & Holland, W. R. (1990). Continuous assimilation of simulated Geosat altimetric sea level into an eddy-resolving numerical ocean model: 2. Referenced sea level differences. *Journal of Geophysical Research. Oceans*, *95*(C3), 3235–3251. doi:10.1029/JC095iC03p03235

Wyrtki, K., Bennett, E. B., & Rochford, D. J. (1971). Oceanographic Atlas of the International Indian Ocean expedition. Academic Press.

Chapter 14
Why Flight MH370 Has Not Vanished in the Southern Indian Ocean

ABSTRACT

This chapter delivers the final conclusions that were raised due to the critical question of why MH370 has not ended up in the Southern Indian Ocean. In this view, the application of the multiobjective genetic algorithm is implemented to explore the final destination of MH370. The results show that the MH370's last destination is not near the coastal water of Perth, Australia as obtained by using a multiobjective algorithm. In this understanding, the Pareto optimization has allowed the author to see the impact of the Southern Indian Ocean dynamics on the MH370 debris trajectory movements. Moreover, the Pareto front verified that the found fragments do not belong to MH370 fuselage. There is a 95% confidence level that the flight found in Cambodia is not MH370. Finally, MH370 has been hijacked and driven to Diego Garcia as it is a short route from the departure point at International Malaysia Airport, Kuala Lumpur (KLIA).

INTRODUCTION

Significant questions are raised over six years from time to time: where is the MH370 and How did it end up? It is precisely six years to the day when MH370 vanished from the radars while flying from Kuala Lumpur to Beijing, China. Notwithstanding the long passageway of time, the world is immobile clueless on its final vanishing destination. All the advanced search technologies have miscarried to explore the resting place of MH370 fuselage or debris. In this understanding, an extensive multinational exploration initiated immediately after the MH370 went vanishing, and succeeding a particular examination expending state-of-the-art underwater equipment months later correspondingly miscarried to reveal sufficient confirmation to patch simultaneously the mystery puzzle.

DOI: 10.4018/978-1-7998-1920-2.ch014

Previous chapters have proved logically the recovery of some fragments, for instance, wings and fuselage in waters off eastern Africa are not MH370 remains. In this sense, more critical questions are raised: why the MH370 deviated off far from its original route.

These critical questions cannot be answered through conspiracy theories that flooded the internet six years ago. One of these conspiracy theories is the plane was taken over remotely in a bid to foil a hijacking. Besides, MH370 perhaps was landed gently in the water and sunk in mostly one piece, which is consistent with the debris found. However, as proved previously, the debris are not the MH370 fragments.

Another theory is that the search was in the wrong area and that investigators should be looking north of Malaysia – perhaps the plane crashed, or perhaps it landed and was hidden somewhere. However, this theory was rejected by Inmarsat. In July 2018, push back on theory was stated that the batteries and fruit in the plane's cargo had somehow formed an explosive mixture that brought down the plane.

More bizarre speculations comprise a remote cyber-hijacking, the plane being shot down by perhaps the US military, the idea that Russian President Vladimir Putin knows the plane's location, a mystery extra passenger taking control of the plane, a Bermuda Triangle-style area causing the plane's disappearance, and North Korea taking the plane. These bizarre speculations were not logically address as Bermuda Triangle-Style historically has never been shown in the South China Sea. North Korea and Russia have huge collaborations with China and not go further to shoot-down a flight as the majority of passengers included were Chinese. Other farfetched theories are also blamed the CIA, Israel, and aliens – all outdated bogies for conspiracy theorists.

The surprising issues that the Malaysin professors never deliver any logical clues and theory to explain how did MH370 end up near Perth, Australia? In this view, Malaysian oceanographers cannot develop any ocean dynamic model to track the trajectory movement of debris either in the South China Sea or the Indian Ocean. As they rely on the consequences that delivered by Australian scientists.

Even though, an advanced mathematical method of Bayesian for the search of MH370 book was published by Springer, not come up to scratch to determine the precise last destination of MH370.

This chapter is mounded to accomplish the logical debriefing: did the MH370 vanish in the Southern Indian Ocean? The intelligible decision can be drawn back to physical evidence of Inmarsat satellite data and its ambiguities. Likewise, the debris detections in optical satellite data and the effect of Southern Indian Ocean dynamic components, i.e. Rossby wave, velocity potential, significant wave height pattern and surface current patterns can exploit to express the rational answer.

GENETIC ALGORITHM ENCODING FOR MH370 VANISHING

Genetic algorithms (Gas) are constructed on machinery inattentive from population genetics. Local minima and sub-optimal results are avoided through genetic algorithms. To describe how Gas can be implemented for solving the mystery of MH370 optimization, let us assume a population of N individuals p_i. In this view, each p_i is represented by a chromosomal string or (string) of L *allele* values. In other words, *Allele* denotes to the detailed value of a gene's position on the string. For instance, assume the function f identifies over a arrange of the real numbers $[r_1,r_N]$. In this understanding, each $p_i \in [r_1,r_N]$, which represents a candidate solution to conclude either MH370 vanished in the Southern Indian Ocean or not: max $f(p)$ for $p_i \in [r_1,r_N]$. A depiction for respectively p_i could be a binary bit string, implementing standard binary encoding. In this regard, respectively gene position or allele value, constraints on either 0 or 1, which is a function of p_i. In this understanding, the searching for space $[r_1,r_N] \in \mathbb{R}$ to maximize the output of "max $f(p)$ ". In this circumstance, f refers to fitness measure or fitness function and represents the circumstances in which the optimized solutions are arbitrated (Akin and Lagerwerff 1965);Mackay et al.,2003;Smojver and Ivančević, 2010).

The initial p_i is commonly picked up randomly. The moment, the population is clued-up the genetic algorithm's evolutionary set can create. The first phase of the genetic set is the evaluation of the individualization of the populated members. Consistent with the above perspective, for all population members $(1 \leq i \leq N)$, every $f(p_i)$ is weighed.

Using the same sets of data it is also possible to evolve rules to dichotomize either MH370 ended in the Southern Indian Ocean or not. In this concern, a logic genetic algorithm rules are established to conform to a wide-ranging stencil (Figure 1).

Figure 1. A generalized algorithm for investigating ditching of MH370 in the Southern Indian ocean

Rule 1:
{
if flaperon fell > 500 km
Then MH370 ditching > 500 km
Or MH370 = 0 i.e. exploded in air
If Captain Azhari committed suicide
MH370 route > MH370 route to Perth
If the frequency of search =0 i.e. no exist of MH370 fuselage,
}
If probability P <0, then flag suspected case of nonexistent of MH370 in the Southern Indian Ocean.

Rule 1:
{
if flaperon fell > 500 km
Then MH370 ditching > 500 km
Or MH370 = 0 i.e. exploded in air
If Captain Azhari committed suicide
MH370 route > MH370 route to Perth
If the frequency of search =0 i.e. no exist of MH370 fuselage,
}

If probability P <0, then flag suspected case of nonexistent of MH370 in the Southern Indian Ocean.

For rule 1 the various parameters, $p_1, p_2, \ldots, p_N$ and weighted function $W_1, W_2, \ldots, W_N$. The score associated with a rule is generated by counting the probability values for each multiobjective occurrence. In the circumstance, the genetic algorithm generates rules of probability changes which have to create based on the biological operators (Figure 2).

The above-mentioned regulations comprise of several conditional clauses to investigate the vanishing of MH370 in the Southern Indian Ocean. In this view, two parents' rules, are chosen for the crossover, which produces two offspring guidelines through swapping segments of the regulations related to particular attributes. Though, conditions and attributes are deliberated as a single element within the crossover processes.

Figure 2. Two rules are chosen for crossover

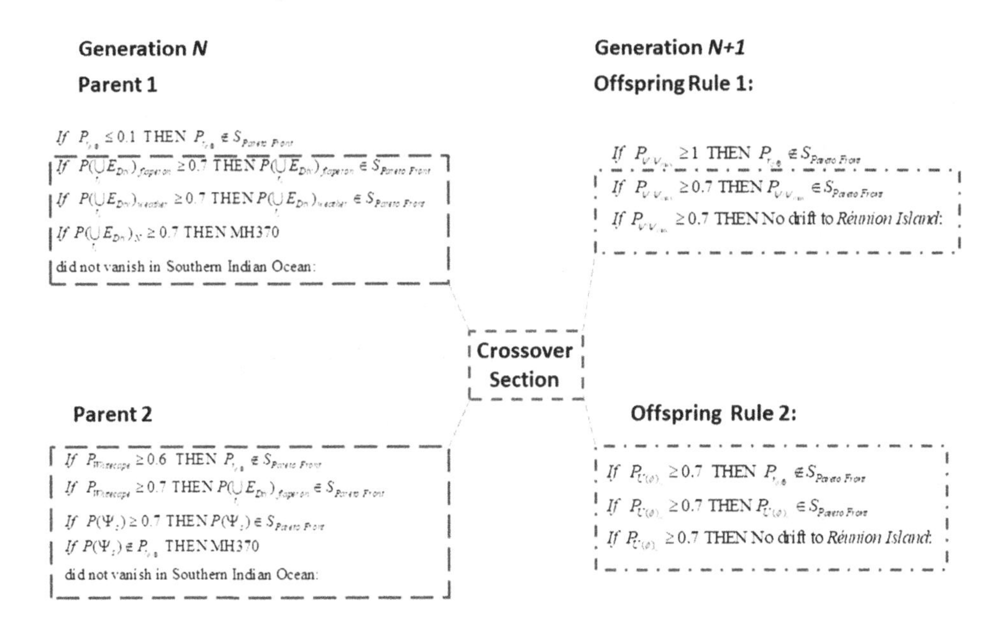

PARETO OPTIMAL SOLUTION FOR MH370 VANISHING

Indicating an approach in the appearance of a couple of targets frequently entails the navigation of trade-offs amongst the objectives. In other words, exclusive options may additionally perform better than others with recognizing specific objectives. For these sorts of problems, decision-makers are frequently fascinated by characterizing the complete set of non-dominated options that is, for, which no other solution performs excellently across all targets.

Consequently, non-dominated solutions are also termed as a *Pareto-optimal*, and the set of objective quantities which, are matched to Pareto-optimal solutions. In other words, it is known as the *Pareto front* (Figure 3). The determination of the Pareto front before decision-making permits decision-makers to determine the optimal solutions, which is based on a considerate of the relevant trade-offs between different objective functions and MH370 suspected search zone. This procedure of optimization is identified as *a posteriori* decision-making. On the other hand, it is considered as consequent to *a priori* decision-making, in which inclinations are quantified before the optimization process(Salman et al., 2007).

In this particular, multi-objective is cogitated. The first is the probability of flaperon existence on the geographical location of 22°S and 112°E (Figure 4). Second, the probability of fuselage of MH370, which could be between 33°S and 40°S and 87°E to 95°E, and the third probability is ocean dynamic impacts on the MH370 debris. These probabilities should satisfy:

1. If maximum flaperon probability $(P_{\max})_{flaperon} \geq 0.6$ and the maximum of individual debris probability $(P_{\max})_{debris} \leq 0$, Then $(P_{\max})_{flaperon} \notin S, (P_{\max})_{debris} \notin S$ where S denotes the Pareto front flow pixel.
2. If $(P_{\max})_{debris} \notin (P_{\max})_{debris} \notin P_{U(\phi)_n}$ then $(P_{\max})_{debris} \notin (P_{\max})_{debris} \notin (P_{\max})_{Réunion\ Island}$.

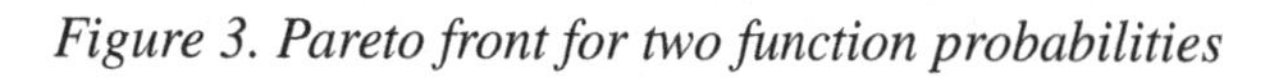

Figure 3. Pareto front for two function probabilities

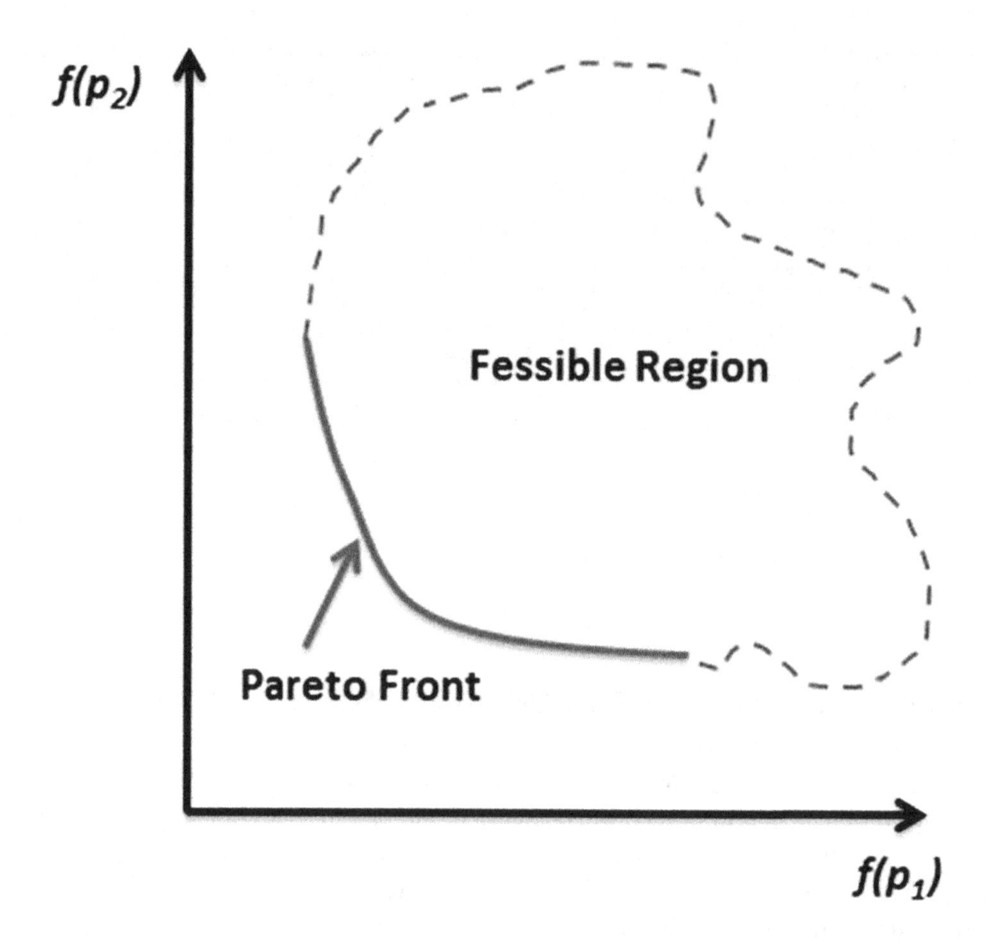

Figure 4. The exact location of flaperon and search area

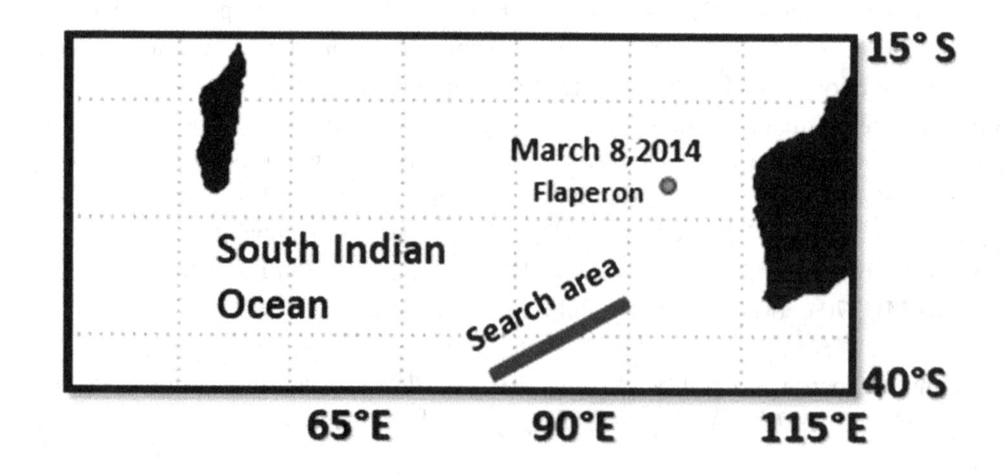

This denotes that the debris or flaperon could not drift towards *Reunion Island* under the impact of current $P_{U(\phi)_n}$ and ocean dynamics.

The high probability of existing flaperon and debris along the search area does not include in the Pareto front (*S*). In fact, both probabilities are located away from the Pareto boundary. In addition, the probability of fuselage existing is zero (Figure 5). Logically, if the flaperon fell down, the MH370 must suffer dynamic instabilities as long under autopilot control. In this regard, the MH370 must chaotically rotate around its flying axes. In this view, it must unsystematically swap in three dimensions: *yaw*, nose left or right about an axis running up and down; *pitch*, nose up or down about an axis running from wing to wing; and *roll*, rotation about an axis running from nose to tail (Chapman et al., 2010). This leads either exploration of flight in the air or ditching obliquely into the ocean with distance must be less than 500 km away from the flaperon drop location. In other words, the MH370 must ditch into water within less than 1 km distance post the flaperon fell down in of 22°S and 112°E. This means the search area cannot be located between 33°S and 40°S and 87°E to 95°E.

Figure 5. Pareto front for MH370 debris flaperon and debris along the search area

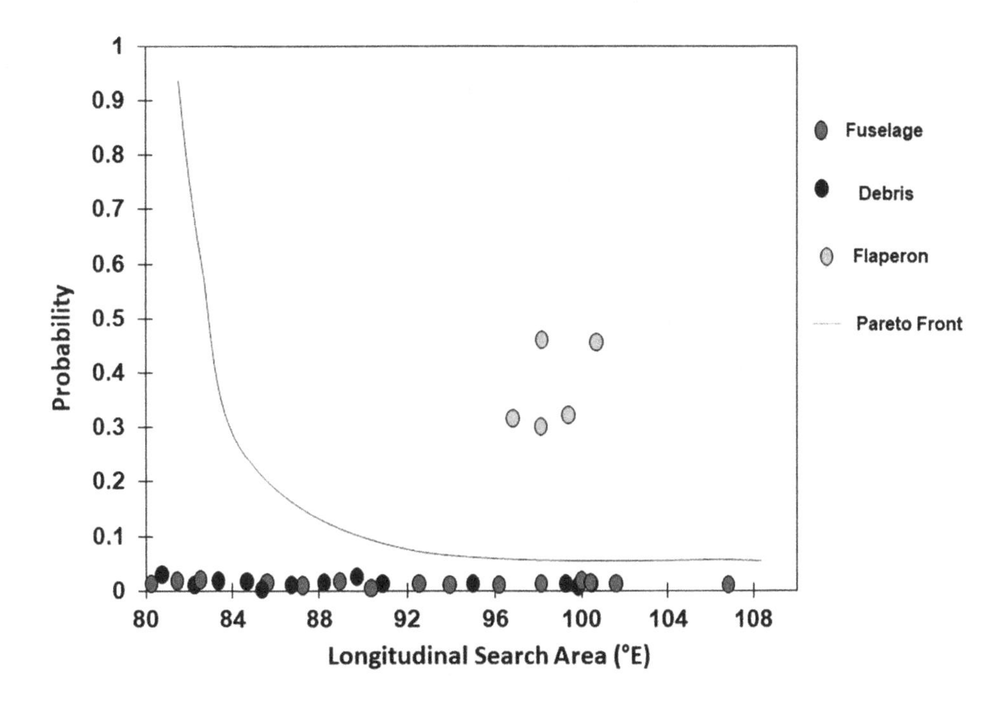

In this circumstance, the plenty of debris must float on the surface of the Indian Ocean and must be well observed. However, most of the debris detected by remote sensing, optical sensors are identified as man-made and wave whitecaps (Figure 6).

Figure 6. 3-D of whitecaps from GF-1 satellite data

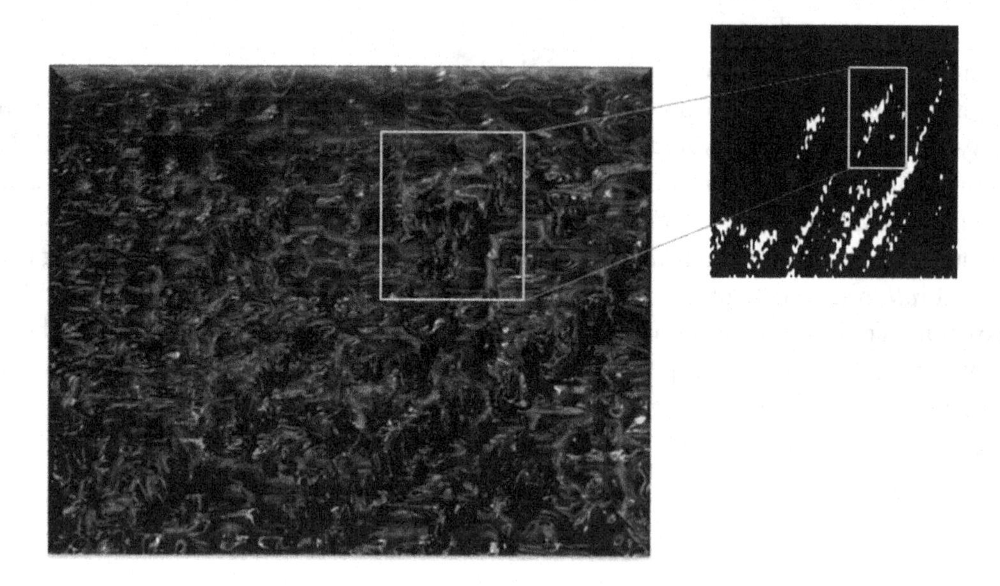

Figure 7. Pareto front for flaperon trajectory movements

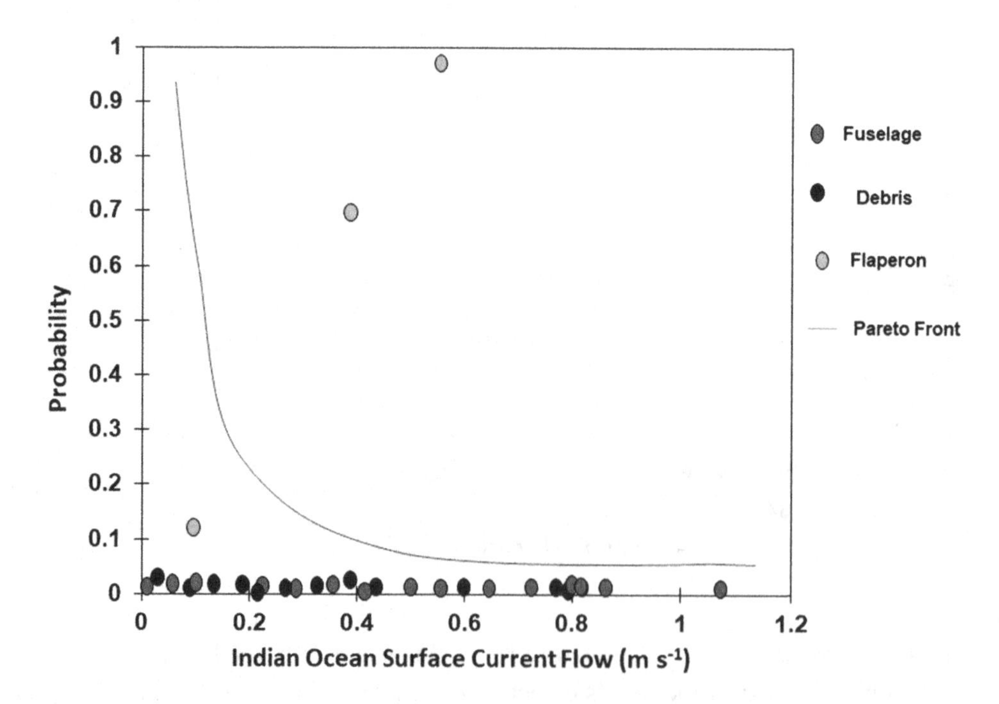

WHY MH370 FLAPERON CANNOT DRIFT TOWARDS RÉUNION ISLAND

The maximum probability of flaperon drifting towards Réunion Island under the impacts of surface current drift does not belong to the Pareto front (Figure 7). Nevertheless, the probability of any debris of fuselage to locate along the Réunion Island is zero. In this understanding, the area where the flaperon fell is dominated by eddy flows (Figure 7). The speed of these eddy flows is approximately 0.6 m s^{-1}. These eddies can stir the flaperon and other debris to a depth of less than 2000 m. Besides, this area is reachable by the Leeuwin Current (Figure 8).

Figure 8. Monthly average of Leeuwin Current

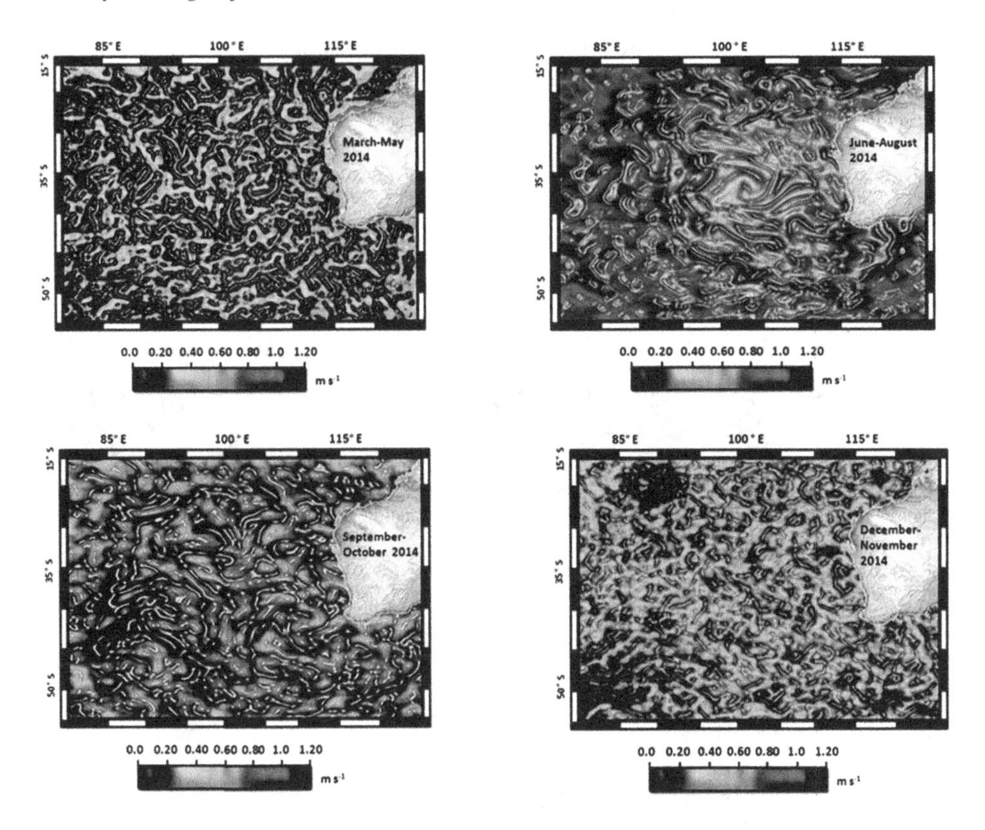

Figure 9. A Pareto front for sinking probability through months

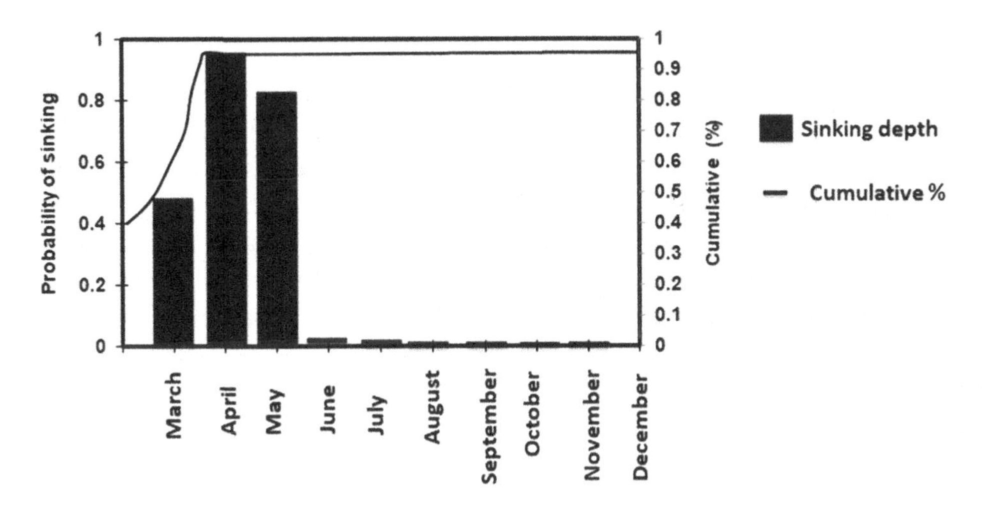

Consequently, the poleward flowing Leeuwin Current along the west coast of Australia is experienced in all the months, however, it spectacles seasonal and spatial variations (Du et al., 2007). The strong flow acquired in April and July with a speed velocity of more than 0.6 m s^{-1}. In this view, the flaperon and debris cannot escape the impacts of the Leeuwin Current in addition to the oscillation of significant wave heights and Rossby waves. This proves that the flaperon and debris must sink with April 2014 below the ocean surface of approximately 1500 m depth with the highest probability of 0.8 (Figure 9).

Figure 10 confirms the sinking possibilities due to shear instability flow occurred due to the Leeuwin Current and offshore ocean current, which they created mesoscale eddy flow through the water column. This can bring the flaperon and debris to the water depth of 1500 m across distance less than 150 km (Rea et al., 1990).

Figure 10. Shear instabilities through eddy flows

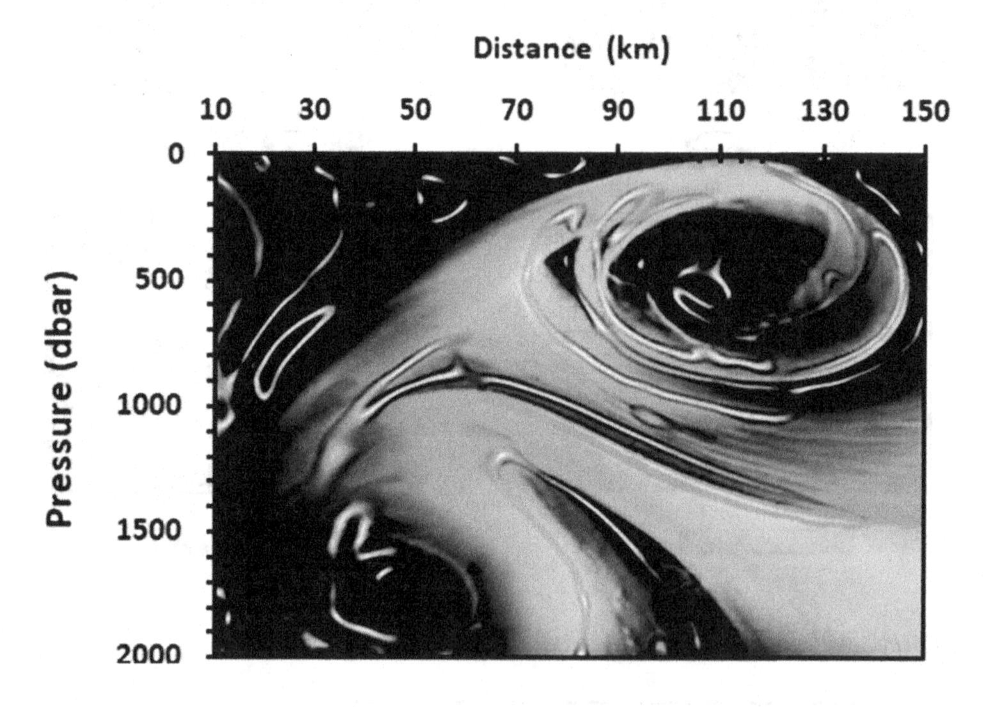

Consequently, if the flaperon escaped from the impact of the Leeuwin Current, it cannot survive under the impacts of eddy shear instability in the coastal waters of Réunion Island. There is evidence of the existence of eddies in the vicinity of the islands. Consequently, it illustrates the high degree of variability of the circulation around Réunion Island (Figure 11). In this circumstance, the flaperon must be trapped in a small-scale of approximately 50 km from the district of Réunion Island. The eddy shear instabilities lead the sinking of flaperon in the water depth of 5500 m offshore of the island (Sundance, 2018).

This proves that the flaperon could not be drifted and located on the coastline of Réunion Island. The eddy shears become more strong along the Mozambique channel. The main circulation feature is the South Equatorial Current (SEC) that is most intense at the surface, but reaches up to 1400 m depth, and forms the northern boundary of the basin-wide anticyclonic Indian subtropical gyre which carries 50–55 Sv westward between 10°S and 20°S. On reaching Madagascar, these cores contribute to the Northeast

Figure 11. Eddy shear instabilities along the Réunion Island

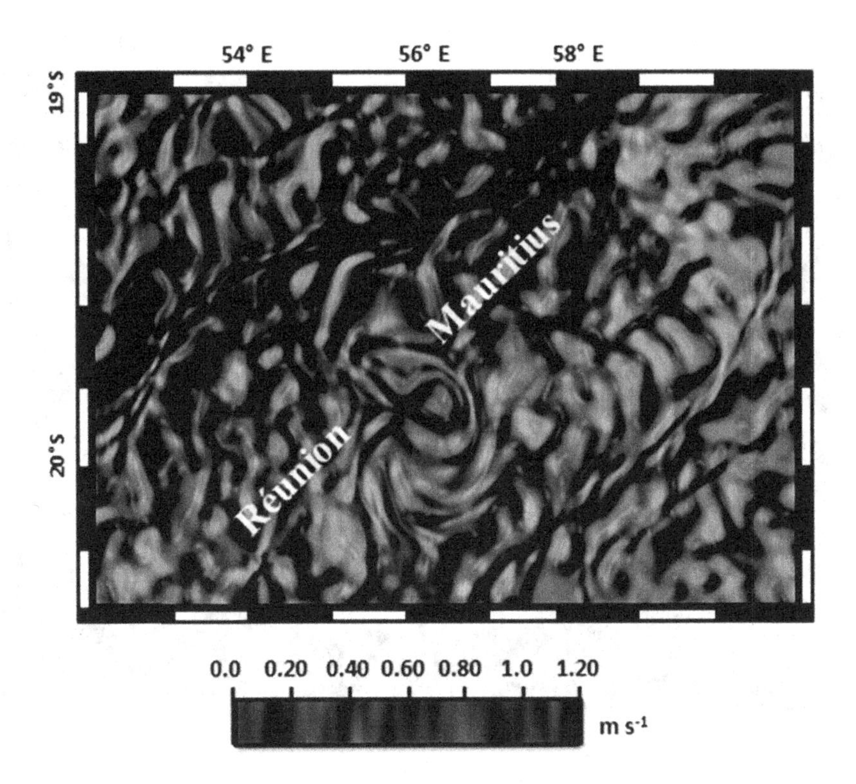

and Southeast Madagascar Currents (NEMC and SEMC) (Frota et al., 2013). The NEMC transports about 30 Sv northward around Madagascar, supplying the Mozambique Channel and the East African Coastal Current. The SEMC transports about 20 Sv southward along with Madagascar. At the southern tip of Madagascar, part of the SEMC splits into a retroflects and feeds a shallow eastward jet identified as the South Indian Ocean Countercurrent (SICC). In other words, the southern Indian Ocean gyre's strongest flows are to the southwest of Madagascar, where they form the Agulhas current (Salman et al., 2007; Picard et al., 2018; Upadhyaya,2018). A relatively strong eddying flow in the Mozambique Channel between Madagascar and Africa connects the gyre's equatorial currents with the vigorous transports of the Agulhas system (Figure 12).

In this view, the smallest dimension of the Mozambique Channel allows the eddies to stir any object until it sinks a bottom. Consequently, The Mozambique Channel feeds the Agulhas Current, one of the strongest currents in the world. The Agulhas Current affects the route of tropical storms and contains warmness towards higher latitudes. In this understanding, if the debris escapes from the eddy flow in the Mozambique Channel, they must be governed by Agulhas Current. In this circumstance, the debris must be drifted to the Southern Atlantic Ocean (Figure 13) under the impact of the fast current of more than 1.8 m s^{-1}.

Figure 12. Gyres along the Mozambique Channel

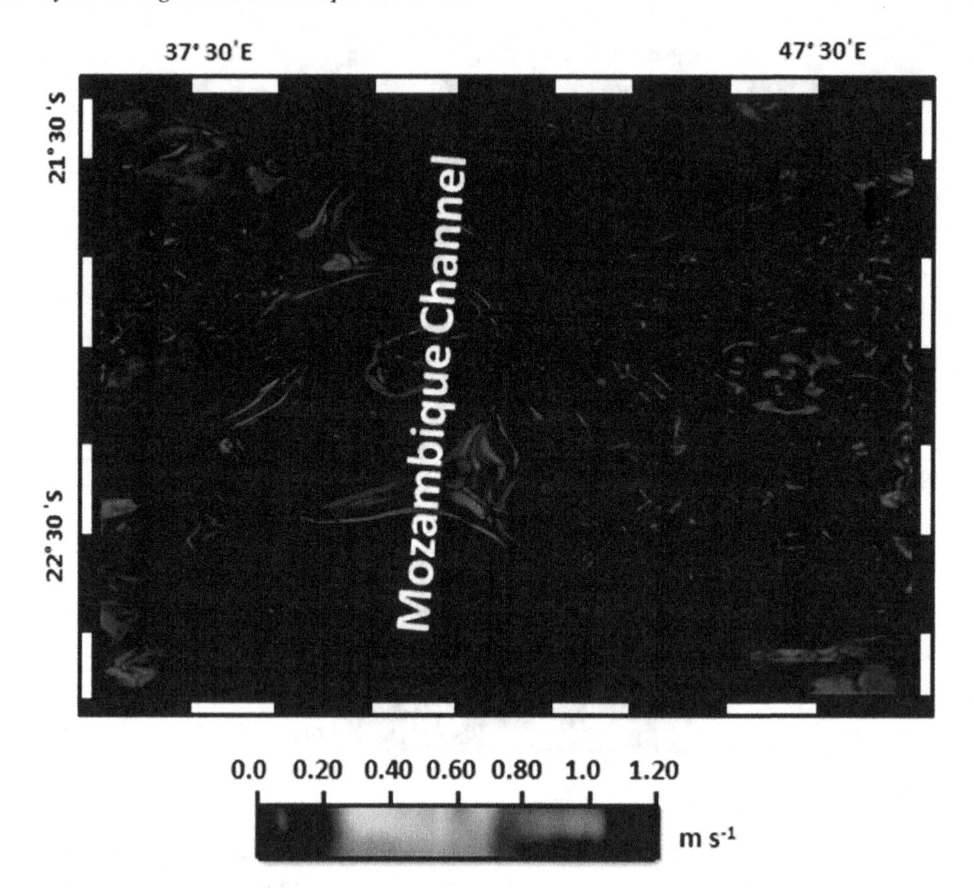

WHY MH370 DID NOT VANISH IN SOUTHERN INDIAN OCEAN

Consistent with the above perspective, the optimal answer regarding the highly technical mystery of MH370 vanishing can deliver. Simply organize the results of the dynamical Southern Indian Ocean against Inmarsat satellite data. In this view, the complicated dynamical Southern Indian Ocean circulation has a great weight against the principle that MH370 flew into the southern Indian Ocean. On the contrary, the Australian authorities deceptively comprehend this confirmation effectively than the media, for oil and gas exploration across the western shelf off Australia (Picard et al., 2018).

Since the previous chapters have proved by using multiobjective genetic algorithm why MH370 was not flown into the southern Indian Ocean, this chapter will conclude the following significant issues:

1. Captain Azhar and his copilot cannot commit suicides as they must fix a short flight route to end their life quickly. In this circumstance, Captian Azhar must vanish the MH370 to the nearest route which should be the Gulf of Thailand.
2. If the MH370 route was ended between the 33°S and 40°S and 87°E to 95°E, the fuel was not enough to carry a flight there.
3. If the flaperon fell down at approximately 22°S to 24°S, the MH370 cannot be ditched between the 33°S and 40°S and 87°E to 95°E i.e., 500 km away from the location of flaperon. Further, as

Figure 13. Agulhas current impact on MH370 debris

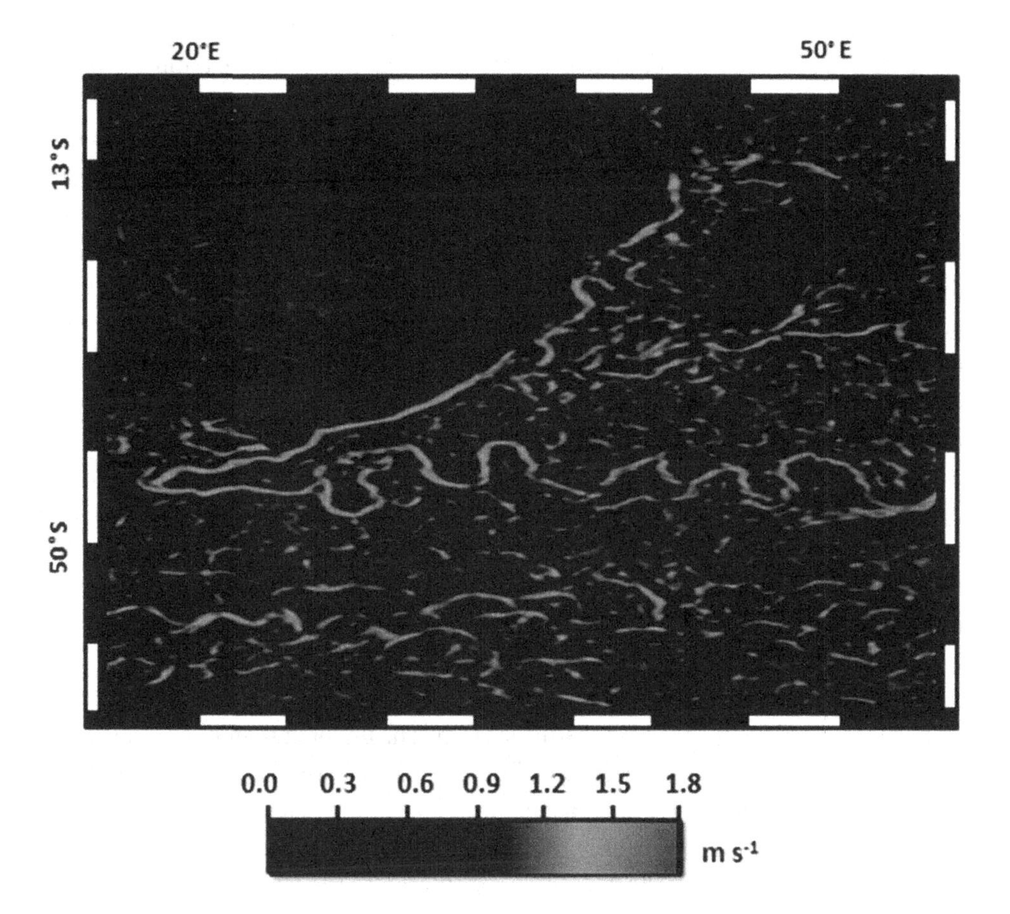

long the flaperon split away from the MH370, the flight must be span and could explode before the debris fell into the ocean.

3. As long the flight plunged into the Indian Ocean without any control, there must be broken into plenty of debris and the fuselage sunk deeply.
4. However, the wreckages are absent along the search zone determined by ATBS between 33°S and 40°S and 87°E to 95°E.
5. Operating Inmarsat communication satellite data provide uncertainties in determining the search area. This proves by the absences of wreckages and fuselage of MH370 debris. However, the Malaysian, Chinese and Australian governments then spent more than $150 million searching this vast, deep abyss without any sign of the MH370 was there. In practice, if the MH370 had turned south, it must have been there physically.
6. The ground radar coverage along the coastal waters of Australia must record some information regarding the route of MH370 especially when approaching the Australian Maritime boundary. Ground radar stations can't be shut down. Australian authorities must monitor its coastal waters over 24 hours due to an illegal immigrant.
7. the inconsistency of operation of the satellite data unit (SDU). In this view, before 18:25, it was turned off and then turned on to reconnect an Inmarsat satellite. In this contrast, the SDU

cannot be reoperated by anyhow. Consequently, this could be evidence that electrical system experts have driven the flight away from the range covers of Inmarsat satellite. In this regard, the aircraft's electrical system must tamper. This requires deeper examination by the experts and investigators.

8. The surface of flaperon was populated by barnacles designated that it had drifted somehow wholly submerged. In this view, the flaperon could not drift over the longer distance of thousands of kilometers till positioned into the French island of La Réunion. The area where the flaperon fell down is dominating by strong eddy shear instabilities due to the existence of the Rossby wave and massive wave power oscillation in addition to Leeuwin Current. These unique features of the dynamic ocean do not allow the flaperon to escape horizontally across the Indian ocean, but they stirred the flaperon till sink at the water depth of 1500 m within April or May 2014.
9. Drift models delivered by an arm of the Australian government named the CSIRO just considered upper surface current movements of the Indian Ocean without considering other dynamic parameters such as wave height, eddies, and Rossby wave, velocity potential, and vorticities. Even if CSIRO only considered upper surface current impacts, the debris cannot be drifted to the French island of La Réunion as it would wholly be submerged according to the above discussions.

These nine reasons create suspicion about the existence of MH370 in the Southern Indian Ocean. The Southern Indian Ocean has a complicated dynamic system. The understanding of this system proves that the flaperon and debris cannot get into La Réunion, Madagascar, and Mozambique channel. The oceanographer experts ignored that if the debris approached the Mozambique channel, it must feed into Agulhas Current. In this circumstance, the search area must be shifted towards the south of the Atlantic ocean.

This can effectively conclude that the MH370 never turned to the south and not vanish in the south of the Indian Ocean.

WHY MH370 SEARCH ZONE IS SIGNIFICANT

It is awkward to expend a massive quantity of millions just to explore the MH370 fuselage. It must countless benefits are elsewhere. The search area is considered as one of the important zones along with west Australia offshore. In this context, the search area is centered in a part of the Broken Ridge Large Igneous Province, which extends north from the Diamantina Escarpment and comprises extensive plains, ridges, rises, and seamounts. The south of the escarpment is a large trough (known as Diamantina Trench) and a series of spreading ridges, fracture zones (e.g., Geelvinck), deep fault valleys, and numerous presumably relict volcanoes (Upadhyaya, 2018).

On the word of Rea et al., (1990), sediments accumulated above the igneous basement until ~40 Ma, producing a sedimentary sequence dominated by chalks and limestones that is between 1 and 1.5 km thick. Furthermore, These provinces comprise iron and magnesium-rich rocks that were erupted from the Kerguelen hotspot between ~130 and ~40 Ma (Picard et al., 2018). In other words, the area is rich by sedimentations, which involve inorganic deposits. In this context, pressure and temperature affect the solubility of these substances. Besides, variations in the turbulence, current speed, and pressure also govern the total mass precipitations (Zhang et al., 2001). In other words, the degree of increment of saturation leads to supersaturated as a function of hydrostatic pressure increases, which equals 5000 dB. In this view, this can generate precipitation of excess salt (Dyer and Graham,2002; Collins and Jordan 2003; Chen et al.,2005). In this regard, the carbonate deposition at a depth exceeding 4000 to 4500 m is beneath the Calcium Compensation Depth (CCD), which is precluded. In this understanding, carbonates are sedimentary rocks that regularly originate in coincidence with shale. Carbonates, nonetheless, are shaped mainly from leftovers of marine life, predominantly shells, and bones, mutual with other minerals. Because of this, they are full of calcium and other compounds that lead to their classification: limestones, which contain calcium carbonate, and dolomites, which contain calcium magnesium carbonate. The spaces between them fused fragments are where oil and gas may be found.

Also, the regularly flat seafloor north of Broken Ridge reveals an old igneous surface covered by > 1.5 km of marine sediment (Akin and Lagerwerff 1965; Mucci and Morse 1983; Gardner et al., 2015). In this regard, are formed from magma within the earth, have larger grains and contain silicate minerals. The MH370 search zone contains the tectonic processes, which contributed to hotspot volcanism and plate tectonics. This could lead to generating igneous reservoirs, which are considered as secondary reservoir targets for oil and gas exploration compared to the more common sandstone and carbonate reservoirs (Surgent,1974; Collins and Jordan 2003; Chen et al.,2005).

The main dominated geological features in the MH370 search area are the existence of ridges, which are mainly spreading fabric or detachment blocks. In this circumstance, the existence of detachment blocks leads to

creating a reservoir into which hydrocarbons will accumulate, and to seal it - such traps for oil and gas are generally found at depths of 2 000 to 4 000 m below the seabed. Conversely, the high hydrostatic pressure over the times also allows cracking of the organic compounds into oil and gas (Collins and Jordan 2003).

Another evidence about regarding oil and gas existing along the MH370 search area is the well stratified surficial sedimentary. It has unit> 300 m thick (> 400 ms TWT) overlies basement is the main feature in the north, the Perth Abyssal Plain, the eastern flanks of Gulden Draak and Batavia Rises, which are dissected by channels and canyons. On the contrary, surficial sediments show chaotic internal bedding due to seafloor mass transport (Du et al., 2007). In this understanding, modern ocean floor sedimentary processes are documented by sediment mass transport features, especially along the northern margin of Broken Ridge, and in pockmarks (the finest-scale features mapped), which are numerous south of Diamantina Trench and appear to record gas and/or fluid discharge from underlying marine sediments (Parnas et al., 1989; Picard et al., 2018).

Needless to say the main reason for expensive sea operation in this area not only to investigate the fuselage of MH370 and its debris but also for oil and gas exploration. As it is the first time in the marine exploration history to host such practical within using advanced and sophisticated instruments.

WHERE IS MH370?

Among the conspiracy theories of MH370 vanishing must be the answer to the most confusing question, which is arising up since March 8, 2018. As long the Captian Zaharie Ahmad Shah did not suicide and scientifically the aeroplane under autopilot control could not fly to the Southern Indian Ocean. Besides, there is no evidence of aliens, sabotage. Missile strike and fire or mechanical failure. However, the two techniques can be used to hijack the MH370 and drove the flight to the hiding place. First is using

Figure 14. Pareto diagram to identify the logical theories beyond the disappearing of MH370

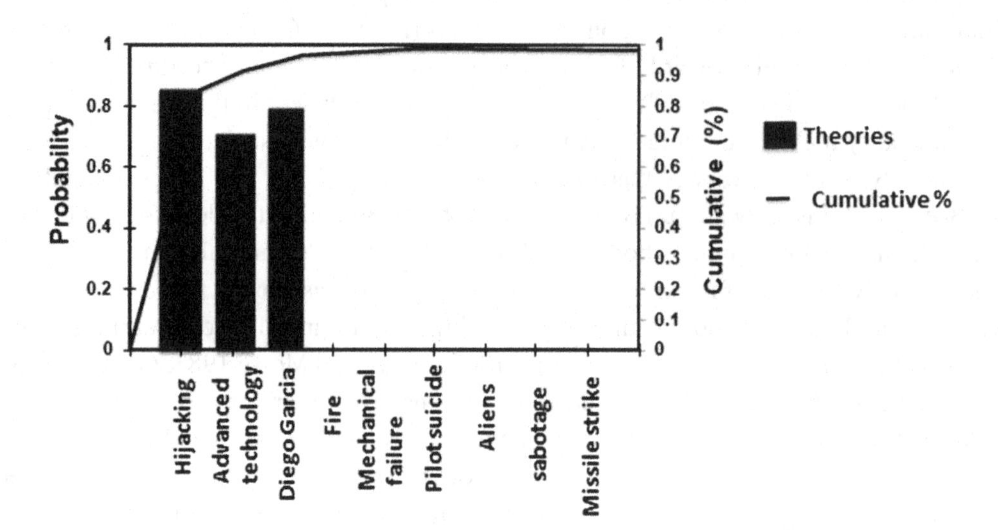

advanced technology to remotely control the flight and second is just internal force, which hijacked the flight. In this view, the MH370 hijacked by the professional crew while to the route to China. This procedure requires at least a runway length of 1 km to land the flight MH370.

In this understanding, MH370 when appeared on radar screen turned right and left and then disappeared, it was hijacked. The professional crews have hijacked the MH370 considered the fuel consumption and also the well-known route to land the flight. They must drive the flight to the nearest route as well as maintain the fuel consumption. Needless to say that the investigators must also concern other the nearest aeroplane routes, which are nearest to Malaysia to find the MH370 fuselage. Lastly, this can be proven by the Pareto diagram, where the three theories can be accompanied by each other: Hijacking, advanced technology, and Diego Garcia (Figure 14). The Pareto diagram declares that the flight MH370 hijacked either by the advanced or conventional method and driven to Diego Garcia with 80% confidence level. Diego Garcia also has a runway of 1 km, which allows the flight to land smoothly.

In this understanding, the flight MH370 hijacked and arrived at Diego Garcia within approximately 4 hours 55 minutes flying nonstop. In other words, MH70 was observed flying lower at 6.15 am on March 8, 2014. This can be confirmed and validate by the hit-rate of 180 (HRI%) (see chapter 13) and the fitness value of 100 (Figure 15). In this regard, Figure 15 illustrates the MH370 route to Diego Garcia. Under this fact, Diego Garcia provides a large runway (Figure 16), which allows the comfortable landing of the MH370.

Conversely, witnesses from the Kuda Huvadhoo corresponded that the aeroplane was roving North to South-East, near the Southern tip of the Maldives – Addu, where Diego Garcia locates. Further, they also observed the extraordinarily deafening noise that the MH370 made when it flew over the island (Picard et al., 2018).

With little to no other air traffic around, the MH370 flight was roving in the midnight towards Diego Garcia. The professional hijackers drove the flight to a low altitude around the immediate beginning of the route to avoid radar detection. In other words, they drove the flight out of high-frequency radar range i.e., a short distance. Nonetheless, they commonly disregard the danger of low-frequency radar,

Figure 15. Simulated MH370 route to Diego Garcia

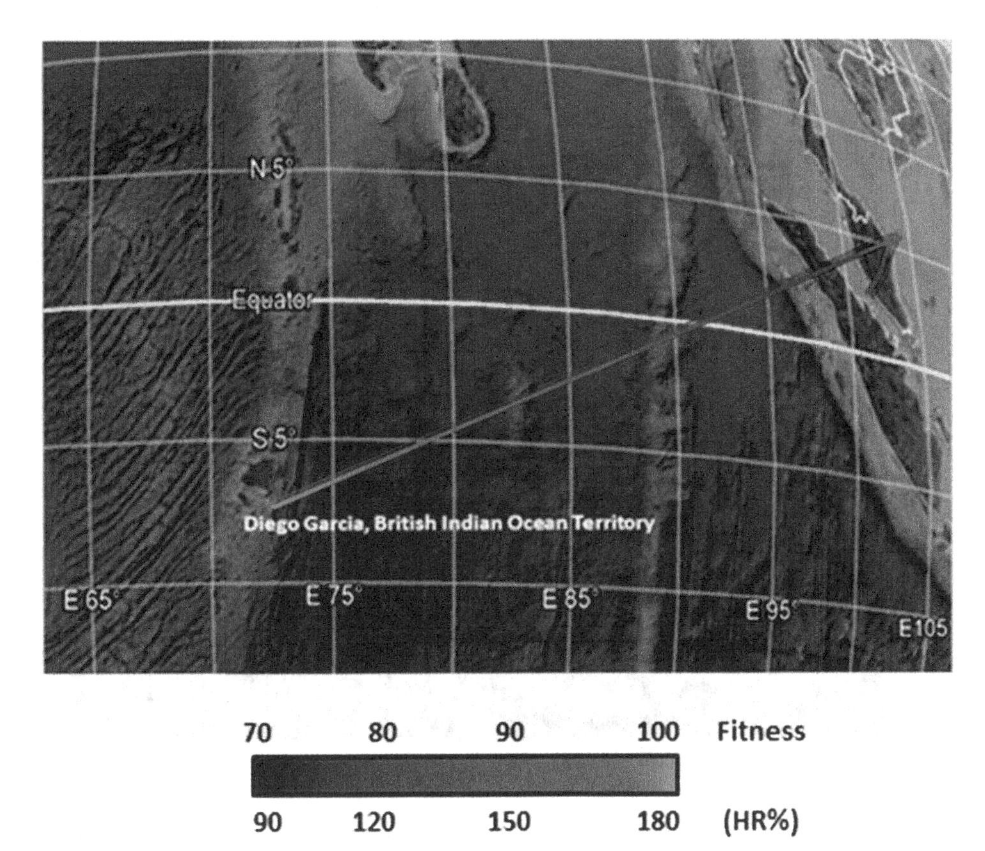

Figure 16. Large runway infrastructure in Diego Garcia

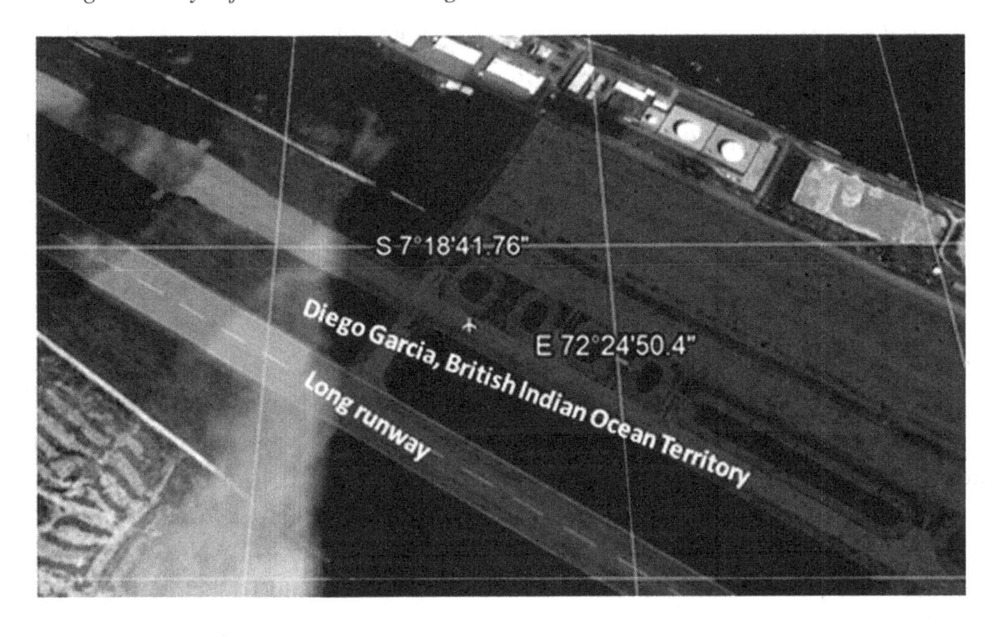

Figure 17. Suspected MH370 in the Jungle of Cambodia

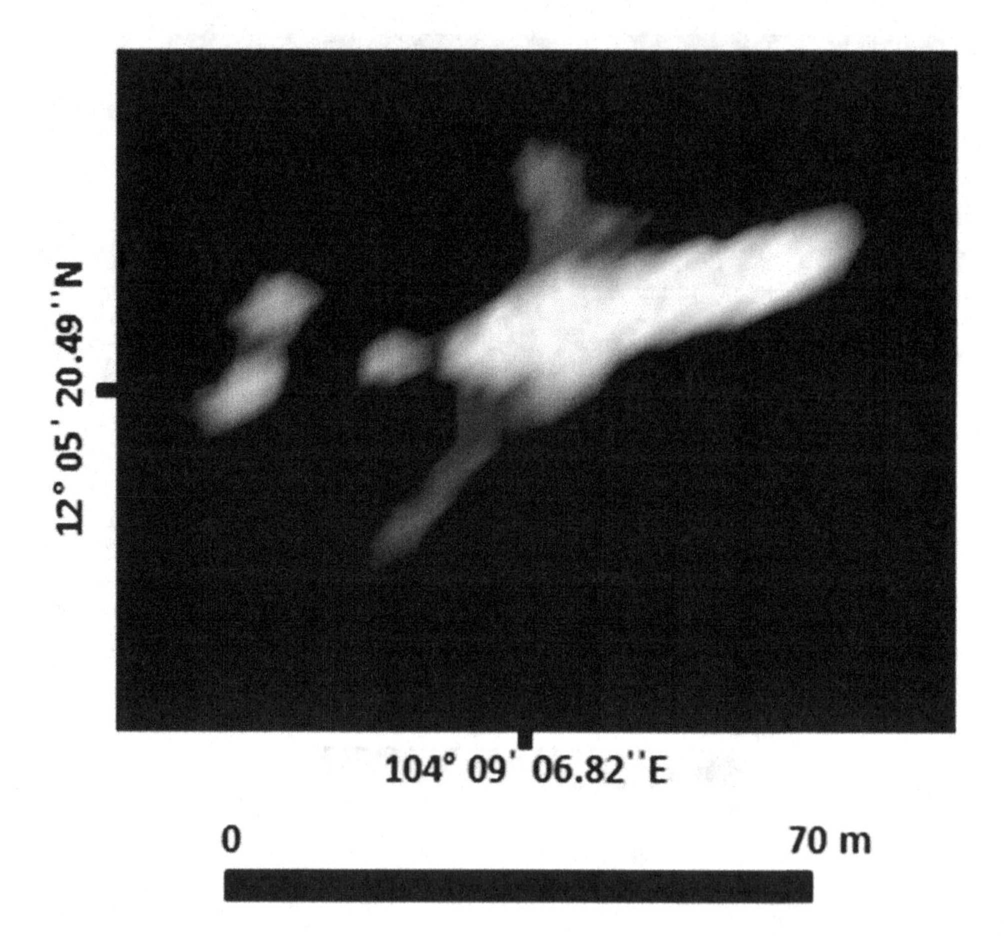

which is sparse and only implemented in very specific zones. In these circumstances, they could do three things to avoid radar detections (i) turned off the transponder to avoid identifying transmission; (ii) they avoided the restricted military radar areas over Malaysia, Sumatra, and the Indian Ocean; and (iii) they also circumvented the routes that shipping channels signify [17-20]. Over open ocean, they drove a flight to fuel efficiency altitude 30,000 ft. Then the journey was secure until landed in Diego Garcia.

WHY THE FLIGHT FOUND IN CAMBODIAN JUNGLES IS NOT MH370

A Chinese satellite Gefon-1 imagines an object look-alike fuselage of flight. It is claimed that its belong to MH370. This flight was observed on the Jungle of Cambodia. Its centre is located at 12° 05' 20.49''N and 104° 09' 06.82''E (Figure 17). The historical LANDSAT data downloaded from the Google Earth, which dated on December 2014 does not show any sign regarding this flight (Figure 18). LANDSAT data have a pixel resolution of 30 m, which allows detecting an object about 70 m.

However, the flight imaged by Gefon-1 cannot be detected earlier in LANDSAT data. In this understanding, there are two possibilities. First, the recorded flight could fly below the satellite during

Figure 18. LANDSAT data acquired in the same area of imaged flight by Geon-1 satellite in December 2015

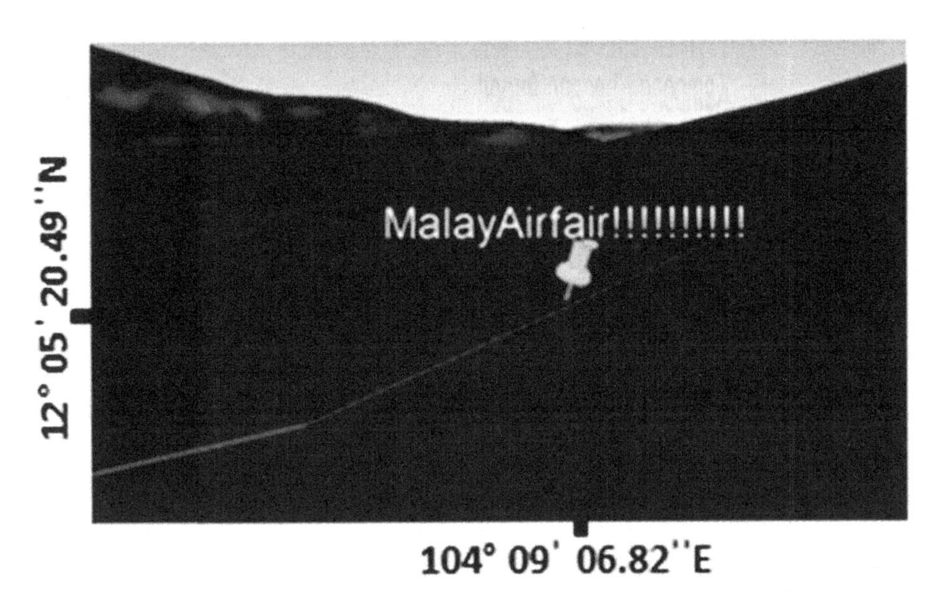

its route to Kampong Chhnang Airport. This airport is the nearest location to imaged flight in Gefon-1 (Figure 19). The distance between a located flight to the airport is approximately 46.50 km. However, the nose of flight in Figure 17 shows the flight is heading somewhere to the east, which indicated that the route is to Kampong Cham Airport. This airport is approximately further far from the flight about 139 km (Figure 20).

Figure 19.

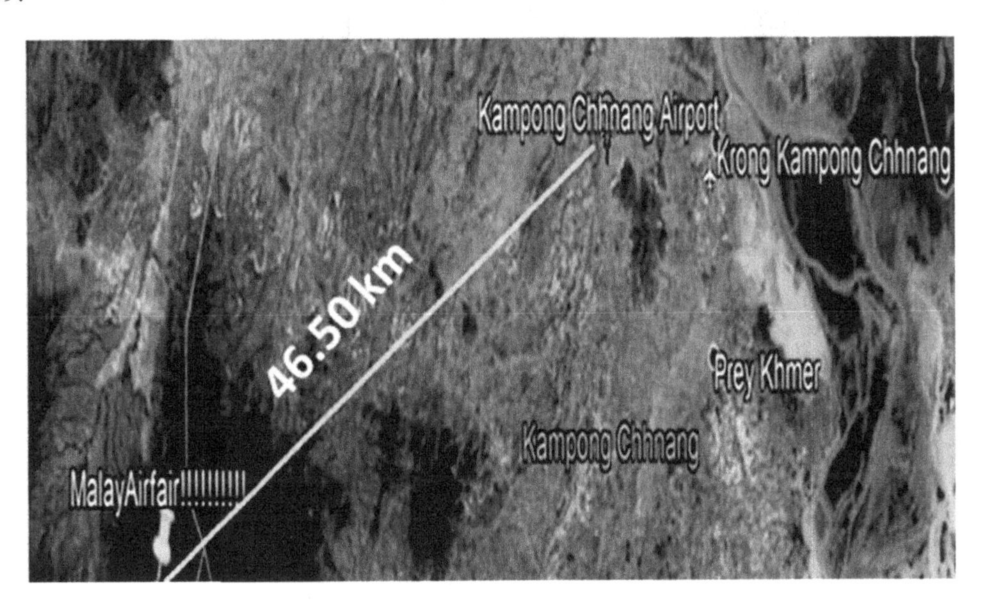

Figure 20. Flight heading to Kampong Cham Airport

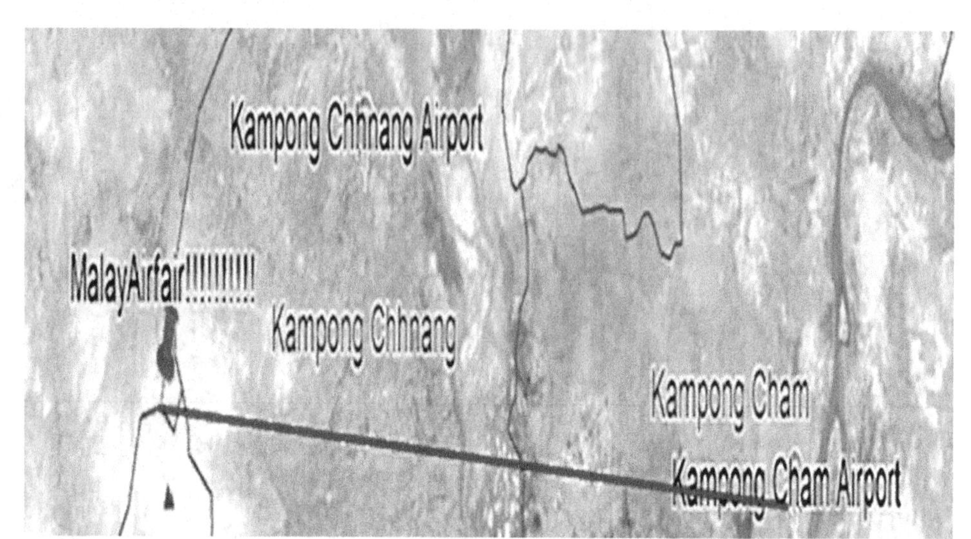

The Pareto optimization diagram spectacles that the flight does not belong to MH370 and the flight headed towards Kampong Cham Airport with a confidence level of 95%, which is higher than the route to the Kampong Chhnang Airport (Figure 21). Further, these hypotheses support that the flight is not belonging to MH370 with a validated probability of 0.95.

Figure 21. Pareto optimization of suspected flight shown by Gefon-1 satellite data

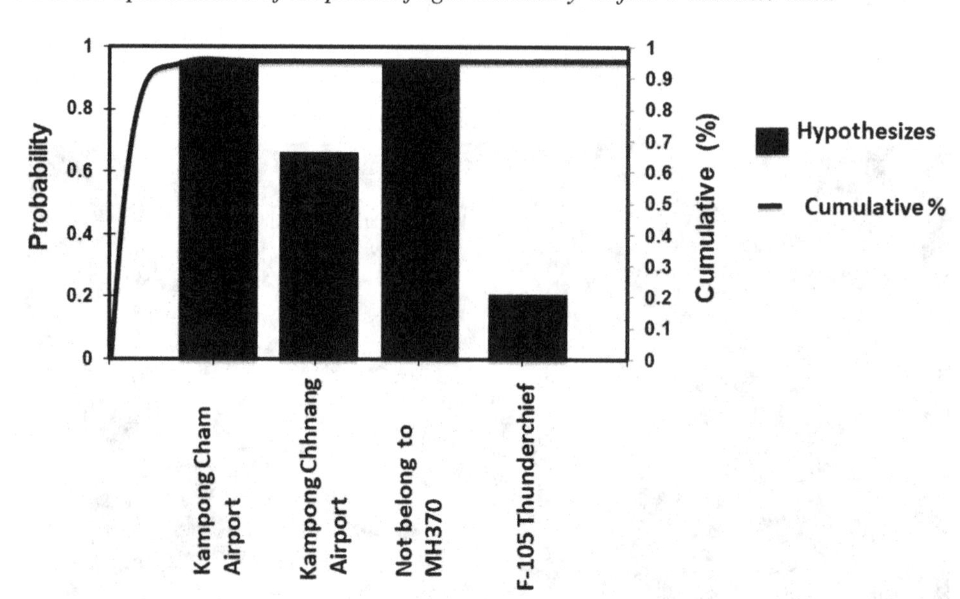

On the contrary, there is a weak possibility, the flight imaged by GF-1 belongs to F-105 Thunderchief a large fighter jet that was the primary attack aircraft of the conflict in Vietnam with a confidence level of 20%. The lack of cockpit cavity advocates this aeroplane smashes the ground vertically in an overturned flat spin. However, F-105 Thunderchief has been just 20 m long, which shorter than Boeing 777. In this circumstance, the flight imaged by GF-1 is precisely heading towards Kampong Cham Airport and the darkness was surrounding its background is just shadow or thermal emission from the fuselage of the flight.

It is worth mentioning that the genetic algorithm based Pareto optimization has been accurately succeeding to investigate the impact of the Southern Indian Ocean dynamics on the MH370 debris trajectory movements. The Pareto principle is also notorious as the 80/20 regulation, the rule of the energetic limited, or the code of factor sparsity, declares that, for countless events as in the case of MH370, roughly 80% of the belongings is derived from 20% of the foundations. In this view. Pareto optimization was developed in this book both concepts in the context of the Southern Indian Ocean dynamic effects and existence debris are shown in satellite remote sensing data.

Furthermore, the most significant differences between the findings of this book and other traditional studies, which are based on Bayesian analysis are summarized below.

An evolutionary algorithm based on genetic algorithm and multiobjective algorithms is a search of points in parallel instead of a single point search. The precise location of MH370 last destination is determined because the route of search is governed by the objective function. Moreover, the evolutionary algorithm involves fitness, which they do not necessitate additional information or knowledge regarding the stated problem. This allows a precise determination of what has been happened to flight MH370. This is also achieved based on the probabilistic transition algorithm, which involves the coding of the evolutionary algorithm. Lastly, evolutionary algorithms, when cast-off with proper contrivances, can muddle through multi-model functions, as a result of several correspondingly valid impending solutions of tracking the MH370 debris and flaperon and even provide its last potential destination as compared to Bayesian algorithms.

Nonetheless, the book cannot answer why the flight hijacked and driven to Diego Garcia. As the book focuses on the logical, scientific explanations based on the existence of flaperon and flight MH370 debris under the impact of the Indian Ocean dynamic system.

CONCLUSION

This chapter reveals the application of the multiobjective genetic algorithm to investigate the last destination of MH370. Based on the multiobjective algorithm, the MH370 last destination is not near the coastal water of Perth, Australia. In this sense, the Pareto optimization has been precisely successive to explore the influence of the Southern Indian Ocean dynamics on the MH370 debris trajectory movements. The Pareto front proved that the fragments do not belong to MH370 fuselage. The chapter is also proved that a 95% confidence level of the flight found in Cambodia is not MH370. In conclusion, MH370 was hijacked and driven to Diego Garcia as it is a short path from the departure point at International Malaysia Airport, Kuala Lumpur.

REFERENCES

Akin, G. W., & Lagerwerff, J. V. (1965). Calcium carbonate equilibria in solutions open to the air. II. Enhanced solubility of CaCO3 in the presence of Mg2+ and SO42−. *Geochimica et Cosmochimica Acta*, *29*(4), 353–360. doi:10.1016/0016-7037(65)90026-8

Chapman, J. W., Nesbit, R. L., Burgin, L. E., Reynolds, D. R., Smith, A. D., Middleton, D. R., & Hill, J. K. (2010). Flight orientation behaviors promote optimal migration trajectories in high-flying insects. *Science*, *327*(5966), 682–685. doi:10.1126cience.1182990 PMID:20133570

Chen, T., Neville, A., & Yuan, M. (2005). Calcium carbonate scale formation—assessing the initial stages of precipitation and deposition. *Journal of Petroleum Science Engineering*, *46*(3), 185–194. doi:10.1016/j.petrol.2004.12.004

Collins, I. R., & Jordan, M. M. (2003). Occurrence, prediction, and prevention of zinc sulfide scale within Gulf Coast and North Sea high-temperature and high-salinity fields. *SPE Production & Facilities*, *18*(03), 200–209. doi:10.2118/84963-PA

Du, S. C., Shi, Z. G., Zang, W., & Chen, K. S. (2007). Using interacting multiple model particle filter to track airborne targets hidden in blind Doppler. *Journal of Zhejiang University. Science A*, *8*(8), 1277–1282. doi:10.1631/jzus.2007.A1277

Dyer, S. J., & Graham, G. M. (2002). The effect of temperature and pressure on oilfield scale formation. *Journal of Petroleum Science Engineering*, *35*(1-2), 95–107. doi:10.1016/S0920-4105(02)00217-6

Frota, T. M. P., Silva, D. R., Aguiar, J. R., Anjos, R. B., & Silva, I. K. V. (2013). Assessment of scale formation in the column of an oil and natural gas producing well: A case study. *Brazilian Journal of Petroleum and Gas*, *7*(1), 15–29. doi:10.5419/bjpg2013-0002

Gardner, R. L., Daczko, N. R., Halpin, J. A., & Whittaker, J. M. (2015). Discovery of a microcontinent (Gulden Draak Knoll) offshore Western Australia: Implications for East Gondwana reconstructions. *Gondwana Research*, *28*(3), 1019–1031. doi:10.1016/j.gr.2014.08.013

Mackay, E. J., Jordan, M. M., & Torabi, F. (2003). Predicting brine mixing deep within the reservoir and its impact on scale control in marginal and deepwater developments. *SPE Production & Facilities*, *18*(03), 210–220. doi:10.2118/85104-PA

Mucci, A., & Morse, J. W. (1983). The incorporation of Mg2+ and Sr2+ into calcite overgrowths: Influences of growth rate and solution composition. *Geochimica et Cosmochimica Acta*, *47*(2), 217–233. doi:10.1016/0016-7037(83)90135-7

Parnas, D. L., Clements, P. C., & Weiss, D. M. (1989). Enhancing reusability with information hiding. In Software reusability: vol. 1, concepts and models (pp. 141-157). doi:10.1145/73103.73109

Picard, K., Brooke, B. P., Harris, P. T., Siwabessy, P. J., Coffin, M. F., Tran, M., Spinoccia, M., Weales, J., Macmillan-Lawler, M., & Sullivan, J. (2018). Malaysia Airlines flight MH370 search data reveal geomorphology and seafloor processes in the remote southeast Indian Ocean. *Marine Geology*, *395*, 301–319. doi:10.1016/j.margeo.2017.10.014

Rea, D. K., Dehn, J., Driscoll, N. W., Farrell, J. W., Janecek, T. R., Owen, R. M., Pospichal, J. J., & Resiwati, P. (1990). Paleoceanography of the eastern Indian Ocean from ODP Leg 121 drilling on Broken Ridge. *Geological Society of America Bulletin, 102*(5), 679–690. doi:10.1130/0016-7606(1990)102<0679:POTEIO>2.3.CO;2

Salman, M., Qabazard, H., & Moshfeghian, M. (2007). Water scaling case studies in a Kuwaiti oil field. *Journal of Petroleum Science Engineering, 55*(1-2), 48–55. doi:10.1016/j.petrol.2006.04.020

Smojver, I., & Ivančević, D. (2010). Numerical simulation of bird strike damage prediction in airplane flap structure. *Composite Structures, 92*(9), 2016–2026. doi:10.1016/j.compstruct.2009.12.006

Sundance. (2018). *MH-370 – Update: Residents Of The Maldives Report Possible March 8th Sighting Of Missing Malaysian Flight…* https://theconservativetreehouse.com/2014/03/18/mh-370-update-residents-of-the-maldives-report-possible-march-8th-sighting-of-missing-malaysian-flight/

Surgent, L. V. Jr. (1974). *Foliage Penetration (FOPEN) Radar Detection of Low Flying Aircraft (No. LWL-TR-74-72).* Army Land Warfare Lab Aberdeen Proving Ground MD.

Upadhyaya, S. (2018). *Maritime Security Cooperation in the Indian Ocean Region: Assessment of India's Maritime Strategy to be the Regional "Net Security Provider"*. Academic Press.

Zhang, Y., Shaw, H., Farquhar, R., & Dawe, R. (2001). The kinetics of carbonate scaling—application for the prediction of downhole carbonate scaling. *Journal of Petroleum Science Engineering, 29*(2), 85–95. doi:10.1016/S0920-4105(00)00095-4

About the Author

Maged Marghany is currently a Professor at Department of Informatics, Faculty of Mathematics and Natural Sciences, Universitas Syiah Kuala Darussalam, Banda Aceh, Indonesia. He is author of, Advanced Remote Sensing Technology for Tsunami Modelling and Forecasting which is published by Routledge Taylor and Francis Group,CRC, Synthetic Aperture Radar Imaging Mechanism for Oil Spills, which is published by Elsevier, and Automatic Detection Algorithms of Oil Spill in Radar Images, which is published by Routledge Taylor and Francis Group, CRC. His research specializes in microwave remote sensing and remote sensing for mineralogy detection and mapping. Previously, he worked as a Deputy Director in Research and Development at the Institute of Geospatial Science and Technology and the Department of Remote Sensing, both at Universiti Teknologi Malaysia. Maged has earned many degrees including a post-doctoral in radar remote sensing from the International Institute for Aerospace Survey and Earth Sciences, a PhD in environmental remote sensing from the Universiti Putra Malaysia, a Master of Science in physical oceanography from the University Pertanian Malaysia, general and special diploma of Education and a Bachelor of Science in physical oceanography from the University of Alexandria in Egypt. Maged has published well over 250 papers in international conferences and journals and is active in International Geoinformatic, and the International Society for Photogrammetry and Remote Sensing (ISPRS).

Index

O

P

R

S

T

U

V

W

Printed in the United States
By Bookmasters